AVIATION EDUCATION AND TRAINING

For my Mom, Cécile Henley
and my sons,
Denis and Eric La France

Aviation Education and Training

Adult Learning Principles and Teaching Strategies

Edited by
IRENE M.A. HENLEY
Associate Professor and Head of Aviation
University of Western Sydney
Australia

ASHGATE

Published by
Ashgate Publishing Limited
Gower House
Croft Road
Aldershot
Hampshire GU11 3HR
England

Ashgate Publishing Company
Suite 420
101 Cherry Street
Burlington, VT 05401-4405
USA

Ashgate website: http://www.ashgate.com

British Library Cataloguing in Publication Data
Aviation education and training : adult learning principles
and teaching strategies. - (Studies in aviation psychology
and human factors)
1.Flight crews - Training of 2.Flight training 3.Adult
learning 4.Aeronautics - Study and teaching 5.Aeronautics -
Human factors
I.Henley, Irene M. A.
629.1'3252'071

Library of Congress Cataloging-in-Publication Data
Aviation education and training : adult learning principles and teaching strategies / edited
by Irene M.A. Henley.
p. cm. -- (Studies in aviation psychology and human factors)
Includes bibliographical references and index.
ISBN 0-7546-1733-5
1. Aeronautics--Study and teaching. 2. Airplanes--Piloting--Study and teaching. 3. Flight
training. 4. Air pilots--Training of. I. Henley, Irene M. A., 1945- II. Series.

TL560.A853 2003
629.13'0071-dc21

2003056039

ISBN 0 7546 1733 5

Printed and bound in Great Britain by MPG Books Ltd, Bodmin, Cornwall

Contents

PART 2 TEACHING STRATEGIES

List of Figures

List of Tables

List of Contributors

Jayne Bye is a Lecturer in Aviation at the University of Western Sydney, Australia where she coordinates the Master of Aviation program and teaches undergraduate and postgraduate subjects in aviation communication and aviation education. She is currently completing her doctoral dissertation on post-compulsory education and training policy at the University of Technology, Sydney. She has experience teaching and consulting across a range of education sectors including post-compulsory school education, vocational training and adult community education. Her current research interests include the education and training of young adults and the area of school to work transition.

Dr Dianne L. Conrad is an Assistant Professor at the University of New Brunswick in Canada where she is responsible for coordinating a provincial, distance delivered instructor development program. Her doctoral dissertation studied community, social presence and engagement in on-line learning. Prior to joining the University of New Brunswick she was the Assistant Director of the Faculty of Extension's Master of Arts in Communications and Technology (MACT) program at the University of Alberta in Canada. Currently the editor of the *Canadian Journal for University Continuing Education*, Dianne is a past-president of the Alberta Association for Continuing Education and served as a board member for a number of provincial and national associations in the fields of education and communications.

Kay H. Ellis is an Air Traffic Control Systems Trainer in Australia. Following a nine year career as a secondary school teacher, she joined the Royal Australian Air Force and trained in air traffic control. She has more than ten years experience as an air traffic controller and on-the-job instructor, and has been involved in developing on-the-job training programs as well as air traffic control familiarisation courses for student pilots. Kay is completing a Master of Aviation at the University of Western Sydney, Australia. Her research area of interest is aviation instruction, particularly on-the-job training.

Jo Harris is a contract employee with Transport Canada, working on several projects: Recreational Aviation Examiner Programs, Personnel Licensing Review, Rewriting of Regulations, and consulting with the Flight Training Standards Division on the Flight Instructor Project. She spent six years as Superintendent of General Aviation for Transport Canada; five years as a Flight Training Standards Inspector and ten years as a Flight Instructor and Chief Ground School Instructor. Jo has been a volunteer with aviation organisations, such as the Ninety Nines (International Organization of Women Pilots), encouraging women pilots to become professionals, and was Governor of Western Canada Ninety Nines for two years. She won the Amelia Earhart Scholarship in 1989 and was on the board of the Alberta Aviation Council furthering the cause of general aviation pilots. She organised and ran the Governor General's Cup Air Rally in 1979 and won the Jean Luc Pépin award for Aviation Safety in 1981.

Dr Irene Henley is an Associate Professor and foundation Head of the Department of Aviation at the University of Western Sydney, Australia where she teaches graduate and undergraduate subjects in aviation education and aviation safety management. A former Transport Canada Flight Training Standards Inspector, Irene holds a Canadian Airline Transport Pilot Licence, a Class I Flight Instructor Rating with aerobatic and float endorsements, an Australian Grade 1 Flight Instructor Rating, and is a Civil Aviation Safety Authority Approved Testing Officer. She has over twenty years of teaching experience as well as over twenty years of experience in civil aviation. Her research interests include flight instructor training and evaluation, student pilot and flight instructor stress, and innovative and effective teaching strategies in aviation education. She has published in all these research areas and has presented papers and workshops at international conferences in the United States, Europe, New Zealand and Australia. She also has conducted numerous seminars and workshops for flight instructors across Australia.

Dr Lynn M. Hunt is the Advisor to the Pro Vice Chancellor (College of Business) Flexible Learning and Teaching at Massey University, New Zealand. In 1986 she began work in the Business College at Massey University, teaching and conducting research in computer-based learning, instructional design and psychology, and adult learning. During this time she became involved in HURDA (Human Resource Development in Aviation) research program in which she developed CALES, a computer-based examination-on-demand system for pilot licensing. Her PhD thesis identified those factors that influence the selection of learning strategies and how students match those learning strategies to task types. Other research interests are learning styles, instructional design, online learning and needs assessment. In addition, she is currently working on a number of projects, including: methodologies in accident investigation, measuring competence and proficiency, developing an online distance education program for the Maldives, air rage and online student self-assessment programs. She worked at the Massey School of Aviation between 1990 and 1999, and continues to teach there.

Patricia McDonell has been an Air Traffic Control Officer in the Royal Australian Air Force for more than twenty two years. In that time she has held ratings at various Australian military and civil airfields. Her career has seen her work as a training officer in a number of air traffic control elements, an air traffic rating authority and senior air traffic controller. Patricia is currently completing a Master of Human Factors (Aviation) at the University of Western Sydney. Her research area of interest is human error within aviation.

Ann Reich is a Lecturer and Course Coordinator of Adult Education courses at the University of Western Sydney, Australia where she teaches undergraduate and postgraduate subjects in adult and vocational education and training. She has written distance education modules in aviation education. She has over twenty years experience in education in community, vocational education and training and higher education contexts. Ann has been a Manager of Research for a state-based industry training board and held positions in state and local government. Her research interests are in workplace learning and vocational education. She is currently completing her doctoral dissertation in workplace learning at the University of Technology, Sydney.

Dr Matthew J.W. Thomas is a Lecturer in Professional Development (On-line Teaching and Learning) within the Flexible Learning Centre at the University of South Australia. He joined the University of South Australia from the School of Aviation at Massey University in New Zealand where he lectured in Aviation Human Factors, Crew Resource Management, Instructional Design and Performance Evaluation. Prior to joining the Massey University's School of Aviation he undertook doctoral research at the University of Adelaide in Australia examining learner interaction and learning outcomes in on-line learning environments. His research activities focus on instructional design in high-risk industry settings, with a particular emphasis on the evaluation of flight crew performance and instructional design strategies for responsive training systems design. Further interest include the use of new information and communication technologies in education and training and the use of new distance learning technologies to enable more flexible modes of instruction.

Dr Mary Ann Turney is an Associate Professor at Arizona State University, USA where she teaches graduate and undergraduate courses in aviation and human factors. She is engaged in research related to adult learning styles, crew resource management and retention of women in aviation. Mary Ann was formerly the director of flight programs at Dowling College in New York, where she administered the flight training program, taught a variety of aviation courses, and trained the college's Precision Flight Team for intercollegiate competition. She has published her research on human factors, crew resource management and gender-related issues, has presented human factors workshops at international conferences in the United States, Europe and Australia and continues to carry out federally-funded research in human factors in collaboration with several colleagues throughout the United States. An active pilot and Gold Seal flight instructor, Mary Ann has given over 1500 hours of instruction. She holds an Airline Transport Pilot Certificate, an instrument and multi-engine instructor certificate and an advanced ground instructor certificate. As an accident prevention counsellor, Mary Ann also taught numerous FAA safety seminars for pilots and flight instructors.

Dr Mark W. Wiggins is a Senior Lecturer in Psychology at the University of Western Sydney, Australia. He has a PhD from the University of Otago in New Zealand, is a Registered Psychologist in New South Wales, and is a member of the Human Factors and Ergonomics Society and of the Australian Psychological Society. Mark is a member of the Human Factors Advisory Group for the Australian Civil Aviation Safety Authority and a qualified pilot. He has been a consultant researcher for the Federal Aviation Administration, the National Aeronautics and Space Administration, the Australian Transportation Safety Board and Airservices Australia. He is the author of a number of refereed journal articles, book chapters, and books including *Assessing Human Factors in the Workplace,* published by Lawrence Erlbaum.

Preface

The main objective in writing this book was to develop a valuable educational resource for practitioners who are responsible for teaching and assessing adult learners in aviation programs in colleges and universities and within all the various segments of the aviation industry. However, this aim cannot be separated from the basic adult learning principles and the major theoretical and practical issues that underpin our practice. Therefore, the focus was on producing a practitioner-oriented book that describes the interaction between theory and practice in an effort to develop practical teaching strategies and materials that are useful in real-world educational settings. Accordingly, the choice of the topics to be included in this book was guided by many years of experience as a practitioner in the field of aviation education, by consultation with other practitioners, and by the desire to include other contributors who, together, would enhance the breadth of the material presented in this book.

As educators, we need to reflect on our practice and occasionally ask ourselves if the principles and the philosophical models guiding our teaching are appropriate for the needs of current professionals within the industry. Do we need to correct our course so as to assure that our efforts lead not only to praiseworthy achievements but also to important practical results? As such, this book will have appeal only if practitioners and trainees have an open mind and are willing to question their own beliefs and practice, and are willing to extend their thinking and experiment with new ideas and new approaches. As Anatole France, French novelist and recipient of the Nobel Prize for Literature, said: "In order that knowledge be properly digested, it must have been swallowed with a good appetite". Therefore, the book aims to provide a basis for reflection, judgement, choice and decision making for the training practitioner or vocational educator in a rapidly changing context, as is the case in the aviation industry.

The fifteen chapters within the book are organised into two sections. The chapters in Part 1 focus on some of the major theories of learning and adult learning principles that underpin our practice as educators in the field of aviation. Part 1 also aims to provide insight into learning strategies that can enhance task accomplishment, and factors that can hamper or facilitate student progress.

Part 2 is concerned with the application of these theories and principles and offers a variety of teaching methods or instructional strategies based on research as well as on principles of instruction gleaned from years of successful teaching experience by practitioners from the academic milieu and from the aviation industry. In short, Part 2 of the book is about engaging in practice. Familiarity with a range of teaching strategies is important for all teachers and trainers since there is no single strategy appropriate for all occasions. The chapters in Part 2 cover a range of useful strategies for the aviation context as well as provide guidelines for choosing the most effective strategies in any teaching situation. While many of the examples refer to pilot training, the basic principles and strategies hold true for other aviation professionals (i.e., flight attendants, air traffic controllers, maintenance engineers, administrative personnel, managers, etc.) engaged in aviation education and training.

Finally, this book is about improving learning through effective training practice. The book is intended not only to enable instructors, trainers, teachers and adult learners to promote more effective learning, it is also intended to promote learning and teaching that is critical in enabling us all as learners to move beyond the necessary improvement in our capacities to learn and use that learning in our journey of life-long learning. It is hoped that this book is just the beginning of a sustained effort to develop strategies that will improve training and education in all the various sectors of the aviation industry.

Irene Henley
Sydney, Australia

Acknowledgements

I am grateful for the infrastructure support provided by the University of Western Sydney in Australia. I am also indebted to Emily Ponniah, Kelly-Ann Gunn and Wendy Peddell for their expertise in word processing and graphics and especially to Kelly-Ann for her patience, dedication and good sense of humour. I am also grateful to Margaret McDonell for her careful inspection of the manuscript and her invaluable advice about formatting and her help in preparing the index. Special thanks go to my friend, Claire Lafrenière, who in spite of her busy schedule took the time to read and provide valuable comments on some of the chapters in the first part of the book. To my sister, Juliette, and my ever-faithful friends, Pauline Piché and Jan Colter, I want to say a heartfelt thank-you for their understanding and unwavering moral support. I would also like to express my most sincere thanks to the contributors who participated so willingly and enthusiastically in the production of the chapters within this book. I realise that they all have busy schedules and have many demands placed on their time. Finally, I would like to express my indebtedness to my friend and valued professional colleague, Mark Wiggins, because without his amiable but persuasive initial nudge and his continued encouragement and inspiration, this book may never have seen the light of day.

Irene Henley

List of Abbreviations

AC	Abstract Conceptualization
AE	Active Experimentation
ANTA	Australian National Training Authority
AQP	Advanced Qualification Program
ASRS	Aviation Safety Reporting System (USA)
ATC	Air Traffic Control
ATPL	Airline Transport Pilot Licence
BSI	British Standards Institute
CAL	Computer-Assisted Learning
CASA	Civil Aviation Safety Authority (Australia)
CB	Cumulonimbus
CBL	Computer-Based Learning
CBT	Competency-Based Training
CBT	Computer-Based Training
CE	Concrete Experience
CLAI	Cognitive Learning of Automated Instruction
CML	Computer-Mediated Learning
CPL	Commercial Pilot Licence
CPLI	Integrated Commercial Pilot Licence
CPL/IR	Commercial Pilot Licence / Instrument Rating
CRM	Crew Resource Management
CTA	Cognitive Task Analysis
CWA	Cognitive Work Analysis
EBL	Experience-Based Learning
FAA	Federal Aviation Administration (USA)
FMS	Flight Management System
ICAI	Intelligent Computer-Assisted Instruction
ICAO	International Civil Aviation Organization
IQ	Intelligence Quotient
IR	Instrument Rating
ITS	Intelligent Tutoring System
JAA	Joint Aviation Authority (Europe)
JAR	JAA Air Regulations

LOFT	Line-Oriented Flight Training
MBTI	Myers-Briggs Personality Type Indicator
MI	Multiple Intelligences
MUSE	Multi-User Simulated Environments
NASA	National Aeronautics and Space Administration (USA)
NTSB	National Transportation Safety Board
NVFR	Night Visual Flight Rules
OJT	On-the-job training
OPS	Operations
PBL	Problem-Based Learning
PPL	Private Pilot Licence
PTS	Practical Test Standards
RO	Reflective Observation
SAL	Student Approaches to Learning
SAT	Systems Approach to Training
SPQ	Study Processes Questionnaire
S-R	Stimulus-Response
S-R-R	Stimulus-Response and Reinforcement Theory
TCAS	Traffic Collision Avoidance Systems
TMT	Transfusion Medical Tutor
VEETAC	Vocational Education, Employment and Training Advisory Committee

Part 1
Learning Principles

1 Learning Theories and their Application to Aviation Education

Jayne Bye and Irene Henley

Introduction

Despite the fact that learning has been the subject of intense research for the last century or more, there is still little agreement about how learning occurs. What we do have, however, is a range of differing explanations about what happens when learning takes place, and these explanations are called learning theories (Merriam & Caffarella, 1999, p. 250). Each of these theories has a set of assumptions about the role of the learner in the process of learning, what motivates people to learn, the importance of the role of the teacher or trainer and the techniques required to facilitate learning. Therefore, looking at the process of learning, and more specifically at how adults learn, is a useful activity for teachers and trainers who wish to improve their skills in facilitating this process with participants in both workplace and classroom training environments.

This chapter focuses on the process of learning and explores a range of explanations of this process from five major psychological orientations: behaviourist, cognitive, humanist, social learning, and constructivist. Traditionally, aviation education (especially flight training) has been based mainly on behaviourist and cognitive learning theories. However, many in the field of aviation education now realise that, in order to understand the use of knowledge and the acquisition of complex psychomotor skills in the aviation domain, we also need to be aware of the role of goals, motivation and affect (emotions, values, feelings) in the learning process.

The first orientation to be discussed is the behaviourist approach. This approach has been the dominant framework for the design of much vocational education and training during the past century and as such provides an important foundation for understanding how this teaching and learning context has been shaped over the years. Indeed, its continued dominance and influence on the form of Competency-Based Training (CBT) in Australia will be examined in terms of what advantages it is still seen to offer teachers and learners in the aviation context.

The cognitive orientation is examined next and the focus of discussion is seen to shift here from the behaviourist interest in teaching techniques to the internal sense-making processes of the learner. An understanding of these processes is seen as useful for educators in helping learners understand and improve their own learning processes. This aspect of Chapter 1 links closely with the discussion of learning styles in Chapter 4.

The third orientation is the humanist approach that provides an interesting contrast to the first two orientations. Humanism's concerns with the autonomy of the adult learner and personal growth through learning makes the learner's needs rather than subject matter or course content central to the learning process. The limitations of this approach in the vocational education and training context are acknowledged; however, useful principles within the orientation are suggested for use in the aviation teaching and learning context.

Social learning and its links to popular workplace and management practices, such as mentoring, are examined in the fourth orientation. This useful approach can be seen to have direct application to workplace learning (on-the-job training) in the aviation context, especially with regard to its focus on self-regulation and on mentoring.

Finally, the constructivist orientation is examined with its interest in the internal learner processes of constructing knowledge as well as in environmental factors. Once again, direct application to vocational education and training in the aviation context can be seen in terms of current research into workplace learning that uses this orientation (e.g., self-directed learning and reflective practice).

Behaviourist Orientation

Overview

The behaviourist approach is one of the oldest scientific approaches to understanding learning (Lefrancois, 1997). Further, it is generally agreed that the behaviourist approach to teaching and learning has influenced much teaching and training at least in the USA, Canada, Great Britain and Australia, for 20 years or so (Pithers, 1998). This influence has its historical basis in psychological research carried out in the early twentieth century. Behaviourist research approached the psychological study of learning and motivation as an objective and experimental science and was popularised in the 1940s and 1950s due to "a wider community belief in the ability of science and engineering to address and solve the world's problems" (Russell, 1999, p. 105). Furthermore, behaviourist techniques promised accountability, quality control and clear outcomes in education and training. Behaviourists claimed to answer important questions teachers and trainers were grappling with, such as "What kinds of feedback promote the best learning? How can material that is to be learned be presented so that it best grabs the learner's attention? How can a lesson be taught so that the principle behind the lesson transfers to new situations?" (Kehoe & Macrae, 1999, p. 211).

Due to the popularity of this approach and the apparent advantages it was seen to offer, the behaviourist approach to learning and teaching came to dominate school education, military training and many vocational training systems (Russell, 1999, p. 105). The reason for its continued influence lies in the fact that accountability, quality control and clear outcomes are still issues of concern for educators in the aviation context today.

Definitions and Basic Tenets of the Behaviourist Orientation

As the name of this orientation suggests, educators working in this approach understand learning as changed *behaviour*. That is, all instruction has as its goal *changed behaviour*. In simple terms, the behaviourist orientation states that we learn by receiving stimuli from our environment that provoke a response. The interplay of stimuli and consequent rewards (or punishments) is seen as the mechanism responsible for the desired behaviour change.

According to Merriam and Caffarella (1999), there are three basic assumptions that behaviourists share about the nature of learning and the processes that cause learning to occur.

1. Behaviourists focus their study and research on observable and measurable behaviour rather than internal thought processes. In fact, one of the earliest behaviourist researchers, John Watson, thought that the study of psychology should "abandon the examination of inaccessible and unobservable mental events" (Tennant, 1997, p. 94). As will be seen below, not all theories of learning agree with this assumption.

2. Behaviourists believe that all behaviour can be shaped by the environment. Therefore, all learning is "determined by the elements in the environment, not by the individual learner" (Merriam & Caffarella, 1999, p. 251). The elements in the environment that have the power or role to shape behaviour are called stimuli. As suggested by the term, these stimuli *stimulate* a response in the learner and this response (if it is the desired one) will be rewarded, therefore providing the conditions for the behaviour to be repeated if the same stimuli are again present in the environment.
3. According to Grippin and Peters (1984), "the principles of contiguity (how close in time two events must be for a bond to be formed) and reinforcement (any means of increasing the likelihood that an event will be repeated) are central to explaining the learning process" (cited in Merriam & Caffarella, 1999, p. 251).

These processes of learning were studied by behaviourists using the scientific method in controlled laboratory settings (Russell, 1999, p. 105). The results of experimentation carried out in this orientation were approaches to learning based on three major conditioning theories — classical conditioning, connectionist conditioning and operant conditioning. The bases of these theories and the contributions of key theorists are discussed briefly below.

Classical conditioning Pavlov is probably the theorist most widely known and associated with conditioning theories. The story of his discovery and experiments are similarly well known.

> While performing a series of experiments on digestion and salivation and dogs, Pavlov and his assistants discovered that after a number of trials that involved placing food in a dog's mouth, he began to salivate at the mere sight of the food, then at the sight of the food dish, and finally even at the sound of the assistants' approaching footsteps. (Gilmer, 1970, p. 298)

Connectionist conditioning Also referred to as the Association Theory, connectionist conditioning was first researched and reported by Edward Lee Thorndike (1874-1949) in 1911. Thorndike's work, sometimes also called the stimulus-response (S-R) theory of learning, has had a major influence on education and training practice this century. "Using animals in controlled experiments, Thorndike noted that through repeated trial-and-error learning, certain connections between sensory impression, or stimuli (S), and subsequent behaviour, or responses (R), are strengthened or weakened by the consequences of behaviour" (Merriam & Caffarella, 1999, p. 251).

Thorndike's work can best be examined by briefly looking at the three laws of learning he proposed and later modified during the course of the first half of the 20th century. Most flight instructors will recognise Thorndike's laws of learning since many flight instructor manuals and guides reflect a behaviourist approach.

- *Law of Effect* — The Law of Effect was introduced by Thorndike in 1911 and was later modified after 1930 (Kehoe & Macrae, 1999). The Law of Effect professed that learning consists of an association between the surrounding cues (stimuli) and one's response to them. In addition, Thorndike indicated that such association would be strengthened only when the cue and the response are followed closely by a reward (stimulus-response and reinforcement theory or S-R-R). However, this reward does not become part of the association. Rather, it acts as a catalyst in forming the association. The result is that when the same cue occurs again, it tends to activate the association and elicit the previously rewarded response (Kehoe & Macrae, 1999, pp. 213-214).

  In the original formulation of the Law of Effect, Thorndike stated that, "if a stimulus leads to a response, which in turn leads to reinforcement, the S-R connection is strengthened. If, on the other hand, a stimulus leads to a response that leads to punishment, the S-R connection is weakened" (cited in Hergenhahn & Olson, 1993, p. 64). However, further testing of this law led Thorndike to a very significant revision after 1930. Thorndike found that punishment, in fact, did not have an impact on the formation of the connection, and therefore, did not have a significant role to play in the learning process (Hergenhahn & Olson, 1993).

- *Law of Readiness* — The Law of Readiness states that "if the organism is ready for the connection, learning is enhanced; otherwise learning is inhibited" (Merriam & Caffarella, 1999, p. 251). Stated simply, this means that:

  1. When someone is ready to perform some act, to do so is satisfying.

2. When someone is ready to perform some act, not to do so is annoying.
3. When someone is not ready to perform some act and is forced to do so, it is annoying.

Therefore, it could generally be said that "interfering with goal-directed behaviour causes frustration and causing someone to do something they do not want to do is also frustrating" (Hergenhahn & Olson, 1993, p. 63).

- *Law of Exercise* — Of the three main laws proposed by Thorndike, the Law of Exercise was the one that he modified the most. Originally, the law had two parts known as the law of use and the law of disuse. Basically, this law attempted to make clear the connection between repetition of a behaviour and learning. The law of use proposed that the connection between a stimulus and a response was strengthened the more the connection was exercised, that is, through repetition. The law of disuse proposed that if the connection was not practised, then the connection between the stimulus and the response would diminish and the thing being learned would be forgotten. However, further experimentation saw Thorndike unable to prove the viability of both the law of use and the law of disuse (Hergenhahn & Olson, 1993).

Operant conditioning Although Skinner's name is most widely connected with operant conditioning, Thorndike's work was also influential in the formulation of this concept of learning. Early experiments using cats in cages explored how the animals could learn to release themselves from the cage by pressing a lever. The incentive for this act was a bowl of food placed in view of the cat outside the cage. Learning was said to occur through the cat making the connection between the act of pressing the lever and the subsequent access to the positive reinforcement of the bowl of food (Kehoe & Macrae, 1999).

Skinner later refined these experiments using rats and pigeons in boxes and examined the connections between desired behaviour and rates of reinforcement. Skinner varied the rate of reinforcement — giving the animals food at fixed rates and at variable intervals. Basically, Skinner discovered that desired behaviour can be elicited even when reinforcement does not occur every time (Gilmer, 1970, pp. 302-303). It is generally agreed that Skinner is responsible for developing behaviourism into a fully fledged theory of learning (Merriam & Caffarella, 1999) and that "his views have had a direct impact on educational theory and teaching practice" (Tennant, 1997, p. 94).

Despite the different emphases placed upon various aspects of the learning process by a number of key theorists in this area, there are some important generalisations that can be made about the behaviourist orientation:

- *View of knowledge* — the behaviourist approach sees knowledge in terms of objective facts. Knowledge is truth and can be known and measured; it is independent of both teacher and learner, and it is the same for all learners. The implications of this belief are that there are clear right and wrong responses in the learning process (Rogers, 1998, p. 97). Therefore, the learning process will involve rewarding right or correct responses and discouraging wrong or incorrect responses.
- *Motivation to learn* — students will be motivated to learn if the learning task is pleasurable or rewarded.
- *Role of the teacher* — the teacher using this orientation is the arbiter of right and wrong responses. He or she knows the correct responses and has an active role in providing the stimulus to elicit the correct responses from the learners. It is, in essence, a teacher-centred approach.
- *Impact on design of teaching and learning* — the focus of this orientation in terms of designing teaching and learning is on structure and outcomes. As Pithers (1998) points out:

> This approach has placed stress on the importance of breaking tasks down into their component steps and specific behaviours and logically structuring them. It has resulted in a focus on structured training and the establishment of behavioural objectives: outcomes that are observable and measurable. It has renewed an emphasis on objective evaluation and assessment of outcomes. (p. 32)

As will be seen, the perceived shortcomings of behaviourist theorising gave rise to a number of alternative learning theories which sought to challenge the dominance that these ideas gained early in the twentieth century.

Current Status of the Behaviourist Orientation

The behaviourist orientation to teaching and learning has been subjected to a wide-ranging critique since its development and popularisation. One focus of the critical response was the way behaviourist theorists tried to link findings from experiments with animals to the behaviour of people. Behaviourists based their claims about the learning process on research conducted on laboratory animals. For many, this was seen as a limited research base and a limited view of the learning process. The notion of conditioned responses to controlled stimuli as the foundation of learning was seen as an insult to educators who saw their role as helping others to be creative and reach their full potential as feeling and thinking human beings (Russell, 1999, pp. 105-106). This critique led to greater interest in the humanist orientation discussed below. More recently, some education theorists have generally rejected the role psychology appears to play in guiding educational practice (Usher & Bryant, 1989; Usher, Bryant & Johnston, 1997). However, the issue of the relationship between theory (including psychological theory) and practice in adult education continues to be debated.

In spite of its limitations, the behaviourist orientation remains a powerful influence on many forms of education and training. As Pithers (1998) points out, Competency-Based Training, as a framework for designing and delivering vocational education and training, with its emphasis on structured programs of learning, behavioural objectives, the teaching of context specific competencies for specified workplace domains, and measurable assessment, has raised the profile of this orientation (Pithers, 1998, pp. 32-33).

Links to Current Practice

Competency-Based Training and behavioural objectives In the discussion above, Competency-Based Training (CBT) was linked to the behaviourist approach. Due to the importance of this framework in current vocational education and training practice, it requires further discussion. As mentioned above, CBT's links to the behaviourist approach can be seen most readily in three of its characterising features:

- focus on structured training;
- the establishment of behavioural objectives; and
- outcomes that are observable and measurable.

A useful question to pose here might be — what are the perceived advantages of such an approach to teaching and learning? One answer to this question concerns the use of behavioural objectives or learning outcomes. Pithers' (1998) inquiry into these aspects of CBT has led him to the following conclusions:

> A clear, unambiguous, planned outcome, especially if it can be specified in a behavioural sense, means that observation of the learner's progress is possible (Fletcher, 1991). What can be observed can usually be measured so that the trainer or assessor can actually measure the attainment of the outcome(s) in an objective manner (Gronlund, 1995). The need for objective, measurable assessment of the changes brought about by learning and of their duration, was what, in part, led to the historical development and adherence to behavioural objectives in training. Behavioural objectives and learning outcomes require that the learning task be analysed so that its key component knowledge and skills can be described. (Pithers, 1998, p. 83)

Thus learning, seen in terms of behaviour change (or skill acquisition in vocational terms) can be clearly assessed and objectively measured. The fact that this system is linked to a framework of qualifications which indicate that a learner is competent to perform work-related activities means that issues of quality control and uniformity of training and assessment are important to this framework. To reiterate Russell's (1999) claim, noted in the overview of this orientation, accountability and quality control are seen as advantageous qualities within an education and training framework in an economic rationalist climate (p. 105). However, when the explanation of the learning process is based solely on observable and measurable behavioural changes, its focus could be too narrowly limited to lower level skills. As a result, higher order competencies, such as the ability to solve problems, to work effectively in teams, to think critically, reflectively and flexibly — which are not easily quantifiable — could be overlooked.

What Does Behaviourism Offer to Instructors in the Aviation Context?

Key understandings gained from behaviourist theories might provide some useful guidelines for aviation instructors in a number of ways. In particular, the findings of behaviourism suggest some practical instructional techniques, such as feedback, the role of practice in learning, and the importance of engaging learners during the learning process.

Feedback Understanding the importance of feedback in the learning process can be traced directly back to the early work of Thorndike and his Law of Effect (outlined above). Originally, Thorndike's experiments were carried out using food as a reward; however, development of his work has seen pieces of information, in the form of feedback, given to learners as the reward. This feedback can be "critical feedback" which indicates the success or failure of the learner's actions or "instructive feedback" which contains the desired response or answer (Kehoe & Macrae, 1999, pp. 214-215). This work provides the basis, therefore, of today's widely used teaching technique of providing feedback and correctives to learners. Indeed, the importance of providing good feedback for learners in a training situation cannot be overstated. A good example of a very specific type of feedback used in some aviation training will illustrate its importance to aviation trainers. This type of feedback is known as knowledge of results feedback and is described by Kehoe and Macrae (1999) below:

> In some learning tasks, it can be difficult for learners to perceive their own actions and/or their outcomes.... If learners are uncertain as to what they have done, it is difficult to benefit from either critical or instructive feedback. However, steps can be taken to help the learner by using what has been called either knowledge of results or augmented feedback (Winstein and Schmidt, 1990).... Where computers are used to teach tracking tasks, for example, flight simulation, knowledge of results can be augmented on the display by showing a trace of the learner's actions. (Kehoe & Macrae, 1999, p. 216)

The role of practice Behaviourist learning experiments and theories also have some important information for trainers regarding the role of practice in learning. As outlined above, with the Law of Exercise, Thorndike was trying to examine the link between the amount of times a stimulus-response learning scenario is repeated and the effectiveness of learning. Although Thorndike felt unable to state directly that there was a clear link between a learned connection and repetition, he did maintain that practice led to improvement in the effectiveness of learning. In fact, subsequent behaviourist experiments have shown that the role of repetitive practice, often exemplified in training scenarios by drill activities, is more complex than at first apparent.

> Learning is often assumed to be a straightforward process in which "practice makes perfect". This popular saying implies that learning occurs at a steady pace over practice sessions, or in other words, each practice session provides an equal increase in learning. In fact, even on fairly simple tasks, learning typically starts out as a slow, laborious process that becomes progressively faster across practice sessions. If one

> were to plot the course of this learning on a graph, it would follow as a S-shaped curve....
>
> The important implication of the S-shaped learning curve is that "patience makes perfect". It is not simply practice, but also patience, that is required by the learner and teacher. This is true for many relatively simple tasks, and it is certainly true for more complex tasks. (Kehoe & Macrae, 1999, pp. 212-213)

Engaging learners Current behaviourist-influenced research also provides useful information to trainers about how best to maximise the learning potential of trainees. In addition to providing feedback to reward learners in the learning context, it also appears to be very important to actively engage learners' attention during the learning process. This finding may sound like a common sense approach to training activities, however, many trainers in vocational contexts make too many assumptions about learner engagement levels. In the aviation training context, for example, it may be assumed that because trainee pilots are paying large amounts of money for training, they will automatically be engaged and highly motivated. In cabin crew safety training, it might be assumed that participants will be actively engaged because of the potentially life-saving impact of the learning. While this might be true, behaviourist research has shown that engaging learners involves more than making assumptions about motivation. Engagement of learners has more to do with making the connection between the learner and what is to be learned more vivid for the participants. In experiments where participants were asked to memorise a set of procedures, participants who had been told to generate their own explanations for each procedure demonstrated a much higher rate of retention than participants who were simply told to memorise the procedures or even those who had been provided explanations (Naikar, 1992 cited in Kehoe & Macrae, 1999). "Hence, by being asked to generate an explanation, the learners appeared to have attended more closely to the relevant material and thereby established stronger associations" (Kehoe & Macrae, 1999, p. 219).

In the aviation context, large amounts of complex content often have to be transmitted to trainees and many procedural drills have to be memorised. Too often it seems trainers give little thought to the most effective ways to prepare trainees for this type of learning. Current behaviourist research gives clear indications of the importance of this in the preparation of training sessions (see O'Hara, 1990).

Cognitive Orientation

Overview

Cognitivist learning theories differ from behaviourist theories in that they are concerned with the mental processes of learning. As its name implies, the cognitive orientation to teaching and learning has *thought* as its focus. Whereas behaviourist researchers are interested in the observable outcomes of learning (demonstrated by changed behaviour), the "major emphasis of cognitive approaches concerns the ways information is processed and stored" (Lefrancois, 1997, p. 157).

This approach gained its important place in educational theorising in the 1960s. Despite the fact that behaviourist approaches certainly dominated education and training at that time, and humanist psychology (discussed below) was beginning to be given more widespread consideration, cognitive psychology's explanations of human responses in learning contexts was gaining ground. It actively challenged behaviourism's claims on the grounds that cognitive psychology did not use animals in experimentation and was, therefore, more likely to make relevant claims about people, and its approach was seen as more scientifically verifiable than humanist theorising (Russell, 1999, pp. 106-107). Indeed, one of the outcomes of the cognitive revolution was a greater emphasis on human cognition and a greater acceptance of research in natural settings. Furthermore, from the 1950s onwards, the development of computers increased general interest in how people process information and, as a result, raised new questions about learning (Foley, 1995, p. 38). Cognitive psychology was seen as able to provide answers to such questions.

Like the behaviourist orientation, the cognitive approach had its origins in research dating back to the early twentieth century. As Foley (1995) notes, "studies by German Gestalt psychologists early this century showed that people, constantly confronted as they are with a mass of stimuli, impose order on it in predictable ways. In other words, people have cognitive frameworks that help them to make sense of the world" (Mayer, 1981 cited in Foley, 1995, p. 38).

There are a number of important implications for approaching learning from the cognitive orientation. First of all, this approach represents a significant shift in the focus of education and training. As Merriam and Caffarella (1999) note, this approach heralds the "shift to the individual" (p. 214). Instead of the focus being upon the content or subject matter to be learned or on the effectiveness of transmission techniques carried out by the trainer, the focus here is on the individual learner and the way he or she responds to the activities in the learning context (p. 214).

Secondly, the learner here is active. Learners "do not passively absorb information but process it in their own ways. It follows from this that it is important for teachers to understand how students think and learn" (Foley, 1995, p. 38). This interest in the processing and storage of information leads cognitivist theorists to ask questions about and consider the importance of:

- paying attention to what is to be learned;
- understanding the relationship between the information being presented and what is already known;
- understanding how learning occurs;
- controlling the rate and quality of learning; and
- being aware that learning has taken place (Ashman & Conway, 1993, p. 33).

Influential Theorists and Their Contributions

A number of influential theorists working within the cognitivist framework have attempted to show links between cognitive activity and effective teaching strategies (Merriam & Caffarella, 1999). For example, Ausubel's interest in how we retain learning led him to explore the nature of meaningful learning as compared to rote learning. Earlier in the century, rote learning was commonly used as a classroom technique, especially with school children. Ausubel's work in the 1960s allowed him to explore this type of learning and led him to the conclusion that "learning is meaningful only when it can be related to concepts that already exist in a person's cognitive structure. Rote learning, on the other hand, does not become linked to a person's cognitive structure and hence is easily forgotten" (Merriam & Caffarella, 1999, p. 254).

Gagné's contribution grew out of his interest in what he considered to be the different types of learning. His work led to the proposition that there are five general classes of learning:

1. intellectual skills,
2. cognitive strategies,
3. information,
4. motor skills, and
5. attitudes (Gagné, Briggs, & Wager, 1988, p. 54).

The implication of this proposition is that for each type of learning, there are corresponding instructional techniques best suited to that type of learning. Several of these key techniques are examined in Part 2 of this book. For example, in Chapter 8, the central role of *demonstration* as a technique for teaching psychomotor skills is discussed. Also, in Chapters 8 and 10, the importance of reflective practice and the value of problem-based learning are explored for their usefulness in the development of critical intellectual skills and attitude changes.

Links to Practice

According to Foley (1995), cognitive approaches encourage educators to shift their attention from developing direct teaching techniques to understanding how learners learn. Once this is understood, the role of the teacher is to facilitate that learning through teaching, course design and encouragement of learners to understand and develop their own learning strategies (Foley, 1995, pp. 40-41).

There is a vast amount of information and ideas for teachers and trainers stemming from this broad orientation. For example, the cognitive orientation is responsible for the research done on:

- learning how to learn strategies (see Chapter 5);
- learning styles (see Chapter 4 for a discussion on Kolb — probably the most influential theorist in this area);
- use of mind maps; and
- instructional design (see Chapter 7).

In summary, the following generalisations can be made about the cognitive orientation:

- *View of knowledge* — cognitivists believe that learners respond to a learning situation according to their own perceptions and interpretation based on their experience. Further, they consider that the internal perceptions of learners are more important than the external response. Therefore, they stress that learning is not limited to what is observable.
- *Motivation to learn* — students will be motivated to learn because of their natural tendency to resolve ambiguity and to solve problems. The motivation to learn stems from the learners themselves. Therefore, the learner's perception of the learning situation is essential.

- *Role of the teacher* — the teacher assists learning by explaining very clearly why the learning task is important (providing a rationale), by enlarging the number and kinds of problems the students need to solve and by encouraging students to develop their own learning strategies. It is essentially a student-centred approach.
- *Impact on design of teaching and learning* — the focus of this approach in terms of designing teaching and learning is on linking new concepts or skills to the student's previous knowledge and experience.

What Does Cognitivism Offer to Instructors in the Aviation Context?

As stated above, workers in the aviation environment are often highly skilled professionals who are required to have a large body of knowledge ready for application in a range of contexts. Trainers with an understanding of cognitive processes, such as memory and information retrieval, learning strategies and learning styles, are better equipped to assist trainees in the type of preparation needed to function in highly skilled and demanding jobs. The importance of understanding learning styles and learning strategies is examined in Chapters 4 and 5; however, some key information about teaching for effective memory retrieval will be explored here.

Memory Most trainers expect participants in training to remember what is covered in each training session. More importantly perhaps, trainers in an aviation environment want their trainees to be able to retrieve, at a later date in the workplace, the skills and information they have learned. However, few trainers give much consideration to *how* the trainees are going to do this often difficult task. Cognitive researchers provide some interesting insights into how the memory works and how trainers can utilise this information in their training sessions.

In summarising the research in this domain, Ashman and Conway (1993) note that researchers have claimed that memory has four components:

- learning — how we gain new information and skills;
- retention — how we store the new knowledge;
- recall — how we remember the knowledge when we need it; and
- recognition — how we determine which information is needed and when (p. 34).

Research has also shown that there are some simple training techniques that can be utilised to make sure trainees maximise learning to facilitate the operation of each of these components of memory. For example,

> … research on long term memory seems to indicate that most of what is forgotten is lost very soon after it is learned and that, by the same token, things that we succeed in remembering days or weeks later may well continue to be available even years afterward. To the extent that this is true, it presents an important challenge for teachers … to teach for long term memory. (Lefrancois, 1997, p. 169)

Clearly this has important implications for trainers who wish to train for long-term memory storage and effective retrieval. Research has shown that key factors in teaching for memory and retrieval include:

- Material that is meaningful to trainees is easier for them to remember. This means that it is important to make links for learners between the knowledge and skills being taught and the job role for which they are training. It should be remembered that trainers understand the "big picture" of the job or training environment, however, the training context usually offers participants only small parts of the picture at one time. This means that they may not have a clear understanding of how the skill or information being taught fits into the larger picture. Therefore, it may not always have immediate meaning for them. It is the role of the teacher to make that link very clear to learners.
- Well-organised material is learned more easily and remembered for longer periods of time. Once again, trainers must think about how the order of material might appear to trainees who do not have the big picture. Sequencing ideas and objectives is an important planning task that is covered in detail in Chapter 7.
- Events that are particularly striking tend to be recalled more easily and more clearly.
- Frequent rehearsal improves long-term recall; and
- Well-utilised, appropriate visual material (instructional aids) has a greater impact on memory than verbal material for many learners (adapted from Lefrancois, 1997, p. 171).

Humanist Orientation

Overview

The humanist orientation came to the fore in the 1970s when philosophers of education and other theorists began to write about the importance of self-autonomy in learning (Pithers, 1998), and about the individual being able to pursue his or her own interests (Maslow, 1970). They stressed the importance of personal freedom in learning (e.g., Rogers, 1969, 1983). Clearly this approach marks a radical departure from both the behaviourist and the cognitive orientations, but it too was a product of historical forces.

It is widely known that the 1960s was a period of great social and intellectual upheaval in many parts of the world. During this period, many questions began to be asked about the true claims of empirical science and about the notion of objectivity in scientific research. In response to these questions, a growing sense of "complexity, uncertainty, instability, the uniqueness of individual response and the conflicts of values" (Schein, 1972 cited in Rogers, 1998, p. 99) pervaded the theorising of human activity.

Once again, the focus of teaching and learning in the humanist orientation can be seen to be quite different from the previous two orientations discussed. Whereas the behaviourist approach focuses on techniques used by the trainer to elicit a change in behaviour and the cognitive approach designs instruction with its focus on the cognitive structures of individual learners, the humanist approach contends that the material on which adult learners exercise their learning skills is less important than the goals that they have set for themselves (Rogers, 1998, p. 100). The individual with his or her own needs is placed at the centre of this orientation rather than curriculum content. Furthermore, the concept of need is broadly conceptualised since a holistic view of the learner and his or her learning needs is taken.

The goals of learning need not be technical learning, skill acquisition or information based. Often these goals of learning are quite vague or broad, for example, personal growth. Thus, the humanist orientation is seen as a student-centred rather than a teacher-centred or subject-centred approach. This is not to say that the teacher has no role in this orientation. The humanist teacher's role is to be warm, empathetic and open, in other words, the role of the teacher is seen as that of a facilitator. These qualities then allow the teacher to put into place favourable conditions where personal learning can flourish (Pithers, 1998, p. 54).

In summary, Pithers (1998) lists the important factors practitioners need to consider about learners when working within this orientation:

- Needs and expectations — the learner's preferences are taken into account;
- Type and level of the learner's past experience;
- The need to facilitate self-direction and autonomy;
- Development of self-evaluation;
- The acquisition of content that has personal meaning;
- Freedom from fear and threat; and
- The provision of a psychological environment that is comfortable and secure (p. 55).

Influential Theorists and Their Contributions

Rogers Without doubt, the most influential humanist theorist is Carl Rogers. In his 1951 book, *Client-Centred Therapy*, Rogers elaborated his theoretical position by outlining nineteen propositions about the human personality. As Lefrancois (1997) notes, gaining an understanding of these principles is crucial for seeing how humanism has impacted upon education and training practices since the 1960s. Six of Rogers' key concepts about personality, as summarised by Lefrancois, are listed below in Table 1.1.

Given these beliefs about the human personality, Rogers had corresponding ideas about the best way to design and implement adult learning activities. According to Rogers (1983), learning for human growth and self-actualisation should have the following characteristics:

1. *Personal involvement*: the affective and cognitive aspects of a person should be involved in the learning event.
2. *Self-initiated*: a sense of discovery must come from within.
3. *Pervasive*: the learning "makes a difference in the behaviour, the attitudes, perhaps even the personality of the learner".
4. *Evaluated by the learner*: the learner can best determine whether the experience is meeting a need.
5. *Essence is meaning*: when experiential learning takes place, its meaning to the learner becomes incorporated into the total experience (p. 20).

Table 1.1 Major characteristics of human personality

Principle	Clarification
1. Our worlds are private; reality is phenomenological.	The significant aspects of reality consist of the world of private experience. Our realities are therefore completely individualistic. They can be *intuited* but not *known* by others.
2. Our reality is immediate, personal consciousness.	Our private experience determines our reality. Our phenomenal field makes up our immediate consciousness.
3. The goal of human existence is self-actualisation.	Each of us has a basic tendency to strive toward becoming a complete, healthy, competent individual through a process characterised by self-government, self-regulation, and autonomy.
4. Behaviour can be understood only from the individual's point of view.	Because behaviour occurs within the context of personal realities, the best way to understand a person's behaviour is by attempting to adopt his or her point of view; hence, humanism emphasises the importance of open communication.
5. We construct our own selves.	We discover who we are on the basis of beliefs and values that we incorporate in our own self-concepts from information provided by people who communicate to us what we are.
6. Behaviours are consistent with notions of self.	In general, we select behaviours that do not contradict who and what we think we are.

Source: Rogers, 1951 cited in Lefrancois, 1997, p. 318.

Knowles — Perhaps the most well known influence of humanism on the education and training of adults has been through the work of Malcolm Knowles. Many of Knowles' (1970) assumptions about the adult learner can be traced back to humanistic psychology. Similarly, one of the key teaching techniques he recommends for use with adult learners, self-directed learning through the use of learning contracts, can be linked to humanism's belief in the uniqueness of each learner and each learner's right to explore his or her individual growth through a learning process centred on one's individual needs (see Chapter 2 for a discussion of Knowles' assumptions about adult learners). Briefly, a few generalisations that can be made regarding the humanist orientation are:

- *View of knowledge* — humanists believe that learning is achieved only if the learner's needs are considered first. Learning is internal to the learner. They also believe that humans have a basic tendency to strive for actualisation, that is, to enhance themselves. Thus, the learning process is seen as a personal act to fulfil one's potential.
- *Motivation to learn* — students will be motivated to learn things which will contribute to their enhancement as autonomous individuals. Adult learners deliberately choose to determine their own direction for their learning. However, adult learners may seek or need advice from others but they prefer to retain responsibility for their learning.
- *Role of the teacher* — the teacher facilitates the learning process and the development of the whole person. Teachers are useful resources who can provide skills and knowledge.
- *Impact on design of teaching and learning* — the focus of this approach is clearly on the learner's individual needs, therefore teachers need to design learning experiences which encourage learners to explore new facts and experiences at their own pace, hence the emphasis on self-directed learning.

What Does Humanism Offer to Instructors in the Aviation Context

A common question among vocational educators and trainers concerns the relevance of the humanist orientation in workplace contexts or in a structured training environment, such as the aviation training context. In such contexts, although learner self-development may be seen as an attractive and even a valuable by-product of training, efficiency and measurable outcomes will always be seen as the essential characteristics of teaching and learning. Pithers (1998) suggests that careful adaptation of the humanist principles as well as a realistic assessment of their level of usage is a reasonable response to using this orientation in vocational education and training.

To effectively implement a humanist approach in vocational training, Pithers (1998) offers the following strategies:

- find out about and work with student needs, desires and expectations;
- establish a warm, friendly, positive environment in which the views of individuals are important and equitably considered;
- be aware of your own controlling teaching style;
- do not neglect the power of facilitation and delegation;
- provide feedback (positive) for people without becoming familiar;

- let people know honest feelings about process and events;
- encourage the development of the learner's self-esteem or confidence, self-direction and autonomy; and
- provide a comfortable physical and social environment, free from fear, in which attitudes and values are considered as well as knowledge and skills (p. 57, also see Chapter 5 for a list of teacher behaviours and attitudes that students find either enhance or hamper learning).

In addition, the humanistic emphasis on the uniqueness of each learner means that "humanistic teachers are especially sensitive to diversity in their classrooms" (Lefrancois, 1997, p. 315) and are culturally sensitive. These are clearly important characteristics of any training context but have particular importance in aviation training contexts whose global span sees personnel and trainers operating in diverse cultural settings and with people of different cultural backgrounds.

Links to Practice

Many of the humanist principles are contained in two of the most widely utilised concepts in adult education, namely the andragogical approach and its related technique of self-directed learning. Both of these ideas and their application to the aviation training context are discussed in Chapter 2 of this book.

Social Learning Orientation

Overview

One of the interesting contributions that the social learning orientation makes to the range of theories on how people learn, and more specifically on how adults learn, is that it recognises the vast array of places in which learning occurs. As the name of the orientation suggests, it sees learning as a social interaction, and as such, this type of interaction is not restricted to a classroom or even any other recognised learning context. In fact, proponents of the social learning approach contend "that much important learning occurs in settings such as workplaces, in shops and everyday activities of a social nature" (Cornford, 1999, p. 73).

Social learning theorists, therefore, focus on the enormous amount of learning that occurs in our daily lives through interactions with those around us. Often we observe other people and model what they do (e.g., a first officer observing the performance of an experienced captain). This is the essence of what is known as social learning (Cornford, 1999, p. 73). In other words, according to social learning theory, while we may learn from receiving rewards and punishments, we also learn from the example of others. In this way, social learning theory emphasises both the person and the surrounding social environment as sources of learning (Saunders, 1999, p. 33). Therefore, it can be seen that the three fundamental ingredients of social learning are:

- the individual;
- the behaviour being observed; and
- the environment.

Each of these variables is seen to contribute to the individual's cognition and behaviour. "In this view of learning and behaviour, modelling is of very great importance and seen as an environmental factor" (Saunders, 1999, p. 77). Cornford (1999) summarises the main points of Bandura's (1977, 1986, 1997) Observational Learning Model that indicates that learning occurs:

> … when an individual sees a model performing some type of skill or action, attempts to learn from this situation and tries to reproduce the modelled behaviour. The model observed may be a real life model, but equally in an information age the model may be a symbolic model. That is to say, the model may be in cartoon form or represented by line drawings… In an age in which television is in almost every home there can be considerable learning from movies or television series which may involve fictional characters as well as real ones. The radio, another common form of media, also provides models which may be imitated. There is considerable evidence that media models have considerable impact … (Cornford, 1999, p. 81)

However, the link between observed modelling behaviour and performance of that behaviour by a learner is not always direct. Other factors may inhibit the learner from performing a particular modelled behaviour. Whether or not this behaviour is performed depends upon our perceptions about the behaviour and its consequences.

Briefly, a few generalisations that can be made regarding social learning are:

- *Motivation to learn* — We are motivated to learn when we see modelled behaviour that appeals to us and accordingly, "some models are more likely than others to be attended to, such as those thought to be competent, powerful, attractive, and so on" (Merriam & Caffarella, 1999, p. 259).
- *Role of the teacher* — In a traditional classroom setting, the teacher can model desirable behaviour for students to attend to and potentially follow. In the workplace, the teacher/model may be given the role of mentor (discussed below). However, given the basic premises of this orientation and its broad view of learning, we are surrounded by "teachers" in our daily lives whose "lessons" we may choose to follow.
- *Impact on design of teaching and learning* — Clearly this orientation supports the idea that much of the learning that we do is *not* achieved by design — it happens as we interact with others in our environment. Having said that, this orientation can be used to support approaches to the design of learning experiences that rely upon interactions with others in an environment, which elicit specific desired behaviours, such as those needed within the workplace.

Links to Practice

Social learning orientation research has resulted in the large amount of information currently available on the technique known as mentoring. According to Cohen (1995), "mentoring offers much to aviation professionals who are committed to assisting adult learners in their pursuit of educational, training, and career goals" (p. 1). This technique has proven popular in recent years in the areas of education and training as well as management and human resource development. Undeniably, mentoring has great value in the development of air traffic controllers, flight attendants, flight instructors, maintenance engineers, and both military and airline pilots. Mentoring is discussed in more detail in Chapter 2 of this book.

Constructivist Orientation

Overview

Constructivism as an approach to teaching and learning has its basis in cognitive science (Blunden, 1997, p. 148). Like the cognitivist orientation described previously, constructivism concludes that there is an innate human drive to make sense of the world. According to this framework for understanding learning, learners actively construct knowledge by integrating new information and experiences into what they have previously come to understand, revising and reinterpreting old knowledge in order to reconcile it with the new (Billett, 1996; Kerka, 1997). Clearly the role of the learner in this process is an active one. In further defining the elements of the constructivist orientation, Blunden (1997) states that:

> the two terms which need to be kept in mind in thinking about constructivist theories of learning are, "construction" and "interaction". Constructivists place different kinds of emphasis on these two influences — one internal and one involving interaction between an individual and the environmental conditions... The term "construction" clearly implies that the individual is building — or constructing — their knowledge through internal mechanisms — cognitive schemes, perhaps ... (p. 148)

Thus, constructivists see the importance of the individual's internal processes of constructing knowledge, but also take into important consideration the environment in which this construction takes place. This notion of the impact of the environment is important to understanding this orientation in terms of vocational training since particular areas of work and specific workplaces may have sets of practices, attitudes, beliefs and activities which impact upon the ways learners and workers construct knowledge about their activities. However, there is no agreement among theorists about which of these processes have the greatest impact on the construction of knowledge (Blunden, 1997).

Briefly, a few generalisations that can be made regarding constructivism are:

- *View of knowledge* — knowledge is actively constructed by the learner in a process whereby knowledge is situated. Learners construct meaning from experience through reflective practice (experiential learning).

- *Motivation to learn* — similar to the cognitivist view, constructivists see learning as the result of the innate drive of the individual to make sense of the world (self-directed learning).
- *Role of the teacher* — given that the constructivist orientation theorises the activities of the learner and not those of the teacher, what can be the role of the teacher within this framework? Kerka (1997) suggests that "teachers facilitate learning by encouraging active inquiry, guiding learners to question their tacit assumptions, and coaching them in the construction process.... A constructivist teacher is more interested in uncovering meanings than in covering prescribed material...." (pp. 1-3). The teacher is seen as a coach who provides guidance that gradually diminishes as learners become more competent.
- *Impact on design of teaching and learning* — Kerka (1997) suggests that strategies, such as cognitive apprenticeship and coaching (discussed in Chapter 2) reflect key tenets of this orientation to learning. Therefore, designing teaching and learning from a constructivist perspective involves the creation of a learning environment that reproduces "the key aspects of communities of practice: authentic activities sequenced in complexity, multiple experiences and examples of knowledge application, access to experts, and a social context in which learners collaborate on knowledge construction" (pp. 1-3).

What Does Constructivism Offer to Instructors in the Aviation Context?

The constructivist type of theorising has uses for vocational teachers since a range of research into workplace learning draws upon this orientation. Billett (1996) is one such researcher who is specifically interested in the type of learning that occurs in a workplace environment. This research focuses on what is known as *situated learning*, attempts to understand how workers learn and, therefore, to suggest ways to improve workplace learning. Obviously, the constructivist orientation has definite applications to the aviation learning environment since most aviation personnel are involved in on-the-job training (see Chapter 14 for a review of on-the-job training in air traffic control). It is easy to see the Crew Resource Management (CRM) trainer, or a training captain or a supervising air traffic controller in the role of a coach, who guides trainees and challenges them to question their decisions and their assumptions.

Conclusion

This chapter introduced some of the major learning theories: behaviourist, cognitivist, humanist, social learning and constructivist. Applications to the aviation learning context were made. However, in relation to the learning theories discussed, it is useful to keep in mind Lefrancois' (1997) point that "most educators do not fall neatly into the humanist, the behaviourist, or the cognitive camps. Most are quite eclectic; they borrow from here and there" (p. 315).

Reflective Exercise

1. How does an understanding of learning theories assist trainers in the aviation context?
2. In what ways do the five theories of learning impact upon teaching and learning in your teaching context?
3. What do you see as the advantages and limitations of each orientation with reference to your teaching environment?

References

Ashman, A.F., & Conway, R.N.F. (1993). *Using cognitive methods in the classroom.* London: Routledge.

Bandura, A. (1977). *Social learning theory.* Englewood Cliffs, NJ: Prentice Hall.

Bandura, A. (1986). *Social foundations of thought and action: A social cognitive theory.* Englewood Cliffs, NJ: Prentice Hall.

Bandura, A. (1997). *Self-efficacy.* New York: W.H. Freeman.

Billett, S. (1996). Towards a model of workplace learning: The learning curriculum. *Studies in Continuing Education 18*(1), 43-58.

Blunden, R. (Ed.). (1997). *Teaching and learning in vocational education and training.* Katoomba, Australia: Social Science Press.

Cohen, N.H. (1995). *Mentoring adult learners: A guide for educators and trainers.* Malabar, FL: Krieger.

Cornford, I. (1999). Social Learning. In J.A. Athanasou (Ed.), *Adult educational psychology* (pp. 73-96). Katoomba, Australia: Social Science Press.

Foley, G. (1995). Teaching adults. In G. Foley (Ed.), *Understanding adult education and training* (pp. 31-53). Sydney, Australia: Allen & Unwin.

Gagné, R.M., Briggs, L.J., & Wager, W.W. (1988). *Principles of instructional design* (3rd ed.). New York: Holt, Rinehart & Winston.

Gilmer, B. (1970). *Psychology* (2nd ed.). New York: Harper & Rowe.
Hergenhahn, B.R., & Olson, M.H. (1993). *An introduction to theories of learning* (4th ed.). Englewood Cliffs, NJ: Prentice-Hall.
Kehoe, E.J., & Macrae, M. (1999). Modern associative theory in teaching and learning. In J.A. Athanasou (Ed.), *Adult educational psychology* (pp. 211-235). Katoomba, Australia: Social Science Press.
Kerka, S. (1997) *Constructivism, workplace learning, and vocational education.* ERIC Digest No. 181. Columbus, OH: ERIC Clearinghouse on Adult, Career, and Vocational Education.
Knowles, M.S. (1970). *The modern practice of adult education: Andragogy versus pedagogy*. New York: Association Press.
Lefrancois, G.R. (1997). *Psychology for teaching* (9th ed.). Belmont, CA: Wadsworth.
Maslow, A.H. (1970). *Motivation and personality* (2nd ed.). New York: Harper-Collins.
Merriam, S.B., & Caffarella, R. (1999). *Learning in adulthood.* San Francisco: Jossey-Bass.
O'Hara, J.M. (1990). The retention of skills acquired through simulator-based training. *Ergonomics, 33,* 1143-1153.
Pithers, R.T. (1998). *Learning through effective training.* Katoomba, Australia: Social Science Press.
Rogers, A. (1998). *Teaching adults* (2nd ed.). Buckingham: Open University Press.
Rogers, C.R. (1969). *Freedom to learn.* Columbus, OH: Charles E. Merrill.
Rogers, C.R. (1983). *Freedom to learn for the 80s*. Columbus, OH: Charles E. Merrill.
Russell, P. (1999). Motivation. In J.A. Athanasou (Ed.), *Adult educational psychology* (pp. 97-127). Katoomba, Australia: Social Science Press.
Saunders, S. (1999). Social psychology of adult learning. In J.A. Athanasou (Ed.), *Adult educational psychology* (pp. 25-71). Katoomba, Australia: Social Science Press.
Tennant, M. (1997). *Psychology and adult learning* (2nd ed.). London: Routledge.
Usher, R., & Bryant, I. (1989). *Adult education as theory, practice and research: The captive triangle.* London: Routledge.
Usher, R., Bryant, I., & Johnston, R. (1997). *Adult education and the postmodern challenge.* London: Routledge.

2 Characteristics of Adult Learners and Adult Learning Principles: Implications for Aviation Education

Jayne Bye and Irene Henley

Introduction

In Chapter 1, five key orientations to research into learning were examined and the role of theory in understanding the nature of learning was explored in some detail. Chapter 2 focuses more specifically on adult learning theories. The literature on adult learning is vast and covers a wide variety of fields, including perspectives from education, psychology and sociology. In attempting to provide a broad coverage of these perspectives, the first part of this chapter concentrates mainly on the characteristics of adult learners and on adult learning principles taken primarily from researchers within the humanist orientation, which emphasises the importance of lifelong learning.

The importance of looking at the practical implications of theory is also explored. Firstly, the concept and practice of mentoring, which grows out of the social learning perspective, is discussed. Key definitions of this process are outlined and the role of the mentor is examined. Next, an extension of the constructivist tradition is presented through an exploration of current research into learning in the workplace. This section brings to the foreground the importance of considering the specific nature of the context in which aviation trainers work. The benefits and limitations of the workplace as a learning context are also critically explored. Finally, the concepts of metacognition and self-regulation are examined as useful elaborations of the cognitivist perspective with implications for teaching and learning in the aviation environment.

Characteristics of Adult Learners

Characteristics of adult learners are often used as the basis for guiding instructional practices. Attributes such as far-sightedness, perspective, maturity, self-control, security, experience and autonomy are often quoted as being fundamental to adulthood and hence suggest ways in which teachers and trainers might view their participants. However, Rogers (1998) notes that these attributes may be interpreted differently among individuals as well as having different meanings in different cultural contexts. He also notes that, despite the fact that many people whom society recognises as having adult status do not display these qualities in their daily lives, these attributes continue to be widely used to define adulthood. Indeed, educational theorists have applied many of these general attributes of adulthood to describe the adult learner.

The Research – What Generalisations Have Been Made?

A vast array of research has been carried out in the area of adult learning in an attempt to provide information to guide adult educators in their practice. The following discussion can attempt to present only the major perspectives covered in this vast literature.

Developmental and Psychological Perspectives

As mentioned in Chapter 1, undoubtedly one of the most influential theorists on the subject of adult learner characteristics has been Malcolm Knowles, an adult educator from the United States. In *The modern practice of adult education: Andragogy versus pedagogy,* Knowles (1970) puts forward the notion of "andragogy" which he describes as "the art and science of helping adults learn" (p. 38).

Knowles (1980) originally based his guidelines for practice on four basic assumptions (he later added a fifth) about the characteristics of adult learners:

1. Their self-concept moves from dependency to independency or self-directedness.
2. They accumulate a reservoir of experience, which is a rich resource and can be used as a basis on which to build future learning.

3. Their readiness to learn is closely associated with the developmental tasks of their social roles.
4. Their time perspective changes from postponed to immediacy of application of knowledge and from subject centredness to problem centredness (pp. 44-45).
5. "Adults are motivated to learn by internal factors rather than external ones" (Knowles and Associates, 1984, pp. 9-12, cited in Merriam & Caffarella, 1999, p. 272).

Knowles' emphasis here is clearly on the importance of the active role taken by adult learners. He aligns this with developmental characteristics of adults, such as the desire to take responsibility for important aspects of their lives, including learning. Knowles (1984, 1990) advocates self-directed learning as the obvious teaching and learning mode with adults because of these characteristics.

However, Knowles' conception of the adult learner has been critiqued by other educationalists (Brookfield, 1986; Merriam, 1993). Merriam, for instance, notes that close scrutiny of these characteristics reveals that they do not describe the full range of adult learners. For example, if adults have no prior experience of an area of knowledge or set of skills, they rarely desire to take complete control of their learning or be self-directing, at least in the first instance. This is certainly true in the initial stages of flight training. In such cases, adults will undoubtedly benefit from direct instruction. Knowles (1984, 1990), due to this scrutiny, reviewed a number of his prescriptions for practice in later editions of his work. However, his work is still very widely utilised and the concept of andragogy and the use of self-directed learning are seen as important contributions to thinking about the characteristics of adult learners and their implications for how teachers and trainers of adults approach their work.

Sociological Perspective

Many studies of the adult learner population have tried to describe participant characteristics in terms of their gender, age, socio-economic position, educational background, kind of employment and marital status. These types of surveys generate information that may be useful for teachers and trainers to consider when they are designing and planning programs for adult learners.

Brookfield (1986) reviews the literature that has, over the years, attempted to describe the characteristics of the "typical" adult learner and states that the generalisation that most research agrees upon is "that the typical adult learner is a relatively affluent, well-educated, white, male, middle-class individual" (p. 5). Yet adult learners who do not fit into this description of the "typical adult learner" are gradually becoming more and more prevalent in adult learning classes and certainly are present in the workplace learning environment.

Integrating the Perspectives

Galbraith (1991) takes an integrated approach to describing the characteristics of the adult learner. In doing so, he focuses on, and celebrates, the range of people we put under the heading of adult learners.

> It is the multifaceted physiological, psychological, sociological, and developmental aspects of the adult learners that contribute to and make a challenging and enriching educational encounter. The various dimensions or characteristics of the adult learner are interrelated, thus one dimension of the individual is affected by every other dimension... Krupp (1982) suggests that understanding this multifaceted characterisation is to understand the adult learner from a holistic perspective. This seems most appropriate, especially when working with adults who have diverse reasons for participating, are at various stages in their adult development, and who possess a multitude of learning styles. (p. 18)

Given the difficulties and pitfalls of definitively describing the characteristics of this vast group of people, Galbraith's approach makes a good deal of sense. Indeed, any attempt to definitively identify the characteristics of adult learners will be necessarily flawed for a number of reasons. The first and most important reason is that trying to describe such a vast group of learners must lead to generalisations so broad as to be barely useful. For example, Jarvis (1995) cites research that suggests that over 90 per cent of adults undertake learning projects (Peters & Gorden, 1974; Tough, 1979). While these statistics may be a product of particular sampling techniques and a very wide definition of "learning projects", it seems that the majority of adults do indeed undertake some type of learning project during their lifetime. Therefore, attempting to describe the characteristics of the general population of adult learners will probably not yield much in the way of useful information for teachers and trainers.

Another problem with generalisations is the tendency for what Rogers (1998) describes as myths about the adult student population to emerge and prevail. For example, a common myth is that all adult learners are highly motivated, organised and goal-oriented in relation to their learning. Yet, it will be seen in later discussion in Chapter 6, that many adult learners feel unsure and highly emotionally challenged by the learning process. Perceptive teachers and trainers will soon realise the lack of truth in such myths and will discard, or at least question the validity of, many of these so called, adult characteristics as guides for practice.

So if this process of identifying the characteristics of adult learners is so flawed, the question for teachers and trainers may become — why bother to generalise at all about adult learners? The answer to this question is an altogether practical one and has to do with the central processes of planning and preparation for teaching and training (see Chapter 7). Rogers (1998) states that thinking about whom our prospective learners might be is an integral part of the planning of any educational program. In fact, as Rogers notes, we tend to unconsciously make assumptions about our learners which, if they remain part of an unconscious process, are never examined or reflected upon in terms of how much truth these assumptions contain (the critical role of the reflective practitioner in education and aviation is addressed in Chapter 8).

The nature of planning and preparation in any type of training or education environment means that generally, prior to meeting the participants, the trainer must make assumptions about the participants in terms of their attributes, needs, and level of current knowledge. This can realistically only be done in a provisional way given that every group of adult learners is different and that also, within any group, there may be a variety of attributes.

Rogers (1998) suggests that it is a good idea for teachers to test out their assumptions about their students in the first (or early) meetings with the participants so that adjustments can be made where appropriate or, indeed, necessary. In this way, the teaching and learning process can become responsive to the characteristics of particular groups of adult learners, within the confines of any particular educational context.

Roles of Teachers in Adult Learning

A group of people involved in a teaching and learning episode have a common task to achieve — *learning*. To accomplish this task, each member of the group carries out a role, including the teacher. Rogers (1998) suggests that the teacher of adults fulfils a range of roles in the process of working with a group of learners. For example, the teacher may be a group leader, a resource person, or a facilitator.

Humanists Focus on the Role of Facilitator

As mentioned in Chapter 1, humanists, such as Carl Rogers and Malcolm Knowles, believe that teachers cannot teach someone directly, teachers can only facilitate learning. *Facilitation* is, indeed, the term most often found in the literature of adult education to describe the preferred approach of many adult educators. Brookfield (1986) suggests that this is the case since educators and trainers of adults wish to distance their approach from that traditionally found in initial schooling, which is generally considered as being less congenial and more rigid. Thus, *teacher* becomes *facilitator* or *resource person*.

Whether or not the negative comparison between school education and adult education is a fair one is difficult to say given the current range of modern teaching practice in schools. However, Brookfield is certainly not alone in focusing on the meaning behind the use of the term facilitator and its construction as a positive role.

Galbraith (1991) discusses the various ways facilitating adult learning has been interpreted. A facilitative approach sees the faciltator's role as that of a guide, mentor, resource person, coach, counsellor, helper, partner or controller (p. xi).

Clearly, facilitation is a complex process, made more complex, in Galbraith's estimation, by the fact that "different approaches are necessary within the teaching and learning process because of the diversity, variability, and varying levels of expectations of the learners, the subject matter presented, and the varying intellectual domain levels of the learners" (p. xi). Galbraith defines the facilitation of adult learning as a transactional process, characterised by collaboration, challenge, reflectivity, action, respect, freedom and equality (p. x).

Facilitation Techniques

Galbraith and Zelenak (1991) describe seven useful techniques designed to develop the facilitation qualities listed above in everyday teaching practice:

1. Discussion
2. Simulation
3. Learning Contracts
4. Inquiry Teams
5. Case Method
6. Critical Incident
7. Mentoring (pp. 103-133).

Clearly, each of these facilitation techniques will not be appropriate for all aspects of aviation instruction. However, three facilitative techniques (discussion, simulation and mentoring) have been selected for further discussion because of their potential usefulness in the aviation context.

Discussion

Discussion has always been a popular technique in adult education and training. This is because it encourages learner participation in the classroom, allows for a range of diverse views to be put forward and allows participants to bring their own knowledge and experiences into the teaching and learning process.

Discussion can take a range of forms in a learning environment, usually depending on the role the facilitator plays. For example, discussion can be closely directed and controlled by the facilitator who may wish to focus on drawing out or constructing very particular knowledge. Alternatively, discussion may be completely directed by the needs and interests of the participants where the facilitator takes less of a directing role, aiding participants to explore new ideas and challenge assumptions as they emerge. The form the discussion takes will depend on the objectives of the learning session.

Brookfield (1986) suggests that an effective discussion is one where participants can prepare themselves by becoming familiar with the issues to be discussed and that the themes or issues should also hold some controversy — encouraging different opinions and views to emerge. Given this potential in the use of discussion, Brookfield also suggests that the participants set some guidelines or rules for participation to create a respectful and tolerant atmosphere.

This technique requires particular skills on the part of the facilitator since most discussion is unpredictable. Again, there are no formulas for successful facilitation of discussion, however, there are some useful guidelines suggested. Galbraith and Zelenak (1991) note that, normally, facilitators will require:

- *Flexibility* — The unpredictable nature of a free discussion means that the facilitator must be ready to accept where the discussion leads, and
- *Sensitivity* — Discussion may result in participants having their opinions, beliefs, and ideas challenged. Therefore, the facilitator must tune into the emotional response of participants. Controversial views and strong feelings may emerge from the discussion and this, although sometimes discomforting, is to be engaged with as it often results in effective learning.

Simulation

According to Galbraith and Zelenak (1991), "a simulation entails setting up situations that approximate reality and then having learners act out the situations. It is most beneficial to learning when followed by discussion and evaluation" (p.108). A limitation of many training environments is the fact that trainees often do not get the opportunity to directly engage with the full range of real life contingencies experienced in their future work environment. There are many reasons for this, such as training resource limitations, ethical considerations and safety issues, to name a few.

For example, just as we would not expect trainee doctors to learn resuscitation techniques on a cardiac-arrested patient, trainee pilots cannot learn emergency procedures in an actual crash landing situation. Simulation is a technique where these crucial skills can be learned in a training environment. Furthermore, simulations that involve such crisis situations provide powerful incentives for participants to not only practice skills but also to reflect on their decisions and their justifications for them (a more detailed discussion on critical thinking and reflective practice can be found in Chapter 8).

The theory behind how and why simulations enhance learning links back to Kolb's (1984) notion of experiential learning (see Chapter 4 for a review of Kolb's model of learning styles). Kolb suggests that learners need to engage in action as well as observation, experimentation *and* reflection for effective and meaningful learning to take place. So, although most individuals will have a preferred learning style, it is important for participants to experience the full learning continuum or cycle.

A simulation exercise should include each of these types of learning. As a direct participant in the simulation, the learner experiences the action directly and may experiment with the application of newly learned skills. As an observer of a simulation, the learner has an opportunity to not only observe but also reflect on the simulated process. This process of reflection is further enhanced in the facilitated debrief of the simulation where interpretations and insights are shared amongst the participants.

The role of the facilitator in simulations is an important and complex one. A skilled facilitator should:

- Prepare the participants;
- Provide clear instructions;
- Check participant understanding of the purpose of the simulation;
- Assess participant tolerance for the complexities of a challenging learning exercise;
- Facilitate participation by all learners; and
- Demonstrate sensitivity to group and individual responses to the simulation.

Simulations are used extensively in aviation training. Lombardo (1985) provides some practical techniques instructors should adopt when teaching in a flight simulator. He suggests that every training period should include the following three sections:

- Preview — to relax and prepare the students or the crewmembers. It should include clear objectives, a thorough description of what will be covered during the session, and a warm up period;
- Training — to cover the objectives. Instructors should start the lesson with easy tasks, gradually increasing the complexity of the problems presented to keep students or crewmembers challenged. However, if performance starts to deteriorate, it normally is an indication that pressure is too great, therefore instructors should simplify the task; and
- Review — thorough debriefing, clarifications provided, and questions answered. It should end on a positive note (adapted from pp. 2 and 4).

Although the list above relates specifically to flight training, the concepts are applicable to any aviation group using simulation (i.e., air traffic controllers, flight attendants, maintenance engineers, dispatchers, etc.).

Mentoring

When describing the social learning orientation presented in Chapter 1, it was noted that one of its key concepts was the notion of learning from behaviour modelled by others. This interest in the learning possibilities which occur in a range of social environments has resulted in the large amount of information currently available on the practice known as mentoring. This technique has proven popular in recent years in the areas of education and training as well as in management and in human resource development, and will be explored here in relation to its potential as an instructional approach in the aviation context. Many senior pilots, qualified flight instructors, and experienced air traffic controllers are expected to mentor newly appointed professionals who are new to flying, instructing, or air traffic controlling, yet few may be fully aware of what the role of a mentor entails.

What is mentoring? Due to the fact that mentoring is a practice that is used in a wide variety of settings, it is probably best described as a concept and practice that evolves to suit the varied situations in which it is used. Therefore, definitions might be difficult. However, a number of key theorists have described the mentoring relationship, each with different emphases. For example, Galbraith (1991) defines mentoring "as a powerful emotional and passionate interaction whereby the mentor and the protégé experience personal, professional, and intellectual growth and development. It is a unique one-to-one teaching and learning method that incorporates the basic elements of the transactional process — collaboration, challenge, critical reflection, and praxis [putting theory into practice]" (p. 126).

Other theorists, such as Cohen (1995) and Daloz (1999), emphasise the dynamic and changing nature of the relationship. Cohen, in particular, acknowledges distinct phases in the relationship between mentor and learner, while Daloz uses the metaphor of a journey to describe the relationship. In this conception of the process, it is the learner who is undertaking the journey and it is the mentor's job to provide "support, challenge, and vision" along the way (Daloz, 1986 cited in Galbraith, 1999, p. 2).

Why use mentoring? In considering the possible use of mentoring in a workplace training context, it is important to assess exactly what the mentoring relationship potentially offers trainees and workers. Given the fact that most areas of the aviation work context have in place a range of formal or off-the-job training arrangements to convey to trainees the complex array of skills and knowledge necessary for employment in the industry, it may well be asked: what can mentoring offer that is not already being taught through other methods?

Modelling behaviour is seen by researchers and social learning theorists as a key element of learning. As noted in Chapter 1, social learning theory emphasises both the person and the surrounding social environment as sources of learning (Saunders, 1999, p. 33). This focus on the effects of the social environment on learning provides an important clue in understanding what mentoring has to offer. Burns (1995) summarises some key considerations below.

> Mentoring can help to bridge gaps in training and education, particularly that conducted in formal institutions which needs to be transferred to the workplace. Important elements of informal learning and application of learning can be addressed in the more informal mentor setting. The needs that adults have to be in control of their learning, the recognition of prior experience, and personal interaction with a skilled colleague cannot be met fully in formal instruction and training.
>
> No matter how detailed the analysis of a job skill on which a training program is based, there are subtle elements of jobs that require interpretation and integration of knowledge and skills. Therefore, mentoring usually not only includes skills but also the norms and mores of the workplace. (pp. 311-312)

Based on the above, it is proposed that mentoring offers an approach to learning that does not necessarily seek to replace more traditional forms of training, rather it very effectively complements them by illuminating the more subtle elements of the nature of the work that learners are being trained to engage in once they are in the workforce. In considering the use of a mentoring program in any work or training environment, these possible benefits must be weighed against the costs of such training. As Burns (1995) notes,

> Mentoring may be costly for organisations. Mentors experience time constraints and may be taken away from their jobs. Mentors need high-level personal skills, as well as work skills. Time is required for a mentor-learner relationship to grow too. Mentors must stress at the outset what they will or will not do for their charges, establishing ground rules and boundaries. (p. 312)

Clearly, these are important issues to consider when assessing the potential benefits of a mentoring program in the aviation training environment.

The mentor's role Cohen (1995) has written in some depth on the subject of mentoring and advocates a set of six behavioural functions which describe the role of the mentor:

1. Relationship Emphasis, to establish *trust*
2. Information Emphasis, to offer tailored *advice*
3. Facilitative Emphasis, to introduce *alternatives*
4. Confrontative Focus, to *challenge*
5. Mentor Model, to *motivate*
6. Mentee Vision, to encourage *initiative* (p. 3).

Cohen (1995) notes, however, that although it is possible to describe the mentor role in terms of the above six components, in fact the mentor's role must "create a synergystic effect which allows the evolving mentoring relationship to become 'larger' than the six separate mentor components. It is this combination or blend which ultimately creates a meaningful mentoring experience for the mentee" (p. 5).

In addition to these functions, Galbraith (1991) adds another level of complexity to the mentoring role by noting the emotional dimension of the role and the relationship which involves development through "elements of challenge and risk taking" (p. 130). Finally, Brockbank and McGill (1998) stress that effective mentoring must incorporate reflective dialogue, which includes astute questioning, "particularly the challenging questions about purpose and intention, as they are likely to uncover discrepancies between espoused theory and theory-in-use" (pp. 260-261).

Qualities of a good mentor Once again, looking to the work of Cohen (1995), it can be seen that there are a number of qualities that a good mentor might be careful to cultivate:

1. The mentor is a role model whose behaviour provides a tangible goal (in terms of appropriate knowledge, skills and attitudes of the workplace) for less experienced workers.
2. The mentor is able to motivate the learner through his or her ability to locate the correct factors that may be inhibiting a learner's success in the work environment.
3. The mentor is able to find the right balance between exercising too much influence over the learner and not providing enough guidance, while respecting the learner's individual path to learning. The mentor is sensitive to finding the right time to offer comments and advice, especially of a challenging or confronting nature. The mentor realises the importance of building trust into the relationship from an early stage, so that when it is appropriate to do so, difficult issues can be tackled without threatening the relationship. The mentor is therefore sensitive to the issue of mentee *readiness* to learn from advice of a challenging nature (adapted from pp. 3-16).

Finally, Burns (1995) offers some important additional advice concerning the issue of which people within an organisation are appropriate mentors for specific individuals. For example, he suggests that mentor choice (for both the mentor and the learner) is an important step in the building of successful relationships and also adds the warning that

> Supervisors are not always the best mentors, as they also may have to assess the person for work performance, pay, or for competency skills. Role conflict could be involved. Potential mentors should have experience and skills directly related to the needs of the learner, be approachable, be given time release, have exemplary attitudes to work, possess sound communication and relationship making skills, and show patience, genuine care and concern. (p. 312)

Learning at Work

Despite the fact that a great deal of the literature dealing with teaching and training skills assumes a formalised and traditional (off-the-job) learning environment as its focus, an important and very current trend in vocational education research focuses much more on the actual workplace as a site of perhaps the most significant learning (see Boud, 1998). This body of research draws largely upon the constructivist orientation, one of the five theoretical orientations covered in Chapter 1. As was noted in the Chapter 1 discussion, constructivists see the importance of the individual's internal processes of constructing knowledge (in this they are similar to the cognitivists) but also take into important consideration the environment in which this construction takes place.

Billett (1996, 1998) is one such researcher who is specifically interested in the type of learning that occurs in a workplace environment. This research focuses on what is known as *situated learning*, and attempts to understand how workers learn and, therefore, to suggest ways to improve workplace learning.

Situated Learning

The concept of situated learning has been explored by a number of theorists and is focused around the contention that "knowledge is created and made meaningful by the context in which it is acquired" (Farmer, Buckmaster, & LeGrand, 1992, p. 46 cited in Kerka, 1997, p. 1). Indeed, Lintern (1995) argues that optimal training benefit in aviation accrues in a situated learning environment, such as Line-Oriented Flight Training (LOFT). That is, a learning environment that duplicates as nearly as possible the situation in which the learning will later need to be demonstrated or applied. Therefore, situated learning "results from undertaking authentic activities guided by expert practitioners situated in a culture of practice" (Billett 1994a cited in Kerka, 1997, p. 1). As McLellan (1996) points out, "this view of knowledge as situated has important implications for our understanding of learning and also for the design of instructional experiences and activities" (p. 6).

In understanding the key processes which result in situated learning, it is useful to look at a model of situated learning. McLellan (1996), drawing on the work of several authors, outlines the following key components of the situated learning model:

Stories In this model, stories are seen as important in the social construction of knowledge (e.g., hangar flying or hangar talk). They provide a method for structuring and remembering information and, when shared by workers in the work environment, provide a means of transferring information to others.

Reflection The ability to stand back from the work and reflect on the events, processes, information and skills evident in the work context is seen as vital for situated learning. Through reflection, learners create an opportunity to gain new insights into the tasks completed and those ahead (the concept of reflective practice is examined in Chapter 8).

Cognitive apprenticeship From the constructivist perspective, cognitive apprenticeship is a key tool through which novices in a work environment gain expertise. As Kerka (1997) explains,

> In cognitive apprenticeship, experts model the strategies and activities needed to solve problems, and learners approximate doing the activity while articulating their thought processes. Experts coach learners with appropriate scaffolds (physical aids and supporting materials), gradually decreasing assistance as, through continued practice, learners internalise the process by constructing their own knowledge base and understanding. (Farmer, Buckmaster, & LeGrand, 1992 cited in Kerka, 1997, p. 2)

Norman (1993 cited in McLellan, 1996) suggests that learning during cognitive apprenticeship will be maximised if teachers:

- provide a high intensity of interaction and feedback;
- have specific goals and established procedures;
- motivate;
- provide a continual feeling of challenge, one that is neither so difficult as to create a sense of hopelessness and frustration nor so easy as to produce boredom;
- provide a sense of direct engagement, producing the feeling of directly experiencing the environment, directly working on the task;
- provide appropriate tools that fit the user and task so well that they aid and do not distract; and
- avoid distractions and disruptions that intervene and destroy the subjective experience (p. 10).

Collaboration Given that most work environments contain groups of workers, collaboration is seen as an important learning tool in the situated learning model. For example, as outlined above, cognitive apprenticeship requires collaboration between the learner and an expert worker. In addition, collaboration is seen as increasingly relevant in "... the emerging age of telecommunications: computer networking and conferencing, groupware, multi-user simulated environments (MUSEs), and other innovative tools for communication and sharing of information. Learners need to be taught collaboration skills in order to work effectively in the age of tomorrow" (Schrage, 1990 cited in McLellan, 1996, p. 10).

Coaching From the description of cognitive apprenticeship outlined above, the process of coaching is clearly very important to the processes of situated learning. McLellan (1996) describes the coaching role as one of "observing students while they carry out a task, providing a 'guide on the side' who intervenes and provides a scaffolding for learning to progress when necessary, but otherwise fades into the background, providing learners with opportunities for initiative and self-directed problem solving: constructive learning" (p. 11). Voll and Beaumont (1997), in a paper entitled "Safety culture on board", stressed the importance of coaching in the development of valuable team skills, such as cohesiveness, assertiveness, effective communication, ownership of role responsibility, understanding the role of self and others, and in increasing safety awareness.

Multiple practice According to the situated learning model, the journey from novice to expert will necessarily involve repeated practice of key skills of the workplace. In this way, "skills are honed and learners become accustomed to deploying them, so that these skills become deeply rooted, and automatically mobilised whenever relevant" (McLellan, 1996, p. 11).

Articulation of learning skills In the situated learning model articulation refers to two important processes. The first involves the novice being exposed to complex skills in an articulated or broken down version so that the simpler skill subsets can be learned effectively prior to being integrated into the complete activity. The second process involves having novice workers articulate their knowledge by describing the processes they are engaging in, and by having them provide reasons for actions taken and decisions made (reflection-in-action). By speaking about these processes and accounting for their actions, novices are said to be better able to understand their learning and perhaps see more clearly where potential deficits might lay (see Chapter 8 for more information on the process of "reflection-in-action" and Chapter 9 for an examination of the process of skill acquisition and the phases involved in acquiring complex psychomotor skills).

Technology Increasingly, technology plays a role in all work environments. In the situated learning model, technology is seen as a tool to support learning. The ability to record workplace activity on video or audio tape or to track performance via computer is seen as a useful way to invite trainees' reflection on both their own performance and the performance of more experienced workers (McLellan, 1996).

It should be obvious that all the key components of situated learning listed above have definite applications when providing on-the-job training to aviation personnel in all the professional groups within the aviation industry (e.g., air traffic controllers, pilots, flight attendants, maintenance engineers, dispatchers, flight instructors).

Benefits of Workplace Learning

According to Billett (1996), the workplace as a learning environment has many distinct advantages. In particular, the workplace offers

- authentic activities;
- expert others; and
- other workers (p. 49).

Each of these aspects of the workplace combine to provide a learning environment where the knowledge required for expert performance can be constructed.

Limitations of Workplace Learning

Despite the fact that the workplace is seen as a rich learning environment by researchers working within the constructivist orientation, it is not viewed in an uncritical light. Billett (1996, 1998), for example, lists a significant range of potential limitations in the workplace as a learning environment:

- *The construction of inappropriate knowledge* — if dangerous work practices are an accepted part of the workplace, then this will impact on the types of knowledge constructed by novices in this environment.
- *Access to authentic activities* — not all workplaces give due

consideration to the types of activities made available to novices. If new workers are consistently only provided with menial tasks or tasks that are not designed to engage them in increasingly complex skills, then effective workplace learning will not occur.

- *Reluctance of experts* — for a variety of reasons, perhaps to do with fear of being supplanted by younger workers, some experts may be unwilling to allow novices to benefit fully from their expertise. In this case, effective workplace learning will once again be thwarted.
- *Access to expertise* — if a workplace does not provide enough of an expert workforce for novices to draw upon, situated learning will not occur.
- *Opaqueness of some knowledge* — in many workplaces, the tasks being carried out are complex to the extent that the conceptual understanding necessary to carry out the task cannot be adequately communicated in the workplace (Billett, 1996, pp. 50-52).

In addition to situated learning, two concepts — metacognition and self-regulation — constitute useful elaborations of the cognitivist perspective which have definite implications for teaching and learning in the aviation learning environment.

Metacognition

Metacognition is a term found throughout the literature in the field of education and refers to the self-monitoring and conscious use of learning strategies (Borkowski, Carr, & Pressley, 1987; Borkowski, Carr, Rellinger, & Pressley, 1990). Stated simply, the term metacognition refers to our knowledge and beliefs about knowing (Lefrancois, 1997, p. 175). According to Ashman and Conway (1993), the term was "introduced in the 1970s to describe the awareness of cognition (that is, a person's understanding of the information-processing involved in complex tasks)" (p. 35). Since then, use of the term has been broadened to encompass a range of cognitive activities which include:

- recognising the need for a strategy;
- evaluating the task requirements;
- searching for the availability of an appropriate strategy within the person's repertoire of strategies;

- executing the strategy; and
- monitoring the effectiveness of the strategy (Campione & Brown, 1978 cited in Ashman & Conway, 1993, p. 35).

Why Are these Concepts Useful to Trainers in Aviation?

Understanding metacognition and the role it plays in learning is important for trainers in vocational contexts because it is a concept that can help trainers plan effective instruction. In any teaching or training situation where trainers expect students to engage with new ideas, information and skills, an important consideration must always be — *how* are the participants going to learn this material? An understanding of metacognition gives trainers insights into the types of strategies learners might be employing in the learning process. This can have a twofold effect.

Firstly, if the trainer employs his or her understanding of the cognitive demands of, say, remembering large amounts of complex material, this will have an impact on how he or she chooses to present the material. Thinking about the type of thinking that learners will do makes for more effective instruction. Secondly, a trainer who is aware of the importance of metacognitive strategies can share ideas about useful strategies with learners. Teaching learners about *thinking* (rather than just teaching them things to think about) can be one of the most useful tools a trainer can introduce to an instructional situation.

> The skills of metacognition allow us to monitor our progress when we try to understand and learn something. They provide us with ways of estimating the effects of our efforts, and they allow us to predict the likelihood of being able to remember the material later. Metacognitive knowledge tells us that there are ways to organise material to make it easier to learn and remember, that some rehearsal and review strategies are more effective for one kind of material than another, and that some forms of learning require the deliberate application of specific strategies whereas others do not. (Lefrancois, 1997, p. 175)

Like most skills, metacognition is not an automatic process but is the result of long-term development. Paris and Winegrad (1990) suggest four approaches that incorporate metacognition into instruction:

1. Direct Explanation — Clear and detailed explanations about what students are learning and about the strategies to be used will help them become more effective learners and develop an awareness of how they learn;
2. Scaffolded Instruction — Involves dialogue between the teacher and the learner in order to provide the learner with enough support and guidance to achieve goals that are difficult to attain or could not be reached without assistance;
3. Cognitive Coaching — Includes mutual dialogue between the teacher and the student, direct explanation, modelling, and motivational encouragement. Students are provided with direct, clear explanations of goals, tactics and strategies, and are encouraged to ask questions to clarify content and the learning activities they are expected to accomplish. Cognitive coaching also ensures that teachers and students have common goals. In addition, it involves ongoing assessment of student performance, therefore the task difficulty can be adjusted to maintain a challenging level.
4. Cooperative Learning — Involves a mixture of instructional practices, including direct explanation, scaffolded instruction, modelling, and group activities. When students work together to complete a task, they often need to provide help to classmates and/or receive assistance from others. In so doing, they develop additional skills, such as improved communication and interpersonal skills. Cooperative learning also helps students restructure and clarify their ideas and, in some cases, it provides additional motivation to students (adapted from pp. 31-41).

Self-Regulation

A key concept that is closely associated with metacogntion and is useful for aviation educators to understand is self-regulation. Indeed, Zimmerman (2000) suggests that "perhaps our most important quality as humans is our capability to self-regulate" (p. 13). Barone, Maddux, and Snyder (1997) define self-regulation as the "automatic and deliberate processes by which people control and direct their cognitions, emotions and behaviours in the pursuit of their goals" (p. 303).

Bandura (1986), on the other hand, views self-regulation as a triadic process involving the interaction of *personal factors* (monitoring and adjusting cognitive and affective states), *behavioural factors* (self-observing and strategically adjusting performance processes) and *environmental factors* (observing and adjusting environmental conditions or outcomes). Zimmerman (2000) explains further that self-regulation "entails not only behavioral skill in self-managing environmental contingencies, but also the knowledge and the sense of personal agency to enact this skill in relevant contexts. Self-regulation refers to self-generated thoughts, feelings, and actions that are planned and cyclically adapted to the attainment of personal goals" (pp. 13-14).

According to Schunk and Zimmerman (1998), the cyclical phases of self-regulation include forethought, performance and self-reflection (p. 3). These self-regulatory processes are interrelated and cyclically sustained.

Forethought refers to the intentions that precede efforts to act and includes:

- a task analysis (i.e., goal setting and strategic planning), and
- motivational self-beliefs (i.e., self-efficacy, outcome expectations, intrinsic interest/value, and goal orientation).

Performance includes self-control, which in turn includes:

- self-instruction (e.g., verbalising or describing how to proceed while one performs the task),
- imagery (e.g., forming mental pictures or visualisation),
- attention focusing (e.g., concentration, avoiding distractions),
- task strategies (e.g., reducing a task to its essentials and reorganising the parts meaningfully), and
- self-observation.

Self-reflection involves a person's response to the experience and includes:

- self-judgement (i.e., self-evaluation and causal attribution), and
- self-reaction (i.e., self-satisfaction, adaptive or defensive behaviours) (adapted from Zimmerman, 2000, p. 16).

Reflection on one's performance, in turn, influences one's forethought regarding subsequent performances, thus completing the cyclical process of self-regulation.

Zimmerman (2000) describes self-regulation as cyclical because the feedback from prior performance is used to make adjustment while performing the current task. He notes that adjustments are needed because personal, behavioural and environmental factors or conditions are constantly changing while one is learning or performing a task and these adjustments must be continuously observed and monitored (reflection-in-action) using three self-oriented feedback loops: behavioural self-regulation, environmental self-regulation and covert self-regulation (p. 14).

Behavioural self-regulation involves self-monitoring and intentionally adjusting performance, whereas *environmental self-regulation* means observing and adjusting environmental conditions or outcomes. For example, environmental structuring involves eliminating distractions and avoiding ruminating on past errors or mistakes. *Covert self-regulation* refers to monitoring and adjusting cognitive and affective states, such as using visualisation for remembering or relaxation techniques to cope with stress (Zimmerman, 2000, p. 14). Zimmerman also underlines that these triadic feedback loops are *proactive* as well as *reactive*. That is, in their attainment of personal goals, highly self-regulated individuals increase "performance discrepancies by raising goals and seeking more challenging tasks" (p. 14).

Environmental and motivational factors also affect students' decisions on how they selectively activate, alter, and sustain their learning practices in home and school contexts (their immediate environment) that vary in structure and content (Zimmerman, 1988). Self-consequences, on the other hand, refer to self-regulated learning strategies directed at improving students' motivation (Zimmerman & Martinez-Pons, 1986).

According to Schunk and Ertmer (2000), self-regulation in a learning environment "refers to self-generated thoughts, feelings, and actions, that are planned and systematically adapted as needed to affect one's learning and motivation" (p. 631). The authors explain that self-regulation

> comprises such processes as setting goals for learning, attending to and concentrating on instruction, using effective strategies to organize, code, and rehearse information to be remembered, establishing a productive work environment, using resources effectively, monitoring performance, managing time effectively, seeking assistance when needed, holding positive beliefs about one's capabilities, the value of learning, the factors influencing learning, and the anticipated outcomes of actions, experiencing pride and satisfaction with one's efforts. (p. 631)

Self-regulation is a useful concept for trainers to understand because, in the teaching and learning environment, we are in constant contact with people who are striving to meet a range of challenging goals. If trainers understand something about the processes individuals go through in their attempt to meet these goals, they may be more effectively positioned to assist learners in their endeavours.

According to Winne (1997), we are all self-regulated to various extents and it would be inaccurate to speak of someone as being devoid of self-regulatory skills. What distinguishes effective from ineffective self-regulation is the quality and the quantity of a person's self-regulatory processes (Zimmerman, 2000). Zimmerman (1986) found that self-regulated learners are metacognitively, motivationally, and behaviourally active participants in their own learning process. Zimmerman and Martinez-Pons (1988) explain that

> in terms of metacognitve processes, self-regulated learners plan, organize, self-instruct, and self-evaluate at various stages during the acquisition process. From a motivational vantage, self-regulated learners perceive themselves as self-efficacious, autonomous, and intrinsically motivated. In terms of behavior, self-regulated learners select, structure, and even create social and physical environments that optimize acquisition. (p. 284)

The importance of self-regulation within the aviation context cannot be overstated. For example, Reason (2001) notes that there are "at least two vital components underpinning both high reliability organisations and individual excellence:

- a mindset that expects unpleasant surprises,
- the flexibility to deploy different modes of adaptation in different circumstances" (p. 29).

When weighed against the definitions provided above, it should be clear that Reason's "vital components" are elements of self-regulation. That is, safe or *excellent* operators are self-regulated, reflective practitioners who "have an unmatched capacity to adapt and adjust to the surprises thrown up by a dynamic and uncertain world" (Reason, 2001, p. 28). According to Reason, only those people who have invested a considerable amount of preparatory effort through studying (cognitive readiness), practicing, and critically reflecting (pre-prepared response) in the pre-crisis period will be able to adapt and compensate in a timely manner so as to avoid a major catastrophe or at least mitigate the effects of a major crisis (p. 30). As such, situational awareness can also be seen as an element of self-regulation. In fact, research recently conducted by Beaumont (1998, 1999, 2000) has revealed that self-regulation appears to be an essential element in the successful management of flight.

Self-Regulated Learning Strategies

To further understand the role trainers play in this process, it is perhaps useful to examine more closely some of the self-regulated learning strategies suggested in the literature. Zimmerman and Martinez-Pons (1986) proposed a model of self-regulated learning comprising fourteen self-regulated learning strategies including:

- self-evaluation;
- organising and transforming;
- goal setting and planning;
- seeking information;
- keeping records and self-monitoring;
- environmental structuring;
- self-consequences;
- rehearsing and memorising;
- seeking peer, teacher or adult assistance; and
- reviewing notes, tests, or textbooks (cited in Zimmerman, 1988, p. 284).

According to Barone, Maddux, and Snyder (1997), research indicates that self-regulation involves the following five components:

1. goals or standards, what we are trying to accomplish;
2. self-monitoring, or observing one's behaviour and the impact of that behaviour on progress towards a goal;
3. feedback, or information about progress toward a goal that people either gather themselves or that is provided by others;
4. self-evaluation, or affectively charged judgements about progress toward a goal; and
5. corrective behaviour, or behavioural responses to goal-related

information, typically attempts to move oneself toward one's goal more effectively (p. 278).

In particular, substantial research on how people set goals and respond to feedback about their progress and attempts to meet these goals provides clues to teachers and trainers about how best to facilitate this process. For example, Locke and Latham's (1990) work on goal theory explores the characteristics of goals and goal-related feedback that best facilitate performance, particularly in the workplace. Also, Bandura's (1986, 1989) work on self-efficacy theory has explored "the influence of self-efficacy beliefs on behavioural choice, persistence, and affective reactions to perceived success or failure" (Barone, Maddux, & Snyder, 1997, p. 279). Clearly, having an understanding of these processes and the outcomes of current research can help trainers, especially in relation to understanding the effectiveness of different types of feedback and how learners react in the face of challenging goals.

Goals

One of the practical implications of research about self-regulation concerns the process of goal setting in an education and training environment, which is one of the roles facilitators of adult learning may undertake in partnership with individual learners. In fact, Boekaerts and Niemivirta (2000) stress that goal setting is an essential aspect of self-regulated learning. Research on the process of goal setting tells us several key facts.

Barone, Maddux, and Snyder (1997) found that goals that are specific, difficult and proximal (i.e., close in time) have the greatest impact on behaviour. The authors note that "when we commit ourselves to goals that are both specific and difficult, we perform better than when we have no goals, when our goals are specific but easy, or when our goals are vague (e.g., 'Do your best'…)" (p. 289). Locke (1991) and Locke and Latham (1990) point out that regardless of how specific, difficult, and proximal goals are, they will not influence motivation and behaviour unless learners receive formative feedback concerning how well they are progressing towards their goals. Locke and Latham (1990) found that when learners have goals and are also provided with feedback about their progress toward their goals, they perform better than when they have only goals or only performance feedback.

However, less is known about the impact of goals and feedback on cognitively complex tasks and, in fact, goal setting, according to Kanher (1994), can have a detrimental effect in the early stages of learning a new skill (cited in Barone, Maddux, & Snyder, 1997, p. 290). Barone, Maddux, and Snyder explain that "this effect seems to occur because goal setting initiates self-regulatory activity, which takes attention away from the task at hand. When starting to learn a new skill, having a performance goal induces self-evaluation, which increases negative affect and reduces cognitive capacity for learning the new skill" (p. 290).

According to Zimmerman (2000), there are several key self-motivational beliefs that underlie the forethought processes of goal setting and strategic planning. These self-motivation beliefs include self-efficacy, outcome expectations, intrinsic interest or valuing and goal orientation (p. 17). Zimmerman also notes that

> goals can reciprocally affect self-efficacy beliefs. Self-regulated learners feel self-efficacious in part because they have adopted hierarchical process goals for themselves whose progressive mastery provides them with immediate satisfaction rather than requiring them to suspend any sense of success until the final outcome goal is attained. (p. 18)

Research by Schunk and Swartz (1993) and by Zimmerman and Kitsantas (1997) has shown that process goal attainment can become intrinsically motivating in its own right and can even outweigh attainment of outcome goals. According to Zimmerman (2000), self-efficacious learners "see outcome rewards merely as milestones in a lifelong mastery process" (p. 18).

Self-Efficacy

As mentioned above, effective self-regulation depends on feeling self-efficacious about using one's skills to achieve the desired goals (Bandura, 1986, 1997; Zimmerman, 2000). Self-efficacy refers to an individual's perceptions or judgements (one's personal belief) about his or her ability to carry out specific activities (for example to learn or perform a task) at a required level of skill (i.e., having the means to learn or perform effectively). Zimmerman specifies that self-efficacy "refers to beliefs about one's capabilities to organize and implement actions necessary to attain designated performance of skill for specific tasks" (p. 14). Furthermore, self-regulatory efficacy influences the use of effective learning strategies (Schunk & Schwartz, 1993), the ability to resist adverse peer pressure (Bandura, Barbaranelli, Caprara, & Pastorelli, 1996), goal setting (Zimmerman & Bandura, 1994), time management (Britton & Tessor, 1991), self-monitoring (Bouffard-Bouchard, Parent, & Larivee, 1991), and self-evaluation (Zimmerman & Bandura, 1994).

In summary, research shows that beliefs in self-efficacy influence self-regulation in a number of ways:

1. Self-efficacy beliefs influence goal setting because people who are highly self-efficacious in a specific domain of achievement will set higher goals for themselves than those with lower self-efficacy (Barone, Maddux, & Snyder, 1997, p. 290).
2. Self-efficacy beliefs also influence one's choice of goal-directed activities, one's expenditure of effort, and one's persistence in the face of challenge and obstacles (Bandura, 1986; Locke & Latham, 1990). According to Barone, Maddux, and Snyder (1997), "those with a strong sense of self-efficacy increase their efforts to master a challenge when obstacles arise" (p. 290).
3. Self-efficacy in solving problems and making decisions also influences a person's efficiency and effectiveness of problem solving and decision making. Barone, Maddux, and Snyder (1997) noted that "faced with complex decision making tasks, people who have confidence in their ability to solve problems use their cognitive resources more effectively than people who doubt their cognitive and intellectual skills" (p. 291).
4. Barone, Maddux, and Snyder (1997) add that "the stronger our beliefs in our abilities and in the eventual attainability of the goal, the more adamantly we will persevere in the face of challenge (e.g., by remaining task diagnostic rather then becoming self-diagnostic). Perseverance, more often than not, leads to success; and success usually strengthens self-efficacy beliefs, which in turn encourage us to set additional, and perhaps even more challenging goals" (p. 292).
5. Zimmerman and Ringle (1981) also note that successes raise self-efficacy beliefs while failures lower them. Learners often receive persuasive information from teachers, peers, or others suggesting that they are capable or incapable of performing a task. Such information can raise or lower self-efficacy beliefs (Schunk & Ertmer, 2000).
6. Research by Green (1991) and by Sarason, Sarason, and Pierce (1990) on goal theory has also shown that "challenging goals and performance

feedback produce better performance, at least on simple tasks. However, on cognitively complex tasks, self-evaluation may affect self-regulation differently. Dissatisfaction with performance on a cognitively complex task may cause people to dwell on personal shortcomings. Such preoccupation with the self may interfere with complex performance by diverting attention from critical task relevant cues" (cited in Barone, Maddux, & Snyder, 1997, p. 293). Indeed, self-regulation also depends on self-beliefs and affective reactions, such as doubts and fears about specific performance contexts or tasks, or about one's ability to successfully complete the task (Zimmerman, 1995).

Clearly, this is important information for trainers in terms of how learners are treated in the training environment. Within the aviation context, research examining the association between student pilot stress and flight instructor training found that "negative" flight training practices impact upon the confidence of trainees and, therefore, upon their sense of self-efficacy, self-reliance and goal setting (Henley, 1987).

Henley's (1987) research revealed that a major source of distress for student pilots who experienced instructor-induced stress (e.g., sarcastic remarks, faulty expectations, intolerance of error, lack of positive feedback, etc.) was a loss of confidence in their ability to fly. As a result, these students began to doubt their ability to master the task and became apprehensive about their capability to think clearly or to cope adequately with the situation (see Chapter 6 for a more detailed examination of the effects of negative and positive attitudes and behaviours of teachers and instructors on student learning, performance and decision making).

Conclusion

This chapter introduced some of the major findings of the research into how adults learn and examined some of the characteristics of adult learners. The difficulty in describing the broad group known as "adult learners" was discussed. However, the point was made that if teachers and trainers consider a range of perspectives from the literature and then choose which aspects of the information apply to their own participants, some use may be made of these learning principles. Indeed, the perceptive practitioner will assess each situation in terms of participant characteristics, preferred learning styles and potential barriers to learning and respond appropriately to ensure effective teaching and learning take place. A more detailed examination of these issues can be found in Chapters 4 and 6, which specifically address learning styles and barriers to learning respectively.

Underpinning the discussion on adult learning principles was the question of how teachers and trainers of adults might use theory in an area of work that is very practically oriented. Theories, by definition, are general and de-contextualised, while teaching and training are always context specific. Four concept: mentoring, situated learning, metacognition, and self-regulation were explored as extensions of the theoretical learning orientations discussed in Chapter 1. Each of these areas of theory and practice were seen to have direct relevance for the aviation training environment. The usefulness of mentoring and situated learning as practices and concepts lies in their close consideration of the importance of the workplace context. In this way, they force education and training practitioners to consider and assess theory and research into learning in terms of real-world practice issues. Similarly, further exploration of the cognitivist orientation in this chapter through a discussion of metacognition and self-regulation forces trainers to think about training from the perspective of the learner. Having an understanding of the cognitive processes in which learners may be engaged is bound to make us all more sensitive, and therefore more effective, educators.

Reflective Exercise

1. Why is it important for teachers and trainers to consider the characteristics of adult learners? What are the limitations of making these types of generalisations?
2. Paris and Winograd (1990) suggest that there are four approaches that incorporate metacognition in instruction: direct explanation, scaffolded instruction, cognitive coaching, and cooperative learning. Discuss how each approach could be used in aviation instruction.
3. Why is it important for aviation educators and trainers to have an understanding of self-regulation?

Suggested Further Reading

Boekaerts, M., & Niemivirta, M. (2000). Self-regulated learning. In M. Boekaerts, P.R. Pintrich, & M. Zeidner (Eds.), *Handbook of self-regulation* (pp. 417-450).

San Diego, CA: Academic Press.

Boud, D. (Ed.). (1998). *Current issues and new agendas in workplace learning.* Leabrook, Australia: NCVER.

Cohen, N.H. (1995). *Mentoring adult learners: A guide for educators and trainers.* Malabar, FL: Krieger.

Galbraith, M.W. (Ed.) (1999). *Adult learning methods: A guide for effective instruction* (2nd ed.). Malabar, FL: Krieger.

Rogers, A. (1998). *Teaching adults* (2nd ed.). Buckingham: Open University Press.

Zimmerman, B.J., & Kitsantas, A. (1997). Developmental phases of self-regulation: Shifting from process goals to outcome goals. *Journal of Educational Psychology, 89,* 29-36.

References

Ashman, A.F. & Conway, R.N.F. (1993). *Using cognitive methods in the classroom.* London: Routledge.

Bandura, A. (1986). *Social foundation of thought and action: A social cognitive theory.* Englewood Cliffs, NJ: Prentice-Hall.

Bandura, A. (1989). Regulation of cognitive processes through perceived self-efficacy. *Developmental Psychology, 25,* 729-735.

Bandura, A. (1997). *Self-efficacy: The exercise of control.* New York: W.H. Freeman.

Bandura, A., Barbaranelli, C., Caprara, G.V., & Pastorelli, C. (1996). Mechanisms of moral disengagement in the exercise of moral agency. *Journal of Personality and Social Psychology, 71,* 364-374.

Barone, D.F., Maddux, J.E., & Snyder, C.R. (1997). *Social cognitive psychology: History and current domains.* New York: Plenum Press.

Beaumont, G. (1998). Situational awareness or metacognition? In B.J. Hayward & A.R. Lowe (Eds.), *Aviation resource management* (pp. 359-368). Aldershot, UK: Ashgate.

Beaumont, G. (1999). A construct of pilot performance. In R. Jensen, B. Cox, J.D. Callister, & R. Lavis (Eds.), *The Tenth International Symposium on Aviation Psychology* (pp. 1197-1202). Columbus, OH: Ohio State University.

Beaumont, G. (2000, November). *Information management at the command level.* Paper presented at the 5th Australian Aviation Psychology Symposium, Manly, Australia.

Billett, S. (1996). Towards a model of workplace learning: The learning curriculum. *Studies in Continuing Education 18*(1), 43-58.

Billett, S. (1998). Understanding workplace learning: Cognitive and sociocultural perspectives. In D. Boud (Ed.), *Current issues and new agendas in workplace learning* (pp. 47-68). Leabrook, Australia: NCVER.

Boekaerts, M., & Niemivirta, M. (2000). Self-regulated learning. In M. Boekaerts, P.R. Pintrich, & M. Zeidner (Eds.), *Handbook of self-regulation* (pp. 417-450).

San Diego, CA: Academic Press.

Borkowski, J.G., Carr, M., & Pressley, M. (1987). Spontaneous strategy use: Perspectives from metacognition theory. *Intelligence, 11,* 61-75.

Borkowski, J.G., Carr, M., Rellinger, E., & Pressley, M. (1990). Self-regulated cognition: Interdependence of metacognition, attributions, and self-esteem. In B.F. Jones & L. Idols (Eds.), *Dimensions of thinking and cognitive instruction* (pp. 53-92). Hillsdale, NJ: Lawrence Erlbaum.

Boud, D. (Ed.). (1998). *Current issues and new agendas in workplace learning.* Leabrook, Australia: NCVER.

Bouffard-Bouchard, T., Parent, S., & Larivee, W. (1991). Influence of self-efficacy on self-regulation and performance among junior and senior high-school age students. *International Journal of Behavioural Development, 14,* 153-164.

Britton, B.K., & Tessor, A. (1991). Effects of time management practices on college grades. *Journal of Educational Psychology, 83,* 405-410.

Brockbank, A., & McGill, I. (1998). *Facilitating reflective learning in higher education.* Buckingham, UK: Society for Research into Higher Education and Open University Press.

Brookfield, S. (1986). *Understanding and facilitating adult learning.* San Francisco: Jossey-Bass.

Burns, R. (1995). *The adult learner at work. A comprehensive guide to the context, psychology and methods of learning for the workplace.* Sydney, Australia: Business & Professional Publishing.

Cohen, N.H. (1995). *Mentoring adult learners: A guide for educators and trainers.* Malabar, FL: Krieger.

Daloz, L.A.P. (1999). Mentorship. In M.W. Galbraith (Ed.), *Adult learning methods: A guide for effective instruction* (2nd ed.) (pp. 353-372). Malabar, FL: Krieger.

Galbraith, M.W. (Ed.). (1991). *Facilitating adult learning: A transactional process.* Malabar, FL: Krieger.

Galbraith, M.W. (Ed.) (1999). *Adult learning methods: A guide for effective instruction* (2nd ed.). Malabar, FL: Krieger.

Galbraith, M.W., & Zelenak, B.S. (1991). Adult learning methods and techniques. In M.W. Galbraith (Ed.), *Facilitating adult learning: A transactional process* (pp. 103-133). Malabar, FL: Krieger.

Henley, I. (1987). *The association between student pilot stress and flight instructor training: A study of perceptions.* Unpublished master's thesis, University of Manitoba, Winnipeg, Manitoba, Canada.

Jarvis, P. (1995). *Adult and continuing education: Theory and practice.* London: Routledge.

Kerka, S. (1997) *Constructivism, workplace learning, and vocational education.* ERIC Digest No. 181. Columbus: ERIC Clearinghouse on Adult, Career, and Vocational Education.

Knowles, M.S. (1970). *The modern practice of adult education: Andragogy versus pedagogy.* New York: Association Press.

Knowles, M.S. (1980). *The modern practice of adult education: From pedagogy to andragogy*. New York: Cambridge University Press.

Knowles, M.S. (1984). *The modern practice of adult education: From pedagogy to andragogy* (2nd ed.). New York: Cambridge University Press.

Knowles, M.S. (1990). *The adult learner: A neglected species* (4th ed.). Houston, TX: Gulf.

Knowles, M.S., & Associates. (1984). *Andragogy in action: Applying modern principles of adult learning*. San Francisco: Jossey-Bass.

Kolb, D. (1984). *Experiential learning: Experience as the source of learning and development*. Englewood Cliffs, NJ: Prentice Hall.

Lefrancois, G.R. (1997). *Psychology for teaching* (9th ed.). Belmont, CA: Wadsworth.

Lintern, G. (1995). Formal instruction: The challenge from situated cognition. *International Journal of Aviation Psychology, 5*, 327-350.

Locke, E.A. (1991). Goal theory vs. control theory: Contrasting approaches to understanding work motivation. *Motivation and Emotion, 15*, 9-28.

Locke, E.A., & Latham, G.P. (1990). *A theory of goal setting and task performance*. Englewood Cliffs, NJ: Prentice-Hall.

Lombardo, D.A. (1985, September). The role of the instructor in flight crew simulator training. *Simulation Newsletter, 1*(1), pp. 2, 4.

McLellan, H. (Ed.).(1996) *Situated learning perspectives*. Englewood Cliffs, NJ: Educational Technology Publications.

Merriam, S.B. (1993, Spring). *An update on adult learning theory*. San Francisco: Jossey Bass.

Merriam, S.B., & Caffarella, R. (1999). *Learning in adulthood*. San Francisco: Jossey-Bass.

Paris, S.G., & Winograd, P. (1990). How metacognition can promote academic learning and instruction. In B.F. Jones & L. Idol (Eds.). *Dimensions of thinking and cognitive instruction* (pp. 15-51). Hillsdale, NJ: Lawrence Erlbaum.

Peters, J.M., & Gordon, S. (1974). *Adult learning projects: A study of adult learning in urban and rural Tennessee*. Knoxville: University of Tennessee.

Reason, J. (2001). Heroic compensations: The benign face of the human factor. *Flight Safety Australia, 5*(1), 28-31.

Rogers, A. (1998). *Teaching adults* (2nd ed.). Buckingham: Open University Press.

Saunders, S. (1999). Social psychology of adult learning. In J.A. Athanasou (Ed.), *Adult educational psychology* (pp. 25-71). Katoomba, Australia: Social Science Press.

Schunk, D.H., & Ertmer, P.A. (2000). Self-regulation and academic learning: Self-efficacy enhancing interventions. In M. Boekaerts, P.R. Pintrich, & M. Zeidner (Eds.), *Handbook of self-regulation* (pp. 631-649). San Diego, CA: Academic Press.

Schunk, D.H., & Schwartz, C.W. (1993). Goals and progress feedback: Effects of self-efficacy and writing achievement. *Contemporary Educational Psychology, 18*, 337-354.

Schunk, D.H., & Zimmerman, B.J. (1998). Conclusions and future directions for academic interventions. In D.H. Schunk & B.J. Zimmerman (Eds.), *Self-regulated learning: From teaching to self-reflective practice* (pp. 225-235). New York: Guilford.

Tough, A. (1979). *The adult's learning projects* (2nd ed.). Toronto, Canada: Ontario Institute for Studies in Education.

Voll, U., & Beaumont, G. (1997, June 18-20). *Safety culture on board.* Paper presented at the Air Safety Seminar: The Asia Pacific Challenge. Singapore Aviation Academy, Singapore.

Winne, P.H. (1997). Experimenting to bootstrap self-regulated learning. *Journal of Educational Psychology, 88,* 397-410.

Zimmerman, B.J. (1986). Development of self-regulated learning: Which are the key subprocesses? *Contemporary Educational Psychology, 11,* 307-313.

Zimmerman, B.J. (1988). Developing self-fulfilling cycles of academic regulation: An analysis of exemplary instructional models. In D.H. Schunk & B.J. Zimmerman (Eds.), *Self-regulated learning: From teaching to self-reflective practice* (pp. 1-19). New York: Guilford.

Zimmerman, B.J. (1995). Self-regulation involves more than metacognition: A social cognitive perspective. *Educational Psychologist, 29,* 217-221.

Zimmerman, B.J. (2000). Attaining self-regulation: A social cognitive perspective. In M. Boekaerts, P.R. Pintrich, & M. Zeidner (Eds.), *Handbook of self-regulation* (pp. 13-39). San Diego, CA: Academic Press.

Zimmerman, B.J., & Bandura, A. (1994). Impact of self-regulation influences on writing course attainment. *American Educational Research Journal, 31,* 845-862.

Zimmerman, B.J., & Kitsantas, A. (1997). Developmental phases of self-regulation: Shifting from process goals to outcome goals. *Journal of Educational Psychology, 89,* 29-36.

Zimmerman, B.J., & Martinez-Pons, M. (1986). Development of a structured interview for assessing students' use of self-regulated learning strategies. *American Educational Research Journal, 23,* 614-628.

Zimmerman, B.J., & Martinez-Pons, M. (1988). Construct validation of a strategy model of student self-regulation of learning. *Journal of Educational Psychology, 80*(3), 284-290.

Zimmerman, B.J., & Ringle, J. (1981). Effects of model persistence and statements of confidence on children's self-efficacy and problem solving. *Journal of Educational Psychology, 73,* 485-493.

3 Aviation, Adult Learning and Andragogy: A Canadian Perspective

Dianne L. Conrad and Jo Harris

Come to the edge, he said,
They said: we are afraid.
Come to the edge, he said.
They came,
He pushed them … and they flew.

(Guillaume, quoted in Poonwassie, 2001, p. 153)

Introduction

As recently as the late 1990s, studies of Canadian flight instructors' and inspectors' opinions of their own flight training experiences indicate fairly consistent levels of dissatisfaction with the quality of those learning experiences (Harris & Conrad, 1997; Henley, 1995). The recent Air Transport Association of Canada (2001) publication, *Human Resources Study of Commercial Pilots in Canada*, supports Harris and Conrad's (1997) call for the creation of a "more learner-centred culture at the basic level of instructor training, a culture that adheres to andragogical principles of learner worth, experiential value and genuine interaction" (p. 1212).

While further examining the status of aviation training in Canada in this chapter, we will review the history of Canadian aviation teaching methodologies and examine current practices. In our conclusion, we will suggest a teaching-learning approach that we feel would better serve a clientele of adult learners in an increasingly sophisticated and changing world.

The teaching-learning transaction is always important — and never more so than in the aviation industry. In his classic piece, "Beyond Tribalism: Renaming the Good, the True, and the Beautiful", Daloz (1988) discusses the role of education metaphorically, likening the learning journey to tribal migration, a difficult but necessary journey of growth and hardship.

> When it works, good education enables people to construct a coherent and responsive stance from complex information radiating from a rapidly shifting world … They have in some sense left the tribe and may suffer for that. Yet, in time they will make themselves at home in a larger tribe ... The journey from one home to the next is seldom an easy one, but there is about it a kind of imperative. For some it seems a matter of life or death. (p. 235)

Daloz's quote is a useful starting point for the discussion that follows as it provides an overarching metaphor for the learning journey that is every bit as true in the aviation world as it is in other learning enterprises. We will premise the important truth of "making the journey", as individuals train and learn to fly aircraft, on a discussion of adult learners and what we know about how they learn.

Our discussion will rest on the philosophical foundation of andragogy made popular in 1970 by noted American adult educator Malcolm Knowles (see chapter 1), and echoed by Daloz (1988) years later: "Our work with adult students requires that we help them recognise both the limitations of their own insularity and the scope of their responsibility" (p. 240). For only in bringing aviation students to the apex of this critical self-knowledge, together with the skills necessary to hone their "wings", can aviation instructors hope to successfully and safely fulfil their professional mandates.

Flight Training in Canada: An Historical Overview

In 2000, Canadian flight instructors trained 2168 professional pilots and 6500 recreational pilots in an aviation training system that has been considered world-class for many years. Reflecting a 21st century economy in the western world, flight training in Canada is a substantial business. The task of training competent pilots falls on the shoulders of the 287 flying schools and the flight instructors who teach there.

A look backwards will help to explain the development of today's current training ideology. Historically, training has used a number of methods and procedures that have resulted in varied rates of success. A brief history outlines the stages of aviation training in Canada.

The Experimental Era

In its infancy, the phenomenon of flight was regarded by the general public with awe and skepticism. It was considered adventurous and strange — like the circus coming to town — and folks, a-buzz with excitement, would flock to the field where a small plane had landed to view and discuss this extraordinary thing:

> In the early days of flying, barnstorming pilots would descend on a community, attract a crowd, "test the aptitudes" of potential aviators during a quick circuit around the fields, and then sell "flying lessons" to those with the widest eyes and sufficient cash for the required payments in advance. That night they would slip out to their planes and fly away, thereby gaining the moniker of "fly-by-night operators". Early operators of legitimate flying schools, anxious to disassociate themselves from the rascals, soon publicly declared themselves "fixed-base operators". (Roscoe, 1980, p. 173)

From these nebulous beginnings, flight training in Canada evolved to meet the needs of a changing world and changing technologies. From 1903, when the Wright Brothers first flew a power-driven, heavier-than-air machine, until 1914, when the experiment was put to the test in World War I, the Experimental Era saw aircraft being built and pilots teaching themselves to fly.

> [Early pioneer pilots] were not foolhardy. Most of them did a great deal of taxiing in order to get the feel of the controls before taking to the air. And the numerous short hops that characterised their early efforts were dictated more by the limitations of the pilot than the aeroplane. Each hop was an experience giving treasured information about the use of a control: information to be thought about and dreamed about in lengthy hangar sessions for weeks afterwards. Most of their grief arose from simple mistakes that had little to do with the flying qualities of their aircraft: failure to take off into wind; failure to preserve those weak undercarriages, devoid of shock absorbers, by smoothing the ground. Indeed, so much thought was given to getting up, that little time was left for the problem of getting down again successfully. This, as it turned out, was the more difficult problem of the two. (Main, 1967, p. 16)

"Show and Go" Era

Because most of the early aircraft were single-seaters, instruction was limited to ground explanations of how to fly the aircraft and a flying demonstration with the student observing from the ground before being given a turn at the controls. Although an improvement from the previous experimental, trial-and-error methods of training, it was still a primitive modeling type of teaching that occurred in potentially fatal situations: Not recommended for the faint at heart. The following vignette describes the first flight testing procedure:

> The standard was not excessively high. A candidate was required to take off, do a figure of eight, turn over the judges, and land where he had taken off. He was also called upon to show some knowledge of the aircraft and engine. What more proof of competence could one ask?
>
> But most of the pilots were unlicensed: Nobody bothered to ask, and indeed nobody was officially empowered to ask. Licensing standards for aircraft — a long, tedious and exacting business — had not yet been established: an aeroplane either flew or it didn't. And it was sometimes difficult to evaluate the performance between these extremes. (Main, 1967, p. 18)

The Era of Instructor "Patter"

During World War I, the importance of pilot training became much more urgent. The Canadian government authorised the Royal Flying Corps to set up bases of operation at Camp Borden, Long Branch, Leaside, Armour Heights, Mohawk and Beamsville, Ontario while the University of Toronto cooperated by providing ground training. In all 2500 pilots were trained (Main, 1967).

During the war years, World War I and World War II, training was rapid and disciplined and relied on intimidation and urgency to get the young pilots into the air as soon as possible. The flight instructors were tough young men whose responsibility was to train as many pilots as possible in the least possible time. Instructors were trained to memorise a series of lesson plans (known as the instructor's "patter") and gave them verbatim to students. The urgency to produce pilots who could instantly fly into battle, combined with the primitive nature of those early machines, created an aviation environment that focused on survival, rather than safety and sound teaching principles and strategies.

Gradually, after the war, civilian flying schools adopted the instructor "patter" method. In 1961, the *Flying training manual* and *Flight instructor's guide* published by the Department of Transportation, Civil Aviation Branch gave instructors a text from which to work. Even so, very little changed in the methodology of instruction from 1914 to the mid 1980s, as revealed in Henley's (1987, 1995) research on the quality of flight instructor training in Canada and Australia and as echoed in Elshaw's (1993) chapter on "Preparing better flight instructors":

> Although there has been a growing research literature on instructional techniques, it has had little impact on the practice of flying training. Flight instructor training, when it is formally taught, has changed very little since Major Smith-Barry first codified his principles of flight training in 1916 while serving at the Central Flying School in the UK. He was the first person to insist on standardization in flight training based on the instructor demonstrating the flight manoeuvre while at the same time describing what he was doing — generally known as the instructor's "patter". (p. 253)

Mold (1982), a retired Air Force pilot, who undertook recurrent training in 1982, also attests to this stagnancy: "The training has not changed very much over the years. The exercise, the instructor's 'patter', the technique, are much the same as they were when I first took my flying training back in England in 1946" (p. 6).

The varying anecdotal perspectives that follow, taken from interviews with two Transport Canada flight officials, illustrate the randomness of flight training procedures over the years. The first speaker began flight training with the Ottawa Flying Club in 1948. He logged 15 hours before entering the military:

> There were no preflight briefings and no Ground School. I trained on an Aeronca Champ, the old two-seater tandem, fabric aircraft. The flight training was more "hop in the airplane with the instructor and try it" than anything. I soloed at five hours but did not complete my private pilot's license before going into the military. The military program was much more structured and comprehensive. Ground School and lesson plans were well laid out. Ground School was integrated with the flying lessons which was an excellent idea. I learned very quickly. (personal interview, May 4, 2001)

Another perspective is related by another Transport Canada official, who trained in the 1960s, with the Ottawa Air Cadet Program. He stated: "I was impressed with my training. As students, we were given the basics in ground school and flight training with lots of opportunity to study and practice the flying exercises. Things seemed to be simple and clear when I was training, whereas, today it's a lot more complex and confusing for the new student, I think" (personal interview, May 4, 2001).

Pre-Flight Briefing and Methodology Era

In 1978, the *Flight instructor guide*, commonly known as the "Blue Book", was published by Transport Canada. The Blue Book was complete with lesson plans, advice to instructors, questioning techniques, student motivation, and tips on how to deal with difficult students. It represented, in fact, a conglomeration of information that most instructors had never seen before.

Yet, implementation and acceptance of the Blue Book was slow at first because its introduction was not accompanied by appropriate training. However, combined with the use of the *Flight training manual*, first published in 1979 by Transport Canada, Class 1 flight instructors began to change their methods of teaching new instructors. As a result, the aviation industry began slowly to become aware that there was much to learn about the education of adults from the academic study of the art and science of teaching adults.

Today, we are in the throes of the next step — we know that we can improve aviation training for both students who are learning how to fly and instructors who are learning how to teach others to fly. However, to continue the growth an educational presence in aviation training is essential if we want to improve the quality of flight training in Canada.

Flight Training in Canada: A Current Reality

Vestiges of history's individualistic and militaristic influences are still visible in present flight instruction's rigid disciplines and didactic methodologies. However, in a rapidly changing world surging ahead with technological and digital advancement in every sector, parallel dynamic changes in the nature of aviation have contributed to the recognition of our "humanness" as an important component in an effective training program (Air Transport Association of Canada, 2001). These changes, although slow in coming, are being driven by several forces:

Manpower — in recent years, a shortage of pilots with appropriate experience and training background has resulted from a number of factors: increased demand for air services, high pilot turnover — a traditional concern — and increased training costs. "No longer able to draw upon the military to the extent they once did, airlines have been increasingly turning to the smaller regional carriers for new pilots leaving the latter to scramble for replacements elsewhere" (Air Transport Association of Canada, 2001, p. 1). One response to skill shortages has been to lower hiring criteria. And, as in so many other sectors, an anticipated baby boomer retirement surge in the next few years is projected to lead to yet greater shortages.

Technology — the advent of new technologies in all aspects of the aviation industry — on the ground, in the air, from systems to satellites — demands new skill levels from those who will pilot aircraft of the future. "There will always be a need for traditional 'stick and rudder' skills, but in an era of computers and the glass cockpit, pilots will need to become, in effect, managers — managers of systems, managers of information and yes, even managers of people" (Air Transport Association of Canada, 2001, p. 1).

Regulatory systems — as aviation worldwide responds to the pressures of globalisation, through strategic partnerships and alliances that bring ground and air operators and pilot associations to concomitant levels of operating protocols, Canadian aviation personnel will be required to comply with new licensing requirements and the standardisation of some aspects of training. "When and if adopted, the standards will effectively set a new bar for all pilots around the world interested in flying on the international stage or in job opportunities abroad. Given current efforts in the United States and in Europe to revamp and elevate their skills and training requirement, that bar is likely to be high" (Air Transport Association of Canada, 2001, p. 1).

Safety and performance — In 1989, the International Civil Aviation Organization (ICAO) responded to studies that indicated that human error, at 80%, was the largest single cause of aviation accidents. Research into this safety and performance issue had concluded that even "the remaining 20 percent of accidents almost always involve some human component, such as a maintenance oversight or unreasonable management pressure to fly whatever the weather" (Reason, 1997, p. 223). Contributing factors to human error, further replicated in simulator research experiments, included:

- Preoccupation with minor mechanical problems;
- Inadequate leadership;
- Failure to delegate tasks and assign responsibilities;
- Failure to set priorities;
- Inadequate monitoring;
- Failure to utilize available data; and
- Failure to communicate intent and plans (Wiener, Kanki, & Helmreich, 1993).

In 1994 Transport Canada initiated the Human Factors Project. Air Canada won the contract and a group of experts were selected to write three manuals that were to be used in flight training in Canada. Published in 1997, the manuals introduced a resource for training in "human factors" and regulations soon came into force that made this type of training mandatory for all levels (see Air Canada Human Factors Project Team, 1997). Gaining tremendous importance in the aviation world, human factors is defined as

> the study of how pilot performance is influenced by such issues as the design of cockpits, temperature and altitude, the functioning of the organs of the body, the effects of the emotions and interaction and communication with the other participants in the aviation community, such as crew members and air traffic control personnel. (Air Canada Human Factors Project Team, 1997, p. 2)

Changes to Date

ICAO has set up a program of audits that checks each country against the international standard in order to streamline and promote better safety practices in the aviation industry worldwide. As a result the introduction of Crew Resource Management (CRM) and Human Factors training into most airlines training programs has been designed to maximise pilot performance and improve safety records.

Other changes are underway. For example, Transport Canada is working with other licensing agencies, such as the Federal Aviation Administration (FAA) in the United States, the Joint Aviation Authority (JAA) in Europe, and ICAO to standardise pilot licences and procedures to meet the international standards. Recognising the need to upgrade and update Canadian licences and ratings to meet the changing technology, philosophies and regulations, Transport Canada is currently developing three projects:

1. The Integrated Commercial Pilot Licence (CPLI) course, approved in 1999 for implementation in 2001, is a major effort by Transport Canada, in cooperation with industry, to upgrade, update and streamline the training of commercial pilots. Where once 40 hours of ground school was required for a commercial pilot licence (CPL), the CPLI will require 300 hours of ground training and the new licence, the CPL/IR, which includes training for the instrument rating (IR) as well as for the commercial pilot licence, will require 400 hours of ground training. It is likely that courses of this length will be taught in universities or colleges. The flight training part of the CPLI and the CPL/IR will take place within a carefully planned and scheduled curriculum and will occur only at Flight Training Units where Transport Canada has approved operations and training manuals (for more details see Air Transport Association of Canada, 2001, pp. 216-219 and 236-237, where there is a comparison between FAA, JAA and the Canadian program).
2. The Flight Instructor Project was begun in September 2000, in recognition of the necessity for flight instructors to receive more training to meet the needs of today's students. Addressing this problem, the Air Transport Association of Canada (2001) stressed that "in addition to other adjustments needed to be made by institutions planning to offer the new commercial pilot licence program (CPLI), attention will have to be given to their instructors. Flight instructors will need new skills in order to conduct training effectively in a competency-based approach" (p. 137).

   A group of experienced flight trainers from across the country met with Transport Canada Flight Training personnel to assess the flight instructor training program and make recommendations for change. The group looked at the structure of the present system, at training standards, at teaching practices and at regulations, with a view to improving the quality of flight instruction in Canada. In all, 29 recommendations were put forth and are currently being considered by Transport Canada senior management.
3. The Airline Transport Pilot Licence (ATPL) will provide new standards that will prepare candidates more fully for the "real world" flying of professional pilots. In the past, the ATPL has been a licence based on knowledge and upon the accumulation of experience (hours). The new standards and training procedures will cover many additional areas, some of which are mentioned below. These are timely adjustments and their introduction is validated by the data put forth in the *Human Resources*

Study on Commercial Pilots in Canada (Air Transport Association of Canada, 2001) indicating a general lack of training in the following critical areas:

- Crew Resource Management (CRM) training;
- Flight Management System (FMS) training;
- preparation for operations in High Level Airspace (18,000 feet and above);
- familiarity with Traffic Collision Avoidance Systems (TCAS);
- familiarity with Glass Cockpit (cathode ray) Instrument Display Systems; and
- familiarity with computerised components in a modern cockpit.

As well, pilots who did their training in the 1970s and 1980s report that their commercial training did not prepare them well for the "real world" of professional flying and did not differ substantially from private pilot licence training. They indicated that they would like to have received more training in the following areas:

- turbine engines/jet engines;
- complex airplanes — i.e., propeller systems, pressurisation;
- operational control of a commercial company;
- aircraft loading problems; documentation in log books and flight sheets; and
- standard operating procedures (personal interviews, May 4, 2001).

Impact on Training

Taken together, the impact of technological growth, of responses to performance issues, of an imminent shortage of qualified personnel and of scrutinised regulatory systems attempting bureaucratically to deal with a changing aviation industry, raises new challenges for those whose responsibility it is to train flight personnel. The research presented in the *Human Resource Study of the Commercial Pilot in Canada* (Air Transport Association of Canada, 2001) anticipates a restructured flight training system that leaves behind it the antiquities of its past and, driven by a "spirit of renewal", adopts "new methodologies, new approaches and new thinking" (p. 1).

> Alliances, partnerships, further consolidating and possibly new financial arrangements for students and schools will undoubtedly be pivotal in that effort, as will the need for a new, more professional breed of flight instructors ... As it reaches out for the best of tomorrow, [the flight training community] must be prepared to discard those things no longer in step with the next century of flight. (Air Transport Association of Canada, 2001, p. 2)

Transport Canada officials and industry operators and owners are becoming aware of flight training as a more inclusive learning process that will be necessary to develop the knowledge, skills and professional attitudes in well-trained flight personnel. The industry's contact with educators has contributed to a growing acceptance that it is important to better understand the concept of the "adult" in the learning environment, not simply the "pilot". In pilot training, concentration must be shifted to "whole person learning", the adult learning process, and accompanying motivational theory, if the resultant pilot performance is to meet an increasingly elevated world standard. The introduction of adult education theory can play a vital role in that development.

The Adult Learner: Fact or Fiction?

It has been understood since the initial work of Knowles (1970) that there is a field of adult education, the principles of which underlie the teaching of adults (see Chapter 1). Over the years, Knowles' theories have been challenged substantially; the "father of andragogy" himself acknowledged that the assumptions that comprised his initial theory were really just markers on a linear progression of techniques that ranged, at the one end, from those appropriate to the teaching of children to those at the other end of the spectrum, more suitable to the teaching of adults. Given the controversy around the andragogy debate over the years, the question remains: Are adult learners distinguishable from more youthful learners, those we think of as "traditional" learners? The answer is firmly *yes*.

As adults, there are certain ways in which we want to learn — ways that complement and ease the facts of our maturity, cognitively, socially, physiologically. This is not the place for a discussion of how adults learn and function cognitively, a still much-debated conversation among educators and psychologists (Barer-Stein & Kompf, 2001; Garrison & Archer, 2000;

Mackeracher, 1996; Schön, 1983). Rather, this discussion will focus on adult education knowledge that should direct teaching strategies, information that a practising flight training instructor could adapt to his or her interactions with adult learners.

The adult often approaches his or her learning situation with a great deal of fear and anxiety, coupled at the same time with the desire to integrate new learning into the rich fabric of already-held experiences. It is by virtue of having *lived* so long and achieved so many other goals that adult learners are driven to continue learning and to want to participate in their own learning process as a peer.

Much of the adult education literature stresses that adults should, in fact, be given some responsibility for planning their own curriculum and learning activities, and there are some instances where this approach is both desirable and possible. An example of this kind of learner-led curriculum would be a seniors' art class or a community swimming program. But still within the ranks of adult educators, the voice of pragmatism — spoken most vocally by instructors in the trades and in skill-related curricula — demands a more doctrinaire approach whereby the curriculum is set in accordance with outside criteria, by outside agencies who are also responsible for testing and measurement. The ultimate goal is often the awarding of a credential that is necessary to enter the workplace. A good example of this type of adult education is a journeyman's carpentry course at a technical institute.

Clearly, pilot training requires a set curriculum, a rigid examination process, and a resultant credential that has been meaningfully earned. Within these constraints, *still*, it is important to recognise that adults bring with them, along with their enthusiasm and excitement and fears, a unique set of characteristics that hallmark their learning styles. Following are some of those characteristics:

- Adult learners prefer single-concept courses that focus on the application of the concept to relevant problems. They are problem-centred.
- Adults need to be able to integrate new knowledge with what they already know if they are going to retain and use the new information.
- Information that conflicts sharply with what is already believed to be true is integrated more slowly. The forced re-evaluation of old material creates a state of cognitive dissonance in the learner that leads to discomfort, although a certain amount of cognitive dissonance is a useful learning tool.

- Adults tend to compensate for being slower at some psychomotor skills by trying to be more accurate and making fewer trial-and-error attempts.
- Adults take mistakes personally. Their self-esteem is more likely to be affected.
- Adults require a learning environment that is as psychologically comfortable as possible: safe, caring, open and non-threatening.
- Adults respond best to a learning environment that is as physically comfortable as possible to accommodate their changing physiological needs. Therefore, issues of lighting, seating, movement and sound should all be addressed.
- Adults vary in their learning styles (as does everyone) and appreciate instructors' efforts to vary their teaching styles.
- Adults need, deserve, and expect respect.

In a study in which he examined the learning journey of adult students in non-traditional learning environments, Woods (1994) identified eight stages that comprised their "painful odyssey": beginning with shock and denial, adults progress through stages of resistance, withdrawal, surrender and acceptance. Ultimately, they enjoy the return of confidence and feelings of success and integration as they master new techniques (Conrad & Kanuka, 1999).

Woods' (1994) work was based on adult students learning to adapt to a non-traditional form of learning — the computerised virtual classroom. Once called computer-mediated learning (CML), this type of online learning is now often referred to as e-learning and is currently a popular form of distance education which also still includes print-based methodologies (correspondence) and various other combinations of audio and video technologies. Woods' subjects went through a full range of negative and positive emotional benchmarks as they were exposed to this new type of learning. The eight stages that he documented were:

1. Shock: "I don't believe she is going to make us use the computer for this!"
2. Denial: "No way am I going to do this".
3. Strong emotion: "I can't do this. I'm going to fail. I hate this".
4. Resistance and withdrawal: "I'll find another way — I'll do this, but in my own way".
5. Surrender and acceptance: "OK, I'll have to do it her way if I'm going to pass".

6. Struggle and exploration: "Hmmm. Maybe I can do this. I'll try harder".
7. Return of confidence: "This is getting easier. I'm amazed that I could learn this much".
8. Integration and success: "I did it! This has been a great learning experience!"

In some measure, all adults trek similar difficult learning curves when confronted with new challenges. Aviation training would surely be placed high on the scale of difficult, daunting, and fear-inducing learning ventures — all the more reason to treat adults *as* adults by acknowledging their learning needs, and introducing adult education methodologies into the various types of aviation training (see Chapter 6 for a further discussion on barriers to learning).

Adult Education, Andragogy and Aviation: Placing Andragogy within the Field of Adult Education

The history of adult education as a field of study is long and not without controversy. The notion of a field of "adult education" was popularised in Europe in the nineteenth century and various institutions emerged in England during the tumultuous changes of the Industrial Revolution, designed to bring civility and training to the working masses in areas both social and vocational. Many of the institutions that developed in England at that time were later imported to the United States and Canada as emigration from the British Isles nurtured the settling of Upper and Lower Canada and the Eastern seaboard states (Conrad, 2001). Organisations such as the Mechanics' Institutes, Young Men's Christian Associations, and the Workers' Educational Association were among those that gave rise to the provision of adult education training.

At the same time, in those rapidly-evolving 19th century industrial societies, observation of the continuing educational needs of the "labouring poor" (Draper, 2001, p. 15) brought forth reflections from educators and philosophers alike on the nature of education, specifically contrasting educational opportunities for youth with the lack of meaningful education for mature and working adults. In Denmark, educator, historian, and theologian N.S.F. Grundtvig founded the innovative folk high schools, "schools for life", family-centred institutions where participants would awaken to a "truer and deeper understanding of life, leading to service for a better community and a better nation" (Kulich, 1984, p. 21). In so doing, he sought to incorporate adults' concrete experiences into meaningful "living" learning and renounced the traditional, theoretical, rote learning exercises that typified, at that time, the education of children.

In 1833, the term *andragogy* was coined by Alexander Kapp, a German teacher, in an attempt to describe Plato's education theory; that is, to distinguish this type of teaching of adults from the teaching of children (Draper, 2001, p. 16). Andragogy has been described by Malcolm Knowles, the American educator who popularised the term in his 1970 publication *The modern practice of adult education: Andragogy versus pedagogy*, as "the art and science of helping adults learn" (p. 38).

In Europe, especially, the terms adult education and andragogy have been interchangeably applied over the years. In English-speaking Canada, they have remained somewhat more separate, and it has been understood that the term *adult education* encompasses more broadly the enterprise of adults learning, wherever they learn, however they learn, whatever they learn. Still, within the field, the argument around the definition of andragogy and its relationship to pedagogy continues to occupy the literature (Draper, 2001; Feuer & Geber, 1988; Pratt, 1993). A full discussion of this controversy, interesting to educators historically, socially, and practically, is available in Draper's (2001) chapter "The metamorphoses of andragogy".

However, what is important for our argument, here, is to document the historical affiliation of andragogy — the art and science of helping adults learn — with the philosophical orientation known as humanism (see Chapter 1 for further information). Although the evolution of these concepts was not linear, sequential, or even closely parallel, we recognise that the growth of the andragogical approach to education was nurtured by the social humanistic outcroppings of the 18th century European Enlightenment, the period in which philosophers decried the severity of the mechanised scientific world as it was then understood, and began upholding instead the dignity of mankind. As the Enlightenment's thinkers — Diderot, Rousseau, and Voltaire — explored the autonomy and nobility of the human spirit, the emerging sense of human potential that they proffered became synonymous with new approaches to adult education. Hence, this "concept of learning within a humanizing social context provided the impetus for expansion in usage of both the concepts and the terminology of andragogy during the early to mid-twentieth century" (Draper, 2001, p. 18).

Working within an andragogical framework...

For our purposes, in considering the application of andragogical adult education principles and practices to aviation training, we can consider as a framework Brookfield's six principles of effective adult education as synopsised by Poonwassie (2001):

1. Facilitators must recognise that adults participate in learning on a voluntary basis; the decision may be prompted by external circumstances, such as loss of job, change in family situation, etc. but it is ultimately theirs. Learners can easily withdraw their participation, but facilitators should avoid coercion, bullying, or intimidation; instead, they should utilise participatory and collaborative learning techniques and design activities relevant to the participants' experiences and learning needs.
2. Effective practice is characterised by mutual respect of participants' self worth. Any behaviors or statements that embarrass, belittle, or otherwise abuse others should be avoided. Although challenges and criticisms are important components of facilitation, they must be presented in a helpful and respectful manner which does not take away from the participant's dignity or self-esteem.
3. Facilitation is based on the collaborative process. Facilitators and learners should work together in identifying learning needs, setting objectives, developing curriculum, identifying methodologies, and evaluating progress; exploration, discussion, and renegotiation of these aspects should be an on-going and integral part of the group process.
4. Praxis is key to effective facilitation. Exploration, action, and reflection placed in the context of the participants' past, present and future experiences help them interpret ideas and/or facilitate changes in their consciousness, attitudes, or beliefs.
5. Facilitation aims to encourage critical reflection. An effective facilitator will provide opportunities for learners to understand that their values, beliefs, behaviors and ideologies are culturally constructed and that alternative structures and practices may exist.
6. Facilitation strives towards the development of nurturing, self-directed, and empowered adults who will have the ability to continually re-evaluate, renegotiate and reinvent their personal relationship, work situations, and social structures (p. 148).

...and considering the questions that inevitably arise from the above

As experienced practitioners in the fields of adult education and aviation, we can hear the gasps from those trainers and instructors who are not familiar with the language of adult education, and we imagine instantly the types of indignant outcries that follow:

- I'm here to teach, not facilitate ... why is the emphasis on "facilitation"?
- How is facilitation like or not like teaching?
- I have a set and demanding curriculum; how can I negotiate anything?
- Learners are paying for my expertise; they deserve to have as much of it as I can give.
- What's praxis?
- Why should I be interested in my clients' personal well-being or sense of self-esteem?
- I teach; they learn.
- When safety is at stake, I need to "give" my students everything they need to be safe pilots.

These questions and comments, and many more similar to these, are asked repeatedly of adult educators by educators and trainers in other settings. Vocational instructors in technical institutes, journeymen tradesmen teaching carpentry in apprenticeship programs, computer and technical trainers in large corporations, academic upgrading instructors in remedial learning settings, adult basic education instructors working in prison environments — all are pressed for time as they address the demands of their employers for quantifiable results — for graduates, for completions, for whatever performance indicators have been deemed to adequately reflect the administrative need of the providing institution.

We recognise that, to enable learners to construct a "coherent and responsive stance" from the complex world of training information that constitutes the content of their learning task, the aviation industry must make a place for Daloz's (1998) "good education". The era of the "single pilot tradition" (Helmrich, 1993) is gone; however, to design a training program to meet 21st century needs, "it is important to explore and understand the human condition in learning" (Harris & Conrad, 1997) and to build learning technologies around the ways we know that adults learn.

Praxis

Praxis is a commonly-used term in adult education. Often used to refer to the radical Brazilian educator Paolo Freire's approach to empowering his country's illiterate peasants in their struggle against corporate and political oppressors, it means, simply, "putting theory into practice".

There are several opportunities for praxis to occur in the flight training system, as training occurs both on the ground and in the air and involves both human and machine interaction. To date, for example, the minimum requirements for Flight Instructor Training in Canada comprises 25 hours of ground instruction and 30 hours of flight instruction. The course for this training includes teaching and learning theory, lesson planning and motivational theory and the whole body of knowledge and skill required to become a competent instructor. Added to that the candidate must be able to demonstrate expertise in flying an aircraft.

To produce knowledgeable, highly skilled, and professional instructors in such a limited amount of time is clearly a difficult, but essential, task. According to the Air Transport Association of Canada (2001), "it will require new, more varied, more elaborate curriculums, sophisticated training devices and a new more modern fleet of aircraft. Most importantly it will require new methodologies, new approaches and new thinking" (p. 2).

What follows here are several scenarios that illustrate various areas of aviation instruction. Think about how each situation could be improved with the introduction of adult education techniques, based on the adults learning needs that are described above in the section, "The Adult Learner: Fact or Fiction?" (also see Chapter 1).

Scenario 1: Ground School

Henry is a flight instructor who has been flying with students all day. He is slated to teach ground school in the evening but has not had time to prepare for the lesson on theory of meteorology, although his knowledge of the material is good. Fatigued and rushed, he arrives five minutes late for class. He asks the 20 students if they've done their reading assignment for this lesson. Dan raises his hand and says: "Yes, I've read the information and I have some questions."

"Good," Henry says, "but just hold on to them for now because we'll cover all the material first and you will get your answers as we go along".

Henry then asks the class to open their books to the appropriate page and he begins to speak, moving through the material quickly. He jokes that the students will never understand it on the first go-around, and tells them: "Don't worry; it will come to you eventually".

Henry answers a few questions from the students with a curt "yes" or "no" but gives no explanations and does not encourage discussion. He is feeling pressured to get through the material in the shortest time possible. He is oblivious to the frustration and puzzlement of his students. How could Henry have made this session more productive and successful? Let's analyse Scenario 1.

- The instructor is suffering from fatigue and stress.
- He has a good knowledge base and some instructional skills, but he has not had time to prepare and plan a good learning environment for his students.
- He is reflecting the "hurry up" syndrome. (The hurry up syndrome was identified by McElhatton and Drew in 1993, based on their findings from an Airline Pilots Association investigation and a NASA study. The hurry up syndrome results in pilots' performances being compromised by a perceived or real need to hurry. Chrystian (2000) reports that the hurry up syndrome has been named as a factor in several aviation accidents.
- His students leave tired and discouraged.
- They have unanswered questions.
- Their desire to learn has been squelched by having been told that "they won't be able to learn this stuff".

Scenario 2: Simulator Sessions

Marion is a commercial pilot working on her instrument rating. She is on her second session in the simulator, working on hold entries. Her instructor, Orville, arrives in the simulator room, fires up the simulator and says: "OK, let's get started, Marion!"

Marion is flustered and anxious. She has forgotten how a hold entry works, even though Orville explained it to her during her last lesson, and she is too embarrassed to ask.

As the lesson progresses, Marion has trouble maintaining altitude, airspeed and heading. Her performance in basic skills deteriorates. This is new for her, as she was doing very well previously with aircraft control. She feels guilty that she has not mastered the "hold entry procedures" and dreads being considered incompetent by her instructor. Orville is impatient and unsympathetic.

He barks his next directive: “You are cleared to the Edmonton VOR to maintain 6000. Hold southwest on the 230 degree radial. Expect further clearance at 1800Z”.

There is no answer. Marion does not read back the clearance. Orville stops the simulator: “What’s the matter? Didn’t you hear the clearance?”

“Oh yes,” a shaking Marion replies, “but I can’t seem to fly this thing today and talk at the same time!”

“Well, you’re going to have to do better than that, you know,” replies Orville. “In instrument flying, there are lots of other things you’ll have to be thinking about all the time!”

What could Orville have done to assist Marion in her training session? Let’s analyse Scenario 2:

- The Law of Readiness was not observed (i.e., Marion was not ready to learn this lesson). Orville would probably have discovered Marion’s problem before the lesson started if he had taken the time to talk to Marion prior to getting into the simulator.
- It is essential when teaching a skill to assess where in the skill process the student is and work from there.
- Orville used sarcasm and ridicule in his response to Marion.
- His lack of concern and understanding of her state of mind contributed to her secondary state of fear and stress.
- Marion has not developed the skill of self-evaluation in order to know where she is in the learning cycle. This knowledge would enable her to say, “I don’t know,” or, “I’m not ready for this yet.” Both instructor and student have a responsibility in the learning environment to check and connect with each other about what is happening.

Scenario 3: A Scene in the Office of the Flight Training Unit

Teresa, a private student, has quietly arrived for her flying lesson. Three instructors are talking and laughing in the office. Teresa overhears their conversation:

Instructor 1: “You should have seen Roger’s face when that airplane bounced on the runway! I had to take over, of course, and land, but he was really upset! I told him not to worry — it’s good to have that experience, so he’ll know what to do next time”.

Instructor 2: "My student John was having a bad day too ... couldn't seem to get the steep turns. I kept showing him, but he just said it was all right for me to be perfect, but he couldn't figure out what he was doing wrong".

Then Teresa's instructor, finally noticing her, said, "Oh, hi, Teresa, are you ready to go?"

Teresa is wondering what they say about *her* when she's not around. Let's analyse what happened in Scenario 3.

- A breach of student confidentiality is evident.
- Teresa's trust in her instructor has been badly shaken.
- When she completed her Private Pilot Licence, she related the incident to her instructor and was told that she was too sensitive and that "everything in aviation is public knowledge".

Scenario 4: In-flight Instruction

This is an excerpt from a letter written by a student after obtaining her licence:

> When I was first learning to land, I felt overloaded with everything I had to watch and do. I asked my instructor if, for one landing, he could control the rudders so I could focus more specifically on the end of the runway and what the correct attitude was. His response was: "Negative!" He gave no explanation.
>
> Needless to say, I did my best to do everything required for that landing and obviously, I eventually learned it, but I felt there was no attempt made to meet my learning needs at that time. I felt that if he had controlled the rudders for even one landing as I had requested, it would have allowed me to grasp the attitude and maybe for the next landing I could have tied more things together. Do you know what I mean?
>
> Anyway, that never happened. After completing my licence, I talked to the instructor about this particular incident and he commented that he never did that for students. "They have to learn to do everything".
>
> I realise that students have to learn everything but they don't have to learn everything all at once. Neither do all students learn at the same pace or in the same way. It is my feeling that if instructors received some educational training they would understand that. This would then be reflected in their adaptability to meet students' different learning needs.

What was the disgruntled student in Scenario 4 really saying?

- A popular myth in aviation is that “loading up” (overloading) student pilots is good for them. “Makes ‘em tough”.
- The oft-heard argument, “If he/she can’t take the pressure, what will he/she do in an emergency?” creates negative learning situations.
- Since stress and pressure leads to poor performance, the ability to learn in this situation is severely hampered (Henley, 1986, 1987).
- Students need to be recognised as individuals.

Debriefing: What Can We Learn from the Scenarios Above?

Taken together, the four scenarios above illustrate a considerable number of infringements upon the learning rights of adults as they are understood by practitioners of adult education. The consistent theme that emerges from all scenarios is one of frustration on the part of learners: This is usually a precursor to failure in the learning endeavour. Is there much difference here between Henry and Orville's training methodologies and the earlier applications of the “show and go” era described by Main (1967)?

Adults learn best in supportive, psychologically safe, cautious, respectful, collaborative, and dignified environments (Rogers, 1969). Is it possible that aviation training can be housed in such an environment? We think so. Let’s examine some of the techniques that could contribute to more adult-centred, positive learning experiences.

Self-Evaluation and Reflective Journals

It is common sense in the aviation world that learners will be rigorously examined to determine that they have achieved the necessary standard in order to be licensed. On the way to that goal, a process of self-evaluation could serve as a useful tool, as it does in many other adult learning situations. Providing opportunities for learners to reflect on their actions is a valuable exercise in building self-knowledge, wisdom, and the ability to critically analyse actions, decisions, and situations (Schön, 1983, also see Chapter 8). Sustained reflection can be achieved through the use of reflective journals that span the length and breadth of training. Reflective exercises, where time is set aside at the beginning or at the end of training sessions, to envision or recapture the events of the day, also encourage responsible learning (Henley, Anderson, & Schick, 1999).

Self-evaluation can also occur outside of reflective activities with the use of self-testing paper-and-pencil instruments, or through the use of conversational prompts in a period of debriefing. The central purpose of any of these activities is to encourage self-analysis on the part of the learner in a thoughtful manner. It does not replace the need for the external examination of the learner by the instructor.

Self-Management Plan

Both journal-keeping and self-evaluation could form part of a self-management plan for learners. An effective self-management plan could be as simple as this:

1. What are my goals? (Will all learners' goals simply be, "learn to fly"? Or could they include: "develop confidence in my skill level"; "overcome fear of flying"; "realise lifetime dream ..." Could these insights be valuable both to learner and instructor?)
2. What steps are necessary for me to achieve my goals?
3. What outcomes do I expect from undertaking this learning goal?
    a) knowledge
    b) skills
    c) attitudes
4. What resources will I need to access in order to achieve my goals?
    a) external
    b) internal
5. What will success look like to me?

The actions of thinking about the actions *outside* of the actions are similar to the "reflection-on-action" process in which learners evaluate and situate events after they have occurred. "This re-evaluation can lead to a questioning of what occurred and of the assumptions that may have guided decision making, and sometimes, can result in a new perspective on a practice, issue or situation" (Henley, Turney, & Schick, 2001, p. 2). Encouraging learners to "pre-vision" parts of their experience also serves to foster critical thinking, anticipation, and structuring (see Chapter 8).

Collaborative or Student-Centred Debriefing

There are several ways in which debriefing can be used as a positive, learning tool in students' instruction, all of which are preferable to the techniques that Henry used in his training — techniques that left students feeling unsatisfied, confused, and undervalued. Student-centred approaches to learning are the basis on which andragogy and adult education practices are founded. In debriefing situations, emphasis on students can heighten and strengthen their learning experiences in these ways:

- It allows them to uncover and articulate their own mistakes.
- It allows them to recognise and come to terms with the extent of their own knowledge.
- It encourages them to learn how to analyse a situation or event.
- It reinforces in them the value of their own observations.
- It builds in them the skill to "speak the language" in meaningful ways, thus reinforcing their aviation orientation (Henley, Turney, & Schick, 2001).

The techniques above can fit into two larger categories — one philosophical, one attitudinal. Constructivism and respect are discussed below.

Constructivism

Constructivism is one of the sets of beliefs that guide our thinking — and our actions as teachers — in the practice of helping others to learn. In the constructivist view, the making of meaning encourages us to reshape ourselves as we move forward in the learning journey (see Chapter 1). Constructivism "proposes that learning environments should support multiple perspectives or interpretations of reality — knowledge construction, context-rich, experience-based activities" (Jonassen, 1992, p. 28). Similarly, Steier (1991) contends that the circularity of reflexivity is central to the notion of constructivism, embracing the notions of echo and resonance into the learning dynamic.

As a philosophical construct, constructivism underpins instructional approaches that recognise the richness and uniqueness of not only learners' chronological and social experience but also the different ways that learners experience their worlds. In aviation training, the tasks at hand cannot be denied; a constructivist approach would recognise that each learner brings to the simulator, for example, his or her own capability to "make sense" of the lesson. The constructivist approach would first look for ways to identify those capabilities and then attempt to address each learner's ability to incorporate the learning task into his or her realm of understanding.

In practical terms, that may translate into using different instructions or different examples for different learners; it would mean, as in Marion's case, pacing the training to accommodate her ability to absorb new learning; it means giving learners the opportunity to feed back their learning in some way that gives voice to the way they have understood the material.

Respect

The discussion on respect is a far easier one than the foregoing discussion of constructivism. Most of us are raised with a good notion of what respect is and what it means to respect another individual. In the adult education classroom, respect towards learners by an instructor contributes to the overall sense of safety that adults, always fearful of embarrassment or failure, require for comfortable learning (Cross, 1981; Renner, 1993). One of the most basic ways to engender a climate of respect is simply to raise the issue — to set out behavioural rules, conditions, and expectations for both teacher and learner. In "rubber meets the road" situations such as the cockpit of the airplane, pre-established behavioural understandings accomplish three things: 1) They serve as advance organisers for what may possibly come; 2) They reduce confusion around incidents, should there be any; and 3) They open the door for a framework for incident analysis after the fact. The salient point around the establishment of safety and respect in this regard is that the rules are observed.

Introducing respect into any teaching-learning dynamic subtly addresses the power differential that creeps into many learning environments, and the remnants of militarism in aviation training are likely contributors to such imbalances of power. The opportunity, and often the need, for instructors to critique their students' performances may provide the platform for unhealthy displays of power. Henley, Turney, and Schick (2001) posit that "feedback needs to be non-threatening, constructive and supportive" (p. 6) in order that it allows students to benefit from receiving it. They point to these outcomes in learners:

- A greater frankness and open-mindedness;
- The ability to link theory to practice;
- The ability to search for alternative answers;
- The ability to reflect;

- The ability to challenge perceptions; and
- The eventual progression to higher levels of reflection, resulting in eventual independent learning (p. 6).

The end-goal of respectful interaction between teacher and learner is an enhanced sense of professionalism, greater confidence, and the subsequent ability of learners to better perform in complex environments. Somewhat similarly, in the style of the humanistic psychologist Carl Rogers, Renner (1993) highlighted respectful, learner-centred teaching in this way: "the educational situation that most effectively promotes significant learning is one in which threat to the self of the learner is reduced to a minimum" (p. 27).

Conclusion

Come to the edge, he said,
They said: we are afraid.
Come to the edge, he said.
They came,
He pushed them … and they flew.
(Guillaume, quoted in Poonwassie, 2001, p. 153)

Aviation training exists at the edge: the commitment of those who undertake to fly is high on all fronts. It is costly; it is exacting; it is thrilling. The stakes are high.

The quote above speaks to the natural fear and hesitation that exist in all of us as we enter new frontiers and challenge our boundaries. At some point in our journey, we are "pushed"; that is, we receive stimuli, motivation, encouragement, a helping hand — some external force complements the internal drive that has nudged us along thus far. In this chapter, we have examined the history of and concepts surrounding aviation training in Canada, and we have presented historical training methodologies as "pushing". And, in Guillaume's words, "they flew". They flew in rickety single-seaters; they flew bravely delivering mail over mountain tops; they flew over enormous grey bodies of water; they flew in the names of freedom and democracy, and they fly for pleasure.

In spite of the fact that aviation progress has been steady and constant, the Canadian aviation industry has documented room for improvement and change in order to accommodate anticipated changes in demand for and expectation of flight personnel (Air Transport Association of Canada, 2001). Future training, incorporating the notion of "human factors", will strive to adopt "whole person learning", where pilots are understood to be more human than godlike and where the health and texture of the environment in which aviation occurs is less rigidly dogmatic and more democratic. As a part of this new balance, the nature of the adult will be considered more centrally in training exercises — who are they as learners? How do they learn? How can their learning be enhanced? How can the learning process be improved while at the same time becoming more accessible? And at no time should aviation standards be compromised in the adoption of new methodologies and approaches to training.

Incorporating basic adult education philosophies into aviation training affords learners the opportunity to develop the higher-level skills that foster their ability to think critically, self-evaluate, analyse and reflect both "in-action" and "on-action" (Henley, Turney, & Schick, 2001, also see Chapter 8). Learners who develop these skills demonstrate higher levels of autonomy, confidence and self-worth (Brookfield, 1995; Cross, 1981; Schön, 1983). The "stick and rudder" skills that in a simpler era were the sole hallmarks of aviation expertise — while still necessary — must now exist concomitantly with information management and "knowledge" skills — sets of decision-making managerial tools that recognise the integration of global facts — competition, terrorism, technology — into the industry.

In a changing world of borderless economies, all front-line workers must become "more sophisticated and knowledgeable about why they are doing the things they do" (Couture, 1993, p. 15). If American economist Robert Reich's division of workers into three types — "routine producers, in-person servers, and symbolic analysts" (p. 14) is correct, then those who front-line the aviation industry in commercial and personal transport must step forward with new confidence and new sets of analytical tools. Their enlarged responsibilities must include elevated problem-solving and communication skills. Adult education practices offer aviation the blueprint for 21st century training methodologies.

References

Air Canada Human Factors Project Team (1997). *Human factors for aviation: Basic handbook.* Ottawa: Transport Canada.

Air Transport Association of Canada (2001). *Human resources study of commercial pilots in Canada*. Ottawa, Canada: HRDC.

Barer-Stein, T., & Kompf, M. (2001). (Eds.). *The craft of teaching adults* (3rd ed.). Toronto, Canada: Irwin Publishing.

Brookfield, S.D. (1995). *Becoming a critically reflective teacher.* San Francisco: Jossey-Bass.

Chrystian, J. (2000). *At the scene of the accident: Is pilot age involved in aviation accidents?* Unpublished paper. University of Alberta, Edmonton, Canada.

Conrad, D.L. (2001). The issuc of access in adult education: Privilege and possibility. In D. Poonwassie & A. Poonwassie (Eds.), *Fundamentals of adult education: Issues and practices for lifelong learning* (pp. 204-213) Toronto: Thompson Educational.

Conrad, D.L., & Kanuka, H. (1999, June). Learning in safety and comfort: Towards managing on-line learning transactions. Proceedings of the Association for the Advancement of Computing in Education (AACE) Conference, Seattle, USA.

Couture, P. (1993). Lifelong learning and prosperity. *Learning Magazine, 6*(2), 12-18.

Cross, K.P. (1981). *Adults as learners: Increasing participation and facilitating learning.* San Francisco: Jossey-Bass.

Daloz, L.A. (1988). Beyond tribalism: Renaming the good, the true, and the beautiful. *Adult Education, 38*(4), 234-241.

Draper, J.A. (2001). The metamorphoses of andragogy. In D. Poonwassie & A. Poonwassie (Eds.), *Fundamentals of adult education: Issues and practices for lifelong learning* (pp. 1-30). Toronto: Thompson Educational.

Elshaw, C. (1993). Preparing better flight instructors. In R. Telfer (Ed.), *Aviation instruction and training* (pp. 253-270). Aldershot, England: Ashgate

Feuer, D., & Geber, B. (1988). Uh-Oh ... Second thoughts about adult learning theory. *Training: The Magazine of Human Resource Development, 25*(12), 31-39.

Garrison, R., & Archer, W. (2000). *A transactional perspective on teaching and learning: A framework for adult and higher education.* Amsterdam: Pergamon.

Harris, J., & Conrad, D.L. (1997). Adult learning in the aviation training environment. In R.S. Jensen & L. Rakovan (Eds.), *Proceedings of the Ninth International Symposium of Aviation Psychology* (pp. 1209-1212). Columbus, OH: Ohio State University.

Helmreich, R.L. (1993). Future directions in crew resource management. I*CAO Journal, 48*(7), 8-9.

Henley, I. (1986). Student pilot stress and flight instructor training. *Journal of Educational Administration and Foundations, 1*(1), 16-24.

Henley, I. (1987). *The association between student pilot stress and flight instructor training: A study of perceptions.* Unpublished Master's thesis, University of Manitoba, Winnipeg, Manitoba, Canada.

Henley, I. (1995). *The quality of the development and evaluation of flight instructors in Canada and Australia.* Unpublished doctoral thesis, University of Newcastle, Newcastle, Australia.

Henley, I., Anderson, P., & Schick, D. (1999, April). Teaching student pilots to become reflective practitioners. In D.L. Burke (Ed.), *Proceedings of the 2nd Flight Instructor Conference: Innovations in Flight Instruction* (pp. 55-62). Daytona Beach, FL: Embry-Riddle Aeronautical University.

Henley, I., Turney, M., & Schick, D. (2001). Practical strategies for aviation educators: Reflective practice and student-centred debriefing. In R.S. Jensen (Ed.), *Proceedings of the 11th International Symposium on Aviation Psychology* (CD-ROM). Columbus, OH: Ohio State University.

Jonassen, D.H. (1992). Evaluating constructivist learning. In T.M. Duffy & D.H. Jonassen (Eds.), *Constructivism and the technology of instruction: A conversation* (pp. 45-56). Hillsdale, NJ: Lawrence Erlbaum.

Knowles, M. (1970). *The modern practice of adult education*. Chicago: Follett Press.

Kulich, J. (1984). Grundtvig: Education for life. *Learning, 4*(1), 19-21.

Mackeracher, D. (1996). *Making sense of adult learning*. Toronto: Culture Concepts.

Main, J.R.K. (1967). *Voyageurs of the air*. Ottawa: Queen's Printer.

McElhatton, J., & Drew, C. (1993, September/October). "Hurry-Up" syndrome identified as a causal factor in aviation safety incidents. *Human Factors and Aviation Medicine*, *40*(5), 1-6.

Mold, E. (1982, June). Born again pilot. *Airforce, 6*(2), 6 & 38.

Poonwassie, A. (2001). Facilitating adult education: A practitioner's approach. In D. Poonwassie & A. Poonwassie (Eds.), *Fundamentals of adult education: Issues and practices for lifelong learning* (pp. 147-158). Toronto, Canada: Thompson Educational.

Pratt, D. (1993). Andragogy after twenty-five years. In S.B. Merriam (Ed.), *An update on adult learning theory* (pp. 15-23). San Francisco: Jossey-Bass.

Reason, J. (1997). *Managing the risks of organizational accidents*. Aldershot, UK: Ashgate.

Renner, P. (1993). *The art of teaching adults: How to become an exceptional instructor and facilitator.* Vancouver: Training Associates.

Rogers, C. (1969). *Freedom to learn*. Columbus, OH: Charles E. Merrill.

Roscoe, S.N. (1980). *Aviation psychology.* Ames, Iowa: Iowa State University Press.

Schön, D. (1983). *The reflective practitioner: How professionals think in action.* New York: Basic Books.

Steier, F. (1991). Research as self-reflexivity, self-reflexivity as social process. In F. Steier (Ed.), *Research and reflexivity* (pp. 1-11). London: Sage.

Transport Canada (1998). *Flight instructor guide (TP 975-E).* Ottawa: Supply and Services Canada.

Wiener, E.I., Kanki, B.G., & Helmreich, R.L. (Eds.). (1993). *Cockpit resource management.* San Diego, CA: Academic Press.

Woods, D. (1994). *Problem-based learning: How to gain the most from PBL.* Waterdown, Canada: Donald R. Woods.

4 Learning Styles, Multiple Intelligences and Personality Types

Irene Henley and Jayne Bye

Introduction

Understanding something of the cognitive processes and the nature of adults' preferred learning styles, different forms of intelligence, and personality type theories can have a powerful impact on planning for effective teaching and learning in aviation since research suggests that flight safety and effective industry functioning rely on a range of factors. Undeniably, the operation of a sophisticated multi-crew aircraft is dependent on the efficient and effective performance of a team of highly trained professionals, and a vast amount of research has been conducted on the way people learn and acquire skills which has resulted in many different theories of learning styles, cognitive styles, and thinking styles (Biggs, 1987; Entwistle, 1981; Entwistle & Ramsden, 1983; Filipczak, 1995; Kolb, 1976, 1984). However, there is also mounting evidence to support the view that personality traits are an important factor in incidents and accidents (e.g., the KLM and the Pan American 747s collision at Tenerife in 1977, and the Trident accident at Heathrow in 1972), and that personality plays a significant role in the overall maintenance of flight safety (Foushee & Helmreich, 1988; Hawkins, 1987; Stead, 1995).

This chapter presents an overview of learning styles, multiple forms of intelligence, and personality types as these can provide some useful guidelines for our practice. Firstly, the information processing theories will be examined with the main focus being on Kolb's (1976, 1984) four learning styles and some of the strategies teachers can use to help students become more effective learners. Research into approaches to learning will also be reviewed briefly as an extension of theories of learning styles.

Secondly, a summary of Gardner's (1983, 1993) multiple intelligences theory will be presented. The multiple intelligences model proposed by

Gardner provides a valuable framework for aviation educators since most of the eight intelligences or abilities identified by Gardner have applications within the aviation context. Finally, the Myers-Briggs Personality Type Indicator (MBTI), one of the most widely used personality models in education, will be considered. The MBTI is a useful tool for aviation educators interested in developing team skills because it provides invaluable insight about personality types for group members. If members within a group are able to better understand and appreciate each other's individual differences and unique characteristics, they will be better equipped to function more effectively as a team.

Information Processing Theories — Learning Styles

Merriam and Caffarella (1999) report that "there is no common definition of learning style, nor is there a unified theory on which this work is based" (p. 209). Nevertheless, many of the learning style theories have led to the development of learning style inventories and one of the most popular instruments used with adult learners has been Kolb's (1976, 1999) Learning Style Inventory. Kolb (1984), in his book entitled *Experiential Learning*, observes that "as a result of our hereditary equipment, our particular past experience, and the demands of our present environment, most people develop learning styles that emphasize some learning abilities over others... Each of us in a unique way develops a learning style that has some weak and some strong points" (pp. 76-77). Rogers (1998) adds that "in any adult learning group there will always be people with a range of different learning styles" (p. 110). Similarly, every teacher has a preference for one or more teaching styles based on his or her preferred learning style.

What is a Learning Style?

According to Smith (1982), a learning style can be defined as "the individual's characteristic ways of processing information, feeling, and behaving in learning situations" (p. 24). Merriam and Caffarella (1999) define cognitive styles as "consistencies in information processing that develop in concert with underlying personality traits" (p. 208). Similarly, Kolb (1984) refers to a learning style as the particular way people *perceive* and *process* information and solve problems. Each one of us perceives the world in different ways. Some predominantly *sense and feel*; they prefer to interact directly with the environment (sensory-based experience). Others *think or reason*; they prefer to think things through (they reason about experience). After perceiving information, we process that information. Some *watch and reflect*, while others *jump right in and try things out* (McCarthy, 1982, p. 21).

Why Consider Learning Styles?

Kolb (1999) notes that understanding our learning style can help us become more effective learners and problem solvers, essential attributes in aviation. As teachers, we should also be aware that the way we prefer to learn will affect the way we teach. From the students' perspective, the more they are actively engaged in the learning process, the more effective the learning will be. However, active engagement means different things to different people at different times. At times, we might prefer to listen, think, read, write or draw, while at other times, we want to talk, discuss, create something or perform an activity.

The challenge for teachers is to:

- be aware of their own preferred teaching style;
- be aware of the preferred learning styles of each of the students; and
- optimise the congruence between teaching styles and learning styles, thereby maximising the learning opportunities for the students.

The challenge for students is to:

- be aware of their own preferred learning style;
- let the teacher know of this preference; and
- attempt to engage in the learning process even when a preferred learning style is not being provided for.

Kolb's Learning Style Inventory

Kolb (1976, 1984) designed a two-axis grid by juxtaposing the two dimensions of perceiving and processing information to form a quadrant system that depicts four different learning styles. The vertical axis (perception) separates those who prefer Concrete Experience (CE) or Abstract Conceptualization (AC). On the horizontal axis (processing), Kolb discriminates between Reflective Observation (RO) and Active Experimentation (AE). By looking at the quadrants, it is possible to identify "types" of people as divergers, assimilators, convergers and accommodators (see Figure 4.1). Rogers (1998), in his brief summary of recent thinking on people's preferred learning styles, provides similar categories of learning styles, but refers to type 1 learners (divergers) as reflectors, type 2 learners (assimilators) as theorisers, type 3 learners (convergers) as pragmatists or experimenters, and type 4 learners (accommodators) as activists or active learners (see Figure 4.1).

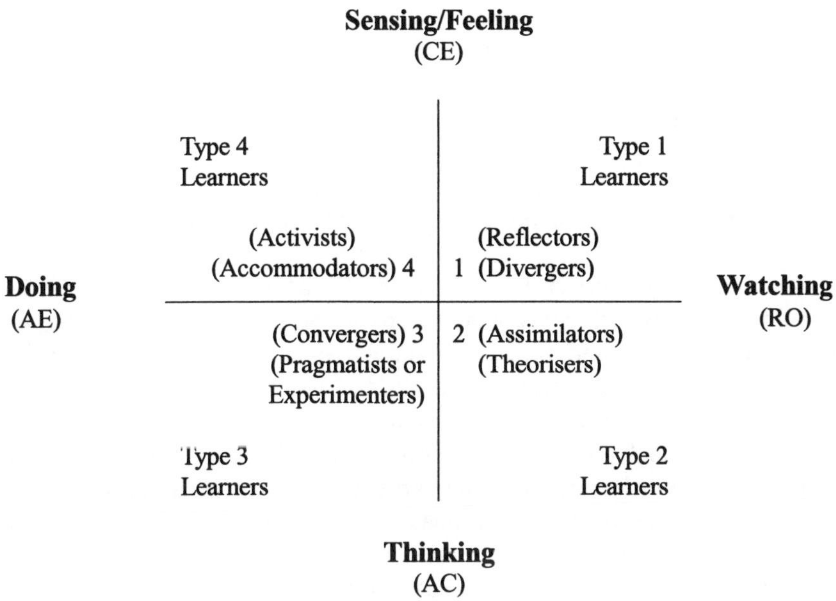

Figure 4.1 Learning styles

Source: Adapted from Kolb, 1976, 1999; McCarthy, 1982; and Rogers, 1998.

Both Kolb (1999) and Rogers (1998) stressed the word *preferred* learning style, since different situations or different tasks may require the use of learning styles that may not be one's preferred learning style. Kolb (1999) further explained that his learning style inventory is based on "Piaget's definition of intelligence as the balance between the process of adapting concepts to fit the external world (accommodation) and the process of fitting observations of the world into existing concepts (assimilation). Convergence and divergence are the two essential processes identified by J.P. Guilford's structure-of-intellect model" (p. 8).

Divergers, as the name coined by Kolb (1976) implies, are divergent thinkers who combine concrete experience and reflective observation. That is, they perceive information concretely and process it reflectively. In a learning situation, divergers (reflectors) normally will not give the first answer or idea that comes into their head; they take the time to reflect first, thus the name reflectors by Rogers (1998). They would rather "wait and see". As a result, divergers usually prefer to learn by listening.

However, divergers also like to share ideas or learning experiences with others because that allows them to gather different views, different perspectives before they make up their own mind. Hence, they function well through social interaction. They normally enjoy discussion group or interactive learning situations. As learners, their primary concern is personal meaning. The strengths of divergers are their ability to view concrete situations from different perspectives, their imaginative ability, and their sensitivity to the feelings of others (Kolb, 1999; McCarthy, 1982; Rogers, 1998).

Assimilators (theorisers) are learners who combine reflective observation and abstract conceptualisation. In other words, they perceive information abstractly and process it reflectively; they learn by thinking through ideas. Assimilators prefer to understand general concepts first, before looking at real issues or cases (i.e., they read the manual before they try to assemble or fix something)! Assimilators normally enjoy the traditional classroom and prefer lectures, reading and exploring analytical models since they are usually more interested in ideas and concepts than in people. As a result, they may be less sympathetic to other people's feelings or subjective judgements. They normally like to analyse ideas or think problems through step by step and often represent ideas or concepts using flow charts or diagrams showing relationships. The strengths of assimilators are their ability to create concepts and models, to understand a wide range of information, and put it into concise, logical form (Kolb, 1999; McCarthy, 1982; Rogers, 1998).

Convergers (pragmatists or experimenters) learn by combining abstract conceptualisation and active experimentation. They perceive information abstractly and process it actively. In simple terms, they need to know how things work; they seek usability and enjoy solving problems. Therefore, teachers should let them try things out, if possible. Convergers like to find practical applications for ideas and theories, thus the name pragmatist. They would rather deal with technical tasks and problems than with social and interpersonal issues (Kolb, 1999).

In a learning situation, convergers prefer to experiment with new ideas. As a result, they enjoy simulations and would rather do assignments that have practical applications, then spend (waste!) time listening to theory lessons. Furthermore, convergers or experimenters will normally not believe that something works until they can try it out for themselves (i.e., "Let's see if this really works for me too"). The strengths of convergers are their ability to solve problems, to find practical applications of ideas, and to make decisions based on their desire to find solutions to questions and problems (Kolb, 1976, 1999; McCarthy, 1982; Rogers, 1998).

Accommodators (activists) combine active experimentation and concrete experience. They perceive information concretely and process it actively. They learn primarily through "hands-on" experience, trial and error or self-discovery approaches. If practical, teachers should allow these learners to explore individual approaches to achieving a learning task. Accommodators are action-oriented people (i.e., they read the manual or listen to explanations/instructions only when everything else fails)! Therefore, they normally will get impatient if a teacher provides lengthy explanations or tells them too many details about the task to be performed.

Accommodators tend to act on "gut" feelings rather than through logical analysis. However, because of their intuitive ability, they can often reach accurate conclusions in the absence of seemingly logical justification. They like to seek hidden possibilities (i.e., What if we did it this way? or What would happen if ...?). They enjoy new challenges and readily adapt to change. They do not mind taking risks. The strengths of accommodators or active learners are their flexibility to adapt to new situations and their ability to carry out plans, to put ideas into action (Kolb, 1999; McCarthy, 1982; Rogers, 1998).

In 1999, Henley administered Kolb's (1999) Learning Style Inventory to close to 450 flight instructors across Australia during national Flight Instructor Seminars conducted by the Civil Aviation Safety Authority Promotion Division. The results revealed that close to 50 percent of flight instructors who attended the seminars were Convergers, slightly over 40 percent were Assimilators, approximately 10 percent were Divergers, while less than 2 percent were Accommodators. This is quite a different distribution when compared to the percentages found in the general population. McCarthy (cited in Burns, 1989) found that only 18 percent of those surveyed within the general population were Convergers, 22 percent were Assimilators, 35 percent were Divergers and 25 percent were Accommodators (p. 43). However, the differences between the flight instructor group and the general population is hardly surprising since people with a particular learning style tend to gravitate toward certain types of careers (Kolb, 1999).

Other educationalists, such as Filipczak (1995) and Karp, Turney and McCurry (1999) propose another model of learning styles, which includes three primary, or dominant, types of learners: *visual learners*, who learn more effectively by reading and looking at pictures, or by visualising how to perform a task; *auditory or aural learners*, who learn best by listening and are more attuned to sounds during task performance; and *kinesthetic or hands-on learners*, who prefer to use their hands or whole body to learn and can feel how to perform a task. Studies by Tannen (1990), Emanuel and Potter (1992), and Turney (1995) have also shown that men and women learn and communicate differently. For example, while most women like to share information and prefer to learn by interacting with others, men generally tend to favour "debate-like" situations (Karp, Turney, & McCurry, 1999; Tannen, 1990). Similarly, Emanuel and Potter (1992) found that women tend to be more participatory in their preferred learning styles, while men tend to be more independent or individualistic. According to Tannen (1990), women are normally more collaborative, they are more inclined to "negotiate" in their communication style, while men, being more competitive, tend to speak to exchange information and establish status in a group.

Notwithstanding the different models of learning styles proposed in the literature, what needs to be stressed is that learning is the process of creating meaning or making sense of "personal" experience (Mezirow, 1981). However, for that to happen, learners need to develop the necessary skills to critically reflect on their learning experiences, and objectively analyse those experiences, so that the insight gained will guide future practice (see Chapter 8 for information on how to help students develop critical reflective skills).

Teaching Strategies to Accommodate Learning Styles

According to McCarthy (1987a, 1996), teachers will be more effective if they use a variety of teaching approaches and techniques in order to accommodate each learning style at least some of the time. Similarly, Burns (1989) notes that "the more the teacher learns to recognise the various ways in which individuals learn, the more flexible he or she becomes in creating effective programs that ensure success for everyone. There's a time for experiencing, a time for lecturing, a time for discovery and a time for exploring new possibilities" (p. 44).

The main strategic approaches that the teacher can adopt to optimise the congruence between teaching styles and learning styles are:

- present a single topic in a number of different ways concurrently;
- present a single topic in a number of different ways sequentially;
- for a series of topics, present each topic in a different way; and
- devise holistic learning activities which incorporate multiple topics and multiple learning styles.

For example, since divergers are primarily interested in personal meaning, teachers should provide reasons or explain why a learning task is important. Assimilators, who seek facts since factual information leads to conceptual understanding, could be given extra reading material or handouts that provide additional information. Teachers should provide practical exercises for convergers, since they want to find out for themselves how things work. Finally, if suitable, accommodators should be allowed to teach themselves (self-discovery), or if they can perform a task well, they could be asked to teach other students who may need assistance.

A complicating factor in this argument promoting congruence between teaching and learning styles is that, if one of the goals of the educational program is to assist the students to *learn how to learn*, the challenge of learning through "non-preferred" styles might be important. Indeed, Kolb (1984, 1999) stresses that if the aim is to develop more effective learners, teachers should not only vary their teaching approaches, but they should encourage students to use other learning styles as well.

Kolb (1984) sees the teacher as cycling students through each of the four quadrants in order to help students become better problem solvers, improve their communication and conflict resolution skills, develop better team skills and critical reflective skills, and as a result, become more effective learners. In short, learners should be encouraged to develop the flexibility to use all four learning styles, depending on the content or the task to be learned. Consequently, teachers need to provide opportunities for students to develop or strengthen learning styles that are not their preferred learning styles. For examples, non-divergers may need to strengthen their reflective ability and their "people skills". They may need to become more sensitive to other people's feelings or listen with an open mind. Non-assimilators may need to become better organised or improve their ability to test theories and ideas. Non-convergers could be encouraged to experiment with new ideas and challenged to make decisions, while non-accommodators may need to develop their team skills, and become more personally involved with people.

Strategies for Developing Less Preferred Learning Styles

Kolb (1999) suggests several ways in which people can strengthen all their learning skills. Teachers could adapt these strategies to assist learners become more effective learners. In summary, teachers could encourage students to:

1. Develop learning and work relationships with people whose learning strengths lie in areas that are different from their own. Of course, in order to do this, learners will have to be aware of their own preferred learning style as well as the learning styles of their classmates or colleagues. Teachers should recognise that this strategy may be more difficult for assimilators, who normally prefer to study and work on their own.
2. Improve the fit between their learning style strengths and the kinds of learning experiences they undertake. This can be done by encouraging students to concentrate on tasks that lie in their areas of learning strengths and rely on classmates or colleagues for tasks that require skills in areas where they have weaknesses. For example, in a group activity an assimilator could gather all the information needed to solve a problem, but rely on a diverger to put forward all the possible options that should be considered, a converger to suggest possible solutions, and an accommodator to recommend an action plan.
3. Practice and develop learning skills in areas that are the opposite of their present strengths. According to Kolb (1999), "this strategy is the most challenging, but it can also be the most rewarding" (p. 10). Teachers can encourage students to become more flexible learners by using learning skills associated with the learning style opposite to their own. However, such risks have to be undertaken in a non-threatening and accepting learning atmosphere, where mistakes will be regarded as challenges for further learning.

Another aspect of cognition related to learning styles is the notion of students' approaches to learning, which includes students' motivation to learn and the learning strategies they use in achieving their goal. Although an important aspect to understanding how people learn, the concept of approaches to learning will only be discussed very briefly here since that topic is covered in depth in Chapter 5.

Approaches to Learning

According to Moore (1991), "an approach to learning is a person's general predisposition towards the acquisition of knowledge, skills, attitudes or values. An approach typically involves both motives and strategies for learning" (p. 1). Learning strategies are various mental operations or techniques that students use to facilitate learning. These learning strategies are normally specific behaviours that are goal-oriented.

Hunt (1997) explains that effective learning proceeds in three stages. Firstly, students assess the learning tasks. Secondly, they decide how they will approach the task, which in turn will determine the quality of the learning outcome. However, in order to select a learning approach, students need to activate prior relevant knowledge and then use specific learning strategies to achieve their goal. In other words, students must first possess metacognitive knowledge (be aware of how they learn) so they can understand when, where, and how to use the various strategies they know. Therefore, students must be taught learning strategies and skills, such as critical reflection, for use in conjunction with factual and conceptual knowledge.

Research focusing on students' approaches to learning has identified three main learning approaches: deep, surface and achieving, each comprising a set of motives and learning strategies (Biggs, 1969; Entwistle & Entwistle, 1970; Marton & Säljö, 1976a, 1976b). Students using a deep approach to learning are intrinsically motivated to understand the material thoroughly or to perform a task as competently as possible. As a result, they inquire widely, discussing and researching the topic extensively in order to discover as much as possible about the subject or the task at hand. However, Coles (1990) found that deep processing students are not as successful as those who elaborate their learning, that is, those who see the interconnections and links between different knowledge areas. This supports Broadbent's (1975) view that remembering is more likely when learners have multiple routes of access to the stored information.

Surface learners, on the other hand, only want to meet the minimum requirements to pass. They predominately learn by rote. They normally memorise the material to be learned and simply regurgitate the information or imitate someone when having to learn to perform a new task. Students who use the achieving approach to learning want to be the best; they want to succeed well and are motivated to earn top grades. They normally are well organised and manage their resources and their time effectively.

Research by Moore and Telfer (1990a) on approaches to learning used by students undertaking flight training found that student pilots' motives and studying strategies affected their ground school results as well as their hours to first solo. The results of Moore and Telfer's research showed that students who used a surface approach to learning received lower grades in ground school, especially in difficult subjects. On the other hand, students who adopted a deep approach to learning were more effective problem solvers and succeeded better in more complex subjects, such as flight planning, navigation, and aircraft performance. Moore and Telfer found that the correlation between the achieving approach and ground school results was not significant, however, when the achieving dimension was combined with the surface approach, some of the negative effects of the surface approach seems to be negated.

In addition, research by Moore and Telfer (1990b) revealed that "a low time to solo is related to a high score on the deep measure" (p. 103). Moore and Telfer (1990a) suggest that in terms of teaching strategies, instructors should encourage surface learners to set personal goals, regularly review material and generate summaries after lectures and study sessions to help them become deeper learners. Furthermore, they stress that the instructional focus should be on encouraging deeper learning approaches (critical inquiry, self-monitoring, self-assessment) because a surface approach to learning was found to be detrimental to learning any of the subjects in *ab initio* (beginning) flight training.

Hunt's (1995, 1997) research into approaches to learning revealed that the tools used by most researchers to determine students' approaches to learning were often too general to provide useful information to students regarding specific areas of their learning behaviour which needed changing or developing. She found that learning approaches inventories simply summarised students' approaches to learning, losing important distinctions between behaviours in different circumstances or when students are faced with different learning tasks.

Hunt's (1995) research revealed that "students see learning tasks as demanding appropriately matched learning strategies, and that the strategies they use are a more complex response than the 'deep' or 'surface' categories popular in many learning questionnaires" (p. 233). She identified seven components representing different functional classes of learning strategies: practice, identify key ideas, relate information to other contexts, remembering information, background filling, project or predict, and group information (see Chapter 5 for a description of each strategy). Hunt (1997) found that "successful learning depends on both *generic factors* such as prior knowledge, non-procrastination and not being concerned with neat, complete study notes and *context specific variables* such as appropriately matching the task with learning strategies" (p. 24).

Theory of Multiple Intelligences

A related yet somewhat different concept to learning styles and approaches to learning is the notion of multiple forms of intelligence. The concept of intelligence has recently become much more complex and multi-faceted than what it was a few decades ago, when it referred mainly to vocabulary and verbal comprehension, numerical reasoning and visual-spatial skills. In the 1980s and 1990s, several theorists broke away from the psychometric tradition of intelligence testing, which they saw as being too narrowly limited to verbal-linguistic and logical abilities, or as being too "academic" and not practical enough to represent aptitudes and skills needed to cope with tasks from everyday life (Gardner, 1983, 1993; Goleman, 1995; Sternberg, 1996). American psychologist Howard Gardner is representative of this group of theorists, who argue that "there is persuasive evidence for the existence of several relatively autonomous human intellectual competencies" (Gardner, 1983, p. 8).

Gardner (1983) defines intelligence as "the ability to solve problems or fashion products that are of consequence in a particular cultural setting or community" (p. 15). He claims that people have "multiple intelligences", not just one, arguing that the concept of a single Intelligence Quotient, or IQ score, to symbolise all that the human mind is capable of puts unnecessary limitations on understanding intellectual endeavour. According to Gardner (1983, 1993, 1997), every individual has different abilities in each of seven different and separate kinds of intelligences:

1. *Verbal-Linguistic Intelligence* involves the mastery of language which includes reading, writing, speaking, debating, creating poetry, doing word puzzles, expressing ideas in words, etc.;
2. *Logical-Mathematical Intelligence* includes the ability to detect patterns, solve mathematical problems, calculate, predict, use logical arguments, reason deductively, research, experiment, analyse, hypothesise, categorise, etc.;
3. *Visual-Spatial Intelligence* involves the ability to create artwork, drawing, painting, creating mental pictures, mapping, making diagrams, solving mazes, designs, photography, navigation, drawing plans, etc.;
4. *Bodily-Kinesthetic Intelligence* refers to the capability to use one's mental abilities to coordinate bodily movements used in playing sports, miming, acting, learning actively, manipulating controls, making things, fixing things, etc.;
5. *Musical Intelligence* involves the sensitivity to pitch, timbre, timing and rhythm, memorising songs and tunes, playing a musical instrument, creating sound effect, etc.;
6. *Interpersonal Intelligence* refers to the ability to understand human behaviour and motivation, to interact socially, to organise and manage others, including cooperation, leadership, negotiation, mediation, empathy and sensitivity to others, etc.; and
7. *Intrapersonal Intelligence* includes self-awareness, self-understanding, goal setting, self-discipline, self-regulation, reflection, analysis of one's own learning and feelings, identifying one's own strengths and weaknesses, using one's experience to understand and explain, etc. (adapted from Gardner, 1993, pp. 8-9 & 17-26).

Recently, Gardner (1997) has added an eighth form of intelligence, which he calls the "naturalistic" intelligence. According to Gardner, the naturalistic intelligence involves the ability to recognise essential distinctions in the natural world (among fauna and flora). This competency is crucial to people who work in nature, such as biologists, horticulturists or zoologists, or to people in hunting and farming cultures.

Gardner (1983) notes that for many years there has been a strongly-held belief in both psychology and education that intelligence was a single underlying general-ability characteristic which accounted for differences in people's behaviour and in their ability to learn. It was assumed that all students could be classified according to how much intelligence (IQ) they had or how "smart" or "bright" they were. However, Gardner's model of multiple intelligences has had an influence in changing this line of thinking among educators worldwide. For many educators the significant question now is not "how smart is this person"? but "how is this person smart"? In other words, "What are the relative strengths of this person across all eight intelligences"?

Gardner (1991, 1993, 1995) contends that most schools teach mainly to the first two intelligences. Therefore, students are not often allowed to use and develop the other intelligences as alternative ways to understand learning material. Most often, verbal-linguistic skills and logical-mathematical skills are equated with "intelligence" and other skills are simply seen as "talents". Traditional tests to assess a person's intellectual ability use mostly verbal and logical-mathematical tasks. Similarly, many selection tests used within the aviation industry employ verbal-linguistic and logical-mathematical tasks, as well as visual-spatial tasks to select pilots or air traffic controllers. However, if we consider the responsibilities and the tasks of the crew of a modern jet airliner, the importance of interpersonal and intrapersonal intelligences should also be as highly valued and assessed during the selection process.

Gardner's model of multiple intelligences accounts for what most experienced teachers have always known: different people have different strengths. While some people are good at writing essays or are good at mathematics, or physics, others will be better at art, handicraft, sport, music, making friends or leadership. The multiple intelligences model presents a useful framework to help teachers identify eight different kinds of intellectual accomplishment. In contrast to traditional views, Gardner's (1993, 1997) view is that all eight areas of competencies should be equally valued and should all be seen as intelligences.

Gardner (1991, 1993) claims that the multiple intelligences do not normally operate independently. Rather, we used our intelligences concurrently with different intelligences complementing each other. For example, a ballet dancer would use several intelligences in performing *Swan Lake* (i.e., musical, bodily-kinesthetic and interpersonal). However, although we possess all eight intelligences, we all differ in the specific profile of intelligence we exhibit. Most of us are talented in two or three areas while we may have little or no abilities in other areas. These relative differences in intellectual strengths can have a significant impact on our lives ranging from the things that interest us through to career choices.

The challenge for teachers is to design worthwhile learning events that acknowledge individual differences and meet the needs of all students while at the same time raising their awareness of their own learning strengths, competence and self-worth. Providing alternative educational activities allows teachers to engage multiple intelligences and thus, increases the number of ways students can learn, and facilitates a deeper understanding of the subject matter. As Gardner (1995) stresses, there is no "single educational approach based on MI [multiple intelligences] theory… [and] educators are in the best position to determine the uses to which MI theory can and should be used" (p. 206).

Personality Theories

Just as research into different learning styles and multiple forms of intelligence has improved our understanding of how people learn and has influenced our approaches to education and training, so too has the growing interest into personality type theory and its impact on learning and multi-crew performance. There is considerable evidence to support the view that personality type can be an influential element in effective teamwork and in aviation safety (Chidester, Helmreich, Gregorich, & Geis, 1991; Hedge, Bruskiewicz, Borman, Hanson, Logan, & Siem, 2000; Helmreich, 1984, 1987; Hörmann & Maschke, 1996; O'Hare & Roscoe, 1990; Prince, Chidester, Bowers, & Cannon-Bowers, 1992; Stead, 1995). Further, accident investigations in the past have identified failures in communication patterns and cockpit teamwork as contributing factors in accidents and incidents (Cooper, White, & Lauber, 1979; Foushee, 1984; Foushee & Helmreich, 1988; Hawkins, 1987; Leedom, 1990).

For example, in its report of an aircraft accident in 1978, where the crew of an airliner was preoccupied with a minor mechanical malfunction and allowed the aircraft to run out of fuel, the National Transportation Safety Board (NTSB, 1979) in the United States of America (USA) indicated that the breakdown in cockpit management and teamwork exemplified a recurring problem. The NTSB (1979) also noted that "admittedly, the stature of a captain and his management style may exert subtle pressure on his crew to conform to his way of thinking. It may hinder interaction and adequate monitoring and force another crewmember to yield his right to express an opinion" (p. 27). A few years later, in its accident report on the jet aircraft that struck the Potomac River bridge shortly after takeoff, the NTSB (1982) indicated that the captain had not heeded the repeated subtle warnings of the copilot, but the NTSB also implied that the copilot had not been sufficiently assertive in his advisories to the captain. Finally, Foushee (1984) provides the example of a copilot who reported the following incident to the Aviation Safety Reporting System (ASRS) in the USA:

> Air traffic control had issued a speed restriction that was repeatedly ignored by the captain. After several attempts [by the copilot] to convey the information, the captain responded by saying, "I'll do what I want". Air traffic control inquired as to why the aircraft had not been slowed, advised the crew that they had almost collided with another aircraft, and issued a new clearance that was also disregarded by the captain despite repeated clarification by the copilot. Following the last advisory from the copilot, the captain responded by telling the copilot to "just look out the damn window". (p. 888)

These three reports are testimonies to the fact that personality type can indeed play a crucial role in crew resource management and in the overall maintenance of flight safely.

Defining Personality Characteristics of Pilots

As is the case with many educational or psychological theories, the study of the personality characteristics of pilots has had a long and controversial history. Personality tests were first used to measure pilots' personality traits during World War 1, and even then, there were diametrically opposite views about the personality characteristics that were considered best suited for the aviator's role. Rippon and Manuel (1918), for example, contended that the ideal pilot was an individual who was "high-spirited" and "easy-going", while Dockeray and Isaacs (1921) portrayed the model pilot as being someone who was quiet and methodical.

Since then, research examining personality profiles of pilots within the military, the airlines and general aviation has yielded interesting and valuable information, but no consensus has been reached on the personality traits needed to be a successful aviator (Alkov & Borowsky, 1980; Ashman & Telfer, 1983; Chidester, Helmreich, Gregorich, & Geis, 1991; Deitz & Johnson, 1991; Fry & Reinhardt, 1969; Lester & Bombaci, 1984; Novello & Youssef, 1974a, 1974b; Picano, 1991; Platenius & Wilde, 1989; Retzlaff & Gibertini, 1987; Sanders & Hoffman, 1975; Stead 1991). Some studies suggest that the persona Tom Wolfe dubbed *The Right Stuff*, that is, an aviator who demonstrates qualities of leadership, cool-headedness, quick judgement and reaction time, remains the personality type that is attracted to aviation.

For example, Novello and Youssef (1974a, 1974b) found that both male and female general aviation pilots deviate in the same way from the general population in the USA. Novello and Youssef (1974b) concluded that:

> It seems that piloting, regardless of the pilot's sex, either requires, attracts, and/or selects out the personality type that has been popularized for so long in song, movie, and verse: a person who is courageous and adventuresome, one who is oriented toward demonstrating competency, skill, and achievement; one who finds pleasure in mastering complex tasks; and one whose manifest sexual orientation is decidedly heterosexual. The results of this study, therefore give credence to what has thus far been lodged in myth and folklore: the pilot as a romantic hero or — heroine. (p. 633)

However, Chidester et al. (1991) stress that such generalisations tend to distort reality since these findings seem to overlook the fact that there is variability in both the general population and the pilot population. Hawkins (1987) also notes that "it must be recognised that different kinds of flying call for an emphasis on different personality traits" (p. 172). Indeed, he adds that "the ideal personality for military missions or emergency relief operations may not be optimum for routine civil flying, and vice versa" (p. 172). Nevertheless, what seems to be widely accepted in recent research is the fact that the great majority of personality tests are poor predictors of successful pilot operational performance (Damos, 1997; Hunter & Burke, 1994; Martinussen, 1995, 1996; Martinussen & Torjussen, 1998). Overall, the results of the research support the notion that no specific personality type is a prerequisite for success within the aviation context.

Despite the uncertainty regarding the personality type that may be best suited to aviation, there is general agreement that an understanding of the basic concepts of personality theories can inform our practice as teachers and is an essential part of crew resource management (CRM) training, since interpersonal behaviour and communication patterns depend to a high degree on personality traits and attitudes of individuals (Chidester et al., 1991; Foushee, 1984; Hunter, 1997; Stead, 1991, 1995). Indeed, according to Stead (1995), "*personality is that which is concerned with a person's typical or preferred way of behaving, thinking and feeling*" (p. 309).

While Hawkins (1987) notes that personality traits and attitudes are closely interwoven and have a fundamental influence on the way people approach life and conduct themselves, he does differentiate between the two concepts. Hawkins agrees with many psychologists who believe that personality traits are innate and acquired very early in life. According to Hawkins, personality traits "are deep-seated characteristics which constitute the essence of a person. They are stable and very resistant to change though certain traits seem to have some tendency to alter in middle age" (p. 172).

Attitudes, on the other hand, describe people's likes and dislikes. An attitude, Hawkins (1987) adds, "can be seen as a learned and rather enduring tendency to respond favourably or unfavourably to people, decisions, organisations, or other objects. The response becomes more or less predictable. An attitude is a predisposition to respond in a certain way; it is not a response or behaviour *per se*" (p. 173). Chidester et al. (1991) agree, stating that "attitudes are defined as learned patterns of responding that are more or less modifiable through training" (pp. 26-27). However, Stead (1995) stresses that both attitudes and beliefs are deeply seated in one's personality. He defines attitude as "*the state of mind which pre-determines and governs our approach to people and situations*" (Qantas CRM, 1989 cited in Stead, 1995, p. 310).

Myers-Briggs Type Indicator (MBTI)

Most personality theorists look at a distribution of personality types along a continuum, for instance, between extroversion and introversion, or between people who display confidence and independence and those who lack self-confidence and may possess feelings of self-doubt or inferiority. Probably the most widely used personality model is the Myers-Briggs Type Indicator (MBTI). One of the reasons for this popularity is possibly because, unlike many personality tests that are designed to identify personality problems, the MBTI describes normalcy. Further, the MBTI is an indicator, not a test. Consequently, it can be a useful tool for understanding one's personality and that of others. Indeed, most educators realise that having an awareness of personality types is of vital importance to our understanding of how people learn and behave, because learning does not just affect the head; the whole person, with his or her feelings, attitudes, values, strengths and weaknesses, is involved in the learning process.

Isabel Myers and Katherine Briggs developed the MBTI in an attempt to sort people according to the different psychological types described by Swiss psychologist Carl Jung. Jung observed patterns and differences in the behaviour of people. He stressed that psychological type preferences were neither good nor bad, instead he considered these differences as gifts. Jung noted that what was important was a person's natural inclination to either "extraversion" or "introversion", combined with the individual's preference for what he called basic psychological functions: thinking, feeling, sensation, intuition, judging and perceiving (Keirsey, 1998).

The Myers-Briggs system uses Jung's pairs of qualities to characterise people according to sixteen *preferred* personality types (Myers, McCaulley, Quenk, & Hammer, 1998). Each of these personality types has a primary orientation toward the world, which in turn affects their ability to learn, to make decision, and to work. The MBTI comprises four dichotomous dimensions: extroversion versus introversion, sensing versus intuition, thinking versus feeling, and judgement versus perception. Obviously, people display all these characteristics, but with varying intensity. Each person tends to prefer, practice, and become more proficient in one of each of the bipolar scales.

- *Extroversion — Introversion*: indicates whether a person prefers social or solitary environments. Extroverts show a preference for the outer world of activity, events and people, while introverts prefer the inner world of thoughts, ideas, memories and feelings. Because they are energised by interactions with the outer world, extroverts are usually seen as outgoing, sociable individuals. They enjoy interacting with others and tend to be expressive and transparent. As a result of their openness, they will often initiate communication. Introverts, on the other hand, are more reserved and reflective, and may communicate only when they are asked. They enjoy time alone and prefer quietness for concentration. Extroverts tend to act and then reflect, while introverts reflect and then act. The strength of extroverts is that they can provide the initial energy in a group, while the power of introverts is the confidence gained from thinking things through, from reflection.
- *Sensing — Intuition*: focuses on whether an individual attaches more importance to the immediate realities of direct experience or prefers inferred meanings and future possibilities of experience. Sensing types prefer to concentrate on present reality; they focus on what works now. They perceive information through their five senses, that is, they are highly observant of things in their immediate environment. They focus on details and facts and need to see the parts and pieces in order to understand the whole. Sensing people respect routine, they like an established way of doing things and work steadily to accomplish a task. They also prefer to handle practical matters. In contrast, intuitive types

seek new challenges and look for future possibilities. They enjoy change, therefore, they may look for variety in handling tasks. They are highly imaginative and inventive, and tend to focus on how things could be improved. However, they need to see the whole picture to understand how the pieces fit together. Intuitive people may work in bursts of energy fueled by enthusiasm interspersed with periods of sagging interest. Sensing individuals usually reach a conclusion in a methodical, step by step fashion, while intuitive types often use a more random approach and may leap to a conclusion quickly based on their inspirations and hunches. In brief, sensing types are more observant than imaginative, while intuitive types are more imaginative than observant.

- *Thinking — Feeling*: reflects on whether someone prefers making judgements or decisions based on logic and objectivity or relies on priorities based on personal and social values. Thinking types are more analytically oriented when making decisions; they identify principles and laws and apply them using logic or reason. As a result, they can be tough-minded or quite impersonal. They are good at putting things in logical order. They seek competence, consistency, justice, truth and honesty. Thinking types focus on goals and tasks, as a result, they usually respond to people's ideas rather than to their feelings. The feeling types, on the other hand, are more people oriented when making decisions; they take interest in the person behind a job or an idea. They are friendly. Feeling types like harmony and will endeavour to make it happen. They weigh values when making decisions and tend to be compassionate and empathetic. They focus on the impact a decision will have on people.
- *Judgment — Perception*: relates to a person's preference for planning and decisiveness or openness and spontaneity with reference to life style. Because they are decisive, judging types usually like to come to closure as quickly as possible. They like to get things settled and finished. Judging types like a lifestyle that is planned and orderly. Therefore, they work best when they can follow a structured plan and usually use lists as agendas for action. They schedule projects so that each step will be done on time. In contrast, perceiving types like to gather as much information as possible or probe for alternatives before making a decision. They like to mull things over for a long time. In fact, they may often have problems making decisions because they enjoy exploring different options. They are flexible, spontaneous and adaptable. They do not mind leaving things open for last-minute

changes. As a result, they often have to rush to get things done at the last minute under pressure of a deadline (adapted from Keirsey, 1998, pp. 331-337; and Myers et al., 1998, p. 6).

In essence, the Myers-Briggs personality theory aims to promote understanding of self and others, facilitate communication, respect other people's uniqueness, and enhance team effectiveness. Clearly, there is much to be gained by being aware and by appreciating individual differences. However, the first step toward valuing the uniqueness of others is to become better acquainted with our own personality traits. The Myers-Briggs system can provide a framework for aviation educators to understand themselves, the students they teach, and the people with whom they interact. The MBTI can also be a useful tool for team building, leadership development and communication facilitation in the aviation training context. It permits group members to recognise how individual differences affect how people think and work together. In addition, group members can become aware of how each personality type in a group can best contribute to the group's successful functioning. The MBTI framework explains some of the similarities and differences among people. Accordingly, educators using this model are encouraged to use a variety of teaching strategies to accommodate all the different personality types, and thus, enhance the effectiveness of training (Chidester et al., 1991).

In the aviation context, research by Kreienkamp (1983), and Kreienkamp and Luessenheide (1985) showed that during the initial phases of training, when student pilots were paired with flight instructors on the proximity of their scores on the extrovert-introvert dichotomy in the MBTI, achievement increased and students could be trained to first solo in a shorter time. Luessenheide (1991) noted that "The critical communication in the environment of a small aircraft during the important first few hours of training is a special situation. If the student and the instructor tend to perceive the world in like fashion and they use similar symbols and metaphors to express themselves, learning is enhanced" (p. 69). However, if the aim is also to encourage students to become more effective learners and to develop useful team skills, this strategy may disadvantage students in the long term. Therefore, in subsequent stages of training, it may be advisable for the students to be exposed to instructors having different personality characteristics to theirs so that they can learn to work effectively with other personality types.

Conclusion

Underpinning the discussion in this chapter was the fact that the perceptive practitioner will assess each situation in terms of participant characteristics, preferred learning styles, approaches to learning, intelligence, and personality type and respond appropriately to ensure effective teaching and learning takes place. However, the danger with any of these personality types and with various learning style inventories is that of assigning learners to particular categories ("putting people in boxes"). Obviously, there is value in knowing as much as possible about learning, about ourselves and about our students or our colleagues, because that information can be helpful in guiding our practice and in explaining individual differences. However, as teachers, it is essential to keep an open mind. The only certainties seem to be that there are always exceptions and we are all constantly changing.

References

Alkov, R.A., & Borowsky, M.S. (1980). A questionnaire study of psychological background factors in U.S. Navy aircraft accidents. *Aviation, Space, and Environmental Medicine, 51,* 860-863.

Ashman, A., & Telfer, R. (1983). Personality profiles of pilots. *Aviation, Space, and Environmental Medicine, 54,* 940-943.

Biggs, J.B. (1969). Coding and cognitive behaviour. *British Journal Educational Psychology, 60*(3), 287-305.

Biggs, J.B. (1987). *Student Approaches to Learning and Studying*. Melbourne: Australian Council for Educational Research.

Broadbent, D.E. (1975). Cognitive psychology and education. *British Journal of Educational Psychology, 45*(2), 162-176.

Burns, S. (1989). There's more than one way to learn. *Australian Wellbeing, 33,* 42-44.

Chidester, T.R., Helmreich, R.L., Gregorich, S.E., & Geis, C.E. (1991). Pilot personality and crew coordination: Implications for training and selection. *The International Journal of Aviation Psychology, 1*(1), 25-44.

Coles, C.R. (1990). Elaborated learning in undergraduate medical education. *Medical Education, 24,* 14-22.

Cooper, G.E., White, M.D., & Lauber, J.K. (Eds.). (1979, June). *Resource management on the flight deck.* (NASA Report No. CP-2120). Moffett Field, CA: NASA-Ames Research Center.

Damos, D.L. (1997). Pilot selection batteries: Shortcomings and perspectives. *The International Journal of Aviation Psychology, 6*(2), 199-209.

Deitz, S.R., & Johnson, W.K., III. (1991). The impaired pilot. In S.R. Deitz & W.E.

Thoms (Eds.), *Pilots, personality, and performance* (pp. 1-23). New York: Quorum Books.

Dockeray, F.C., & Isaacs, S. (1921). Psychological research in aviation in Italy, France, England, and the American Expeditionary Forces. *Journal of Comparative Psychology, 1,* 115-148.

Emanuel, R., & Potter, W. (1992). Do students' style preference differ by grade level, orientation toward college, and academic major. *Research in Higher Education, 33,* 395-414.

Entwistle, N.J. (1981). *Styles of learning and teaching*. London, Wiley.

Entwistle, N.J., & Entwistle, D. (1970). The relationship between personality, study methods and academic performance. *British Journal of Educational Psychology, 40,* 132-143.

Entwistle, N., & Ramsden, P. (1983). *Understanding student learning*. Beckenham, UK: Croom Helm.

Filipczak, B. (1995, March). Different strokes: Learning in the classroom. *Training, 32,* 43-48.

Foushee, H.C. (1984). Dyads and triads at 35,000 feet: Factors affecting group process and aircrew performance. *American Psychologist, 39,* 886-893.

Foushee, H.C., & Helmreich, R.L. (1988). Group interaction and flight crew performance. In E.L. Wiener & D.C. Nagel (Eds.), *Human factors in aviation* (pp. 189-227). San Diego, CA: Academic Press.

Fry, G.E., & Reinhardt, R.F. (1969). Personality characteristics of jet pilots as measured by the Edwards Personal Preference Schedule. *Aerospace Medicine, 40,* 484-486.

Gardner, H. (1983). *Frames of mind: The theory of multiple intelligences.* New York: Basic Books.

Gardner, H. (1991). *The unschooled mind: How children think, and how schools should teach*. New York: Basic Books.

Gardner, H. (1993). *Multiple intelligences: The theory in practice*. New York: Basic Books.

Gardner, H. (1995). Reflections on multiple intelligences: Myths and messages. *Phi Delta Kappan, 77*(3), 200-209.

Gardner, H. (1997). *Extraordinary minds: Portraits of exceptional individuals and an examination of our extraordinariness*. New York: Basic Books.

Goleman, D. (1995). *Emotional intelligence: Why it can matter more than IQ*. New York: Bantam Books.

Hawkins, F.H. (1987). *Human factors in flight*. Aldershot, UK: Gower Technical Press.

Hedge, J.W., Bruskiewicz, K.T., Borman, W.C., Hanson, M.A., Logan, K.K., & Siem, F.M. (2000). Selecting pilots with crew resource management skills. *The International Journal of Aviation Psychology, 10* (4), 377-392.

Helmreich, R.L. (1984). Cockpit management attitudes. *Human Factors, 26,* 583-590.

Helmreich, R.L. (1987). Theory underlying CRM training: Psychological issues in flight crew performance and crew coordination. In H.W. Orlandy & H.C. Foushee (Eds.), *Proceedings of the NASA/MAC workshop on cockpit resource management* (pp. 14-21). Moffett Field, CA: NASA-Ames Research Center.

Hörmann, H.J., & Maschke, P. (1996). On the relationship between personality and job performance in airline pilots. *The International Journal of Aviation Psychology, 6*(2), 171-178.

Hunt, L.M. (1995). Effective learning strategies for *ab initio* pilots. In N. Johnston, R. Fuller, & N. McDonald (Eds.), *Aviation psychology: Training and Selection. Proceedings of the 21st Conference of the European Association for Aviation Psychology (EAAP)* (pp. 228-233). Aldershot, UK: Avebury Aviation.

Hunt, L.M. (1997). Influences on the learning process and learning outcomes: Practical implications for the instructor. In G.J.F. Hunt (Ed.), *Designing instruction for human factors training in aviation* (pp. 17-27). Aldershot, UK: Avebury Aviation.

Hunter, D.R. (1997). Pilot characteristics. In R.A. Telfer & P.J. Moore (Eds.), *Aviation training: Learners, instruction and organization* (pp. 41-53). Hants, UK: Ashgate Publishing.

Hunter, D.R., & Burke, E.F. (1994). Predicting aircraft pilot-training success: A meta-analysis of published research. *The International Journal of Aviation Psychology, 4*(4), 297-313.

Karp, M.R., Turney, M.A., & McCurry, W.K. (1999). Learning style theory and aviation education. In R.S. Jensen, B. Cox, J.D. Callister, & R. Lavis (Eds.), *Proceedings of the 10th International Symposium on Aviation Psychology* (pp. 956-962). Columbia, OH: Ohio State University.

Keirsey, D. (1998). *Please understand me II.* Del Mar, CA: Prometheus Nemesis Book Company.

Kolb, D. (1976). *Learning style inventory.* Boston, MA: McBer.

Kolb, D. (1984). *Experiential learning: Experience as the source of learning and development.* Englewood Cliffs, NJ: Prentice Hall.

Kolb, D. (1999). *Learning style inventory.* Boston, MA: Hay/McBer Training Resource Group.

Kreienkamp, R.A. (1983). *Flight instructor-student pilot perceptive similarity and its effect on flight training time.* Unpublished master's thesis, University of North Dakota, Grand Forks, ND.

Kreienkamp, R.A., & Luessenheide, H.D. (1985). Similarity of personalities of flight instructors and student-pilots: Effect on flight training time. *Psychological Reports, 57,* 465-466.

Leedom, D.K. (1990). *Aircrew coordination training and evaluation for army rotary wing aircrews: Summary of research for fiscal year 1990.* Fort Rucker, AL: Army Research Institute.

Lester, L.F., & Bombaci, D.H. (1984). The relationship between personality and irrational judgment in civil pilots. *Human Factors, 26,* 565-572.

Luessenheide, H.D. (1991). Cockpit communication and initial aviation training. In S.R. Deitz & W.E. Thoms (Eds.), *Pilots, personality, and performance* (pp. 65-70). New York: Quorum Books.

Martinussen, M. (1995). Selecting the right stuff: Problems and challenges in conducting validation research. In R.S. Jensen & L.A. Rakovan (Eds.), *Proceedings of the Eighth International Symposium on Aviation Psychology* (pp. 1117-1121). Columbus, OH: Ohio State University.

Martinussen, M. (1996). Psychological measures as predictors of pilot performance: A meta-analysis. *The International Journal of Aviation Psychology, 6*(1), 1-20.

Martinussen, M., & Torjussen, T. (1998). Pilot selection in the Norwegian Air Force: A validation and meta-analysis of the test battery. *The International Journal of Aviation Psychology, 8*(1), 33-45.

Marton, F., & Säljö, R. (1976a). On qualitative differences in learning I: Outcome and processes. *British Journal of Educational Psychology, 46,* 4-11.

Marton, F., & Säljö, R. (1976b). On qualitative differences in learning II: Outcome as a function of the learner's conception of the task. *British Journal of Educational Psychology, 46,* 115-127.

McCarthy, B. (1982, October). Improving staff development through CBAM and 4Mat™. *Educational Leadership,* 20-25.

McCarthy, B. (1987a). *The 4MAT system: Teaching to learning styles with right/left mode techniques* (2nd ed.). Barrington, IL: Excel.

McCarthy, B. (1987b). *The 4MAT workbook: Guided practice in 4MAT lesson and unit planning*. Barrington, IL: Excel.

McCarthy, B. (1996). *About learning*. Barrington, IL: Excel.

Merriam, S.B., & Caffarella, R. (1999). *Learning in adulthood.* San Francisco: Jossey-Bass.

Mezirow, J. (1981). A critical theory of adult learning and education. *Adult Education, 32,* 3-24.

Moore, P. (1991, August) *Approaches to learning in aviation: Does the difficulty of the topic make a difference.* Paper presented at the Human Factors in South Pacific Aviation Symposium. Palmerston North, New Zealand.

Moore, P., & Telfer, R. (1990a, Spring). Approaches to learning: Relationship with pilot performance. *Journal of Aviation/Aerospace Education and Research, 1*(1), 44-58.

Moore, P., & Telfer, R. (1990b, March). Learning to be a pilot. *Australian Aviation,* 102-103.

Myers, I. B., McCaulley, M.H., Quenk, N.L., & Hammer, A.L. (1998). *MBTI manual: A guide to the development and use of the Myers-Briggs Type Indicator* (3rd ed.). Palo Alto, CA: Consulting Psychologists Press.

National Transportation Safety Board (1979, June). *Aircraft accident report* (NTSB Report No. AAR-82-8) Washington, DC: NTSB Bureau of Accident Investigation.

National Transportation Safety Board (1982, August). *Aircraft accident report* (NTSB Report No. AAR-79-7) Washington, DC: NTSB Bureau of Accident Investigation.

Novello, J.R., & Youssef, Z.I. (1974a). Psycho-social studies in general aviation I: Personality profile of male pilots. *Aerospace Medicine, 45,* 185-188.

Novello, J.R., & Youssef, Z.I. (1974b). Psycho-social studies in general aviation II: Personality profile of female pilots. *Aerospace Medicine, 45,* 630-633.

O'Hare, D., & Roscoe, S.N. (1990). *Flightdeck performance: The human factor.* Ames, IA: Iowa State University Press.

Picano, J.J. (1991, June). Personality types among experienced military pilots. *Aviation, Space, and Environmental Medicine, 62,* 517-520.

Platenius, P.H., & Wilde, G.J.S. (1989). Personal characteristics related to accident histories of Canadian pilots. *Aviation, Space, and Environmental Medicine, 60,* 42-45.

Prince, C., Chidester, T.R., Bowers, C., & Cannon-Bowers, J. (1992). Aircrew coordination: Achieving teamwork in the cockpit. In R.W. Swezey & E. Salas (Eds.), *Teams: Their training and performance* (pp. 329-353). Norwood, NJ: Ablex Publishing.

Retzlaff, P.D., & Gibertini, M. (1987). Air Force pilot personality: Hard data on the right stuff. *Multivariate Behavioral Research, 22,* 383-389.

Rippon, R.S., & Manuel, E.G. (1918, September). The essential characteristics of successful and unsuccessful aviators. *The Lancet,* 411-415.

Rogers, A. (1998). *Teaching adults*. Buckingham: Open University Press.

Sanders, M., & Hoffman, M.A. (1975). Personality aspects of involvement in pilot-error accidents. *Aviation, Space, and Environmental Medicine, 46,* 186-190.

Smith, R.M. (1982). *Learning how to learn: Applied theory for adults.* New York: Cambridge.

Stead, G. (1991). A validation study of the QANTAS pilot selection process. In E. Farmer (Ed.), *Human resource management in aviation: Proceedings of the 28th WEEAP conference* (pp. 3-18). Aldershot, UK: Avebury Technical.

Stead, G. (1995). Personality on the flight deck. In N. McDonald, N. Johnston, & R. Fuller (Eds.), *Applications of psychology to the aviation system. Proceedings of the 21st Conference of the European Association of Aviation Psychology (EAAP)* (pp. 309-314). Aldershot, UK: Avebury Aviation.

Sternberg, R.J. (1996). *Successful intelligence: How practical and creative intelligence determine success in life.* New York: Simon & Schuster.

Tannen, D. (1990). *You just don't understand: Women and men in conversation.* New York: Ballentine Books.

Turney, M.A. (1995). Women's learning and leadership styles: Impact on CRM. In N. McDonald, N. Johnston, & R. Fuller (Eds.), *Applications of psychology to the aviation system* (pp. 262-268). Aldershot, UK: Avebury Aviation.

Wiener, E.L., & Nagel, D.C. (Eds.). (1988). *Human factors in aviation.* San Diego, CA: Academic Press.

5 The Role of Learning Strategies in the Learning Process

Lynn M. Hunt

Introduction

The mid to late 1960s witnessed a change in direction for educational research. Whereas the focus had previously been on finding general principles that improved learning for all, educationalists now began to consider the effect of individual differences in learners on learning outcomes (Gagné, 1967; Glaser, 1970).

On this wave of interest in individual differences, a group of researchers examined more closely the relationship between personality traits, such as neuroticism and extroversion, and success in learning (Biggs, 1969; Entwistle & Entwistle, 1970). These researchers found that correlations between personality variables and academic success were low, but statistically significant. More importantly, they discovered that the kind of learning strategies used by students, often influenced by personality variables, were much more strongly related to learning outcomes (Biggs, 1970a, 1970b). The significance of this finding was that the strategies a student uses to learn could be changed, whereas personality and cognitive style variables could not.

At the same time as researchers, such as Biggs (1976), Entwistle (1977) and Marton and Säljö (1976a), were establishing the role of learning strategies in the learning process, two significant shifts in the focus of learning research were taking place. First, there was a move away from experiments based on meaningless or trivial content in laboratory-based experiments to the more complex, real-life learning context of actual classroom content. When it is remembered that meaningful learning is the basis of education, the importance of this shift becomes apparent.

The second trend was to view learning from the student's perspective (Entwistle, 1982). In spite of the claim of many behavioural theorists that they were studying individual differences in learning, in fact their work focused on teacher and organisational control of learning rather than student learning processes (Cronbach, 1967). It was held that improvements would come from better teaching methods, instructional design or more media resources. To the behaviourist, students were regarded as relatively passive participants to be influenced or manipulated by the primary educative agent, the teacher. The new approach turned this perspective on its head. Students were now viewed as principal agents in the learning process. Their behaviour, attitudes and decisions were recognised as having a strong influence on their learning outcomes.

This view of learning with its focus on the learner's role in the learning process is embodied in a broad-based theory of learning called *constructivism*. Seeing learners as central to the learning process is not new (Piaget, 1950), however, since the 1970s it has taken on additional perspectives. More recently, constructivists identify the learner as an *active* agent in the construction of his or her knowledge (Laurillard, 1999; Merrill, 1991; Pines & West, 1986; Reese & Overton, 1970, also see Chapter 1 in this book). This theory advocates that instructors should be less concerned with transmitting content knowledge than with helping students to develop and use alternative strategies for constructing their knowledge structures, defining their own goals, and monitoring and correcting their own performance.

Two main approaches have evolved from this challenge, the intensive small group studies typified by Ference Marton and his co-workers at Göteborg University (Fransson, 1977; Marton & Säljö, 1976a, 1976b; Svensson, 1977), and the more traditional large-scale psychometric approach used by Biggs (1976) and Entwistle and Ramsden (1983). The Göteborg studies used interview and case study techniques to examine learning behaviour. Small sample sizes of between 20-40 students were asked to read a prose extract. Each student would read the extract in the presence of a researcher, and then answer a series of questions to test his or her understanding of the text. Students were then asked to respond to a series of open-ended questions in order to provide an introspective account of how they approached the reading task. The conversation with the researcher was recorded in order to allow a complete analysis of the whole pattern of the students' responses. The two outcomes, level of understanding and the *approach* used, were classified by Marton and Säljö (1976a, 1976b).

The Göteborg studies have consistently demonstrated that the processing level, or type of learning strategies, used by students qualitatively affects learning outcomes. From these studies, the authors concluded that there were two major approaches to learning: *deep* and *surface*. Surface level processors focused on the surface features of the content, that is, specific facts or disconnected pieces of information that they learned by rote. Deep processors, on the other hand, were more concerned with the underlying meaning or significance of the content. The depth of processing used by students while studying was reflected in the quality of post-study responses to questions. Whereas deep processors were more able to give an analysis of the intent of the content, surface processors were more concerned with regurgitating a reproductive conception of the content (Marton & Säljö, 1976a).

Further, Marton and Säljö (1976a, 1976b) argued for understanding learning in the context of the content being learned. The same learning task was seen to be understood differently by different students and this difference in perception influenced what and how students learned. The findings of these studies were much less equivocal than the traditional psychometric studies of the same phenomena in that they produced a clear demonstration of the link between *depth* of processing and the *quality* of learning outcomes. For example, all students who normally used a *deep level* of processing were classified as achieving learning at the highest levels of complexity, whereas the surface processors could only achieve the lowest levels (Marton & Säljö, 1976a, 1976b).

Valuable as these findings appeared to be, the studies suffered from some weaknesses. Firstly, the sample sizes were very small and the homogeneity of the samples (mainly volunteer female Swedish social science students) raised the question of generalisability (Entwistle, 1981). Secondly, the method of data collection and the interpretation were highly subjective.

The approach used by Biggs (1978) and Entwistle, Hanley, and Hounsell (1979) overcame many of the problems of the *qualitative* method described above. These studies were large scale and conducted over a variety of faculties and subject areas. The instruments used were self-report questionnaires that were easily administered and scored.

Despite the differences in methodology, the psychometric approach produced remarkably similar results to the Göteborg studies. Using second order factor analysis, Biggs (1978) and Entwistle, Hanley, and Hounsell (1979) discovered a similar dichotomy to that of Marton and Säljö (1976a, 1976b), namely *surface* and *deep* approaches to learning. In addition, all three research groups identify "motive", (Marton and Säljö's "intent") as a dimension of each learning approach. Further, the psychometric studies found a third learning approach: *achieving*. This achieving approach was found to be qualitatively different to either deep or surface approaches, and was characterised by high achievement motivation and organisational strategies. In a sense, students classified as achieving could also be seen as being *strategic* in that their primary strategy was to discover the rules of the game, and play to win. They would use either deep or surface learning strategies depending on which one they believed would earn them their best marks. They would read all of the set readings, know when assignments were due and always have them completed on time. While they were often characterised as 'model' students, they often had something of a ruthless streak in them. A summary of the motives and strategies associated with each approach is shown in Table 5.1.

Table 5.1 Approaches to learning

Approach	Motive	Strategy
Surface	Instrumental: aim to meet minimum passing requirements.	Reproductive: focuses on bare essential and uses rote learning.
Deep	Intrinsic: to gain competence in specific subjects.	Meaningful: reads widely, interrelates material with prior knowledge.
Achieving	Competitive and ego-enhancing: to obtain highest grades whether or not interested in material.	Organising: effectively organises time and resources, behaves as model student.

Biggs' (1985) *Study Processes Questionnaire* (SPQ) and Entwistle and Ramsden's (1983) *Approaches to Studying Inventory* have been successful in establishing consistent links between approaches to learning and learning outcomes. The study described below builds on this early work by identifying additional influences on learning strategy selection and the impact of the interaction between learning strategy and type of learning task on learning outcome.

The Selection and Use of Learning Strategies

Sample

Two different samples were used in this research. In the first part of the study, 102 first and second year students studying for a Bachelor of Aviation were used to gather qualitative data that summarised the learning process from the students' perspective. This exercise described the *variability* in the students' learning processes and was expressed in the student's own terminology. The data were then used to design questionnaires for large-scale administration to measure the *consistency* between students in these learning processes. In this way, information rich in the experiences of learning reality for individual students was captured in a format that was then administered to the second, much larger, sample of students.

The second sample of 479 participants, who completed the final version of the instrument, was drawn from second year students in accounting, human resource management, engineering technology, and management. The classes were addressed a week before the administration of the questionnaire and had the purpose and objectives of the study explained to them. Those who volunteered to participate were offered an individual learning profile of their results and a booklet on improving learning. The questionnaire was administered the following week during a class session. It is estimated that close to 100 percent of those who attended the original briefing returned to complete the questionnaire.

Of the 479 respondents, 296 (62%) were male and 181 (38%) were female. The majority of students (74%) were under 25 years of age and of European descent (85.7%). Asians made up 7.7 percent (34) of the sample and Polynesians 6.3 percent (29).

Measures

The first sample of 102 participants was used for a series of intensive instrument studies to explore alternative question types, the context of the questions and the most appropriate timing of questions as they related to specific learning events. Three conclusions were drawn from these studies:

1. students gave the most specific and complete information when asked about learning specific tasks as opposed to how they generally go about learning;
2. students needed to have actually engaged in a learning exercise before they were able to accurately describe how they would go about learning it; and
3. students were unused to reflecting on their own learning, and felt burdened by the cognitive workload required to express these processes. Any instrument used for measuring these reflective processes needed to do so in a manner that minimised the workload and assisted the students in identifying strategies and processes they used.

From these instrument studies, three new measures were devised. The first measured the learning strategies students used to learn a specific learning task. These are defined as any cognitive interaction between the student and the learning content for the purpose of remembering, understanding or improving skills. Students were asked to rate the importance of these strategies in relation to a specific task using a Likert-type scale. The second measure allowed students to describe how they represented the learning task. The third measured those behaviours associated with organising and managing learning resources and opportunities, including time management. These were called study management skills. Students were also asked for biographical data, including their previous year's examination results. These results were used to generate a grade point average, which was used as a measure of prior knowledge.

Results and Discussion

A principal components analysis with varimax rotation was used to explore the three measures described above. The *learning strategy* measure produced seven components that may be considered as classes of strategies that serve different functional purposes in learning (see Table 5.2, also see Hunt, 1995a, 1995b for a more detailed explanation of the results). The individual items within each class are learning strategies that may be used singly, in clusters with other strategies from the same class, or in conjunction with strategies from other classes, depending on the nature of the learning task.

Table 5.2 Classes of learning strategies

Learning Strategies	Description
Practice	*Skill Development:* enables the student to manipulate and apply knowledge. It implies more than just repetitive drill exercises that might normally be associated with practice. The student is practising *using* the information in some way (e.g., students work through examples, find new examples, change formulae into words, explain ideas to self to test understanding).
Identify key ideas	*Comprehension:* describes the different ways in which information can be reduced or condensed to extract key ideas. This also involves the student evaluating the significance of the ideas (e.g., students identify significant points, reduce notes to get a skeleton of main ideas, use headings and subheadings, underline or highlight important ideas, identify key words and phrases).
Relate information to other contexts	*Comprehension:* recognises similarities between the knowledge being learned and other knowledge structures. The student is able to use the "other knowledge" to interpret and understand the new knowledge (e.g., students relate ideas to their own experience, views, beliefs or emotional reactions, apply concepts to real-world situation or in new situations, work out how the ideas are related to each other).
Remember information	*Memorising:* strategies to organise or interact with information to increase its memorability (e.g., students make up acronyms or rhymes, associate new ideas with something easy to remember, change diagrams into words or words into diagrams).
Background filling	*Comprehension:* describes the different methods that a student can use to collect additional material about the information being learned (e.g., students read more widely and collect more information from a variety of sources, find out more about the context in which the ideas were developed, talk to people).
Projecting/ predicting	*Skill Development:* student abstracts meaning from the information and uses it to predict, to project information beyond its immediate or apparent meaning (e.g., students determine trends or patterns, identify or simulate the consequences or results of doing something in a certain way, visualise what is described).
Grouping information	*Memorising*: organises and labels information in a manner that enhances memory rather than understanding (e.g., students chunk information into groups, label groups of information, put information into lists).

For example, being able to write an essay may require that students understand the information first. Two useful classes of strategy for this task would be *identifying key ideas* and *relating information to other contexts*. A task, such as learning a list of dates and events, however, would probably benefit from the classes of strategy that deal with remembering, that is *grouping information* and *remembering information*.

Task Representation

Four components emerged from the *task representation* measure. These described different ways of viewing a learning task from a student's perspective. Whereas a task can be objectively described, a task representation is an individual's subjective appraisal of the learning situation (see Table 5.3).

Table 5.3 Task representation

Task Representation	Description
Declarative	Task requires student to be able to explain or describe the information, but not use or apply it.
Investigative	Students must find much of the information for themselves.
Contained	Task structure and assessment requirements clearly understood.
Fragmented	Task made up of many fragmented, unrelated components.

Study Management Skills

Three components emerged to describe a student's *study management skills*. Collectively these components describe the organisational skills that the student brings to bear on the learning resources available to the student. It is thought that the management of these resources creates conditions for learning (see Table 5.4) that may be more or less favourable.

Table 5.4 Study management skills

Study Management Skills	Description
Planning	The preparation the student makes for study, including time management, goal setting, and availability of information.
Procrastination	The degree to which starting study is delayed or avoided.
Neat, organised study notes	Concern with having complete organised study notes.

The results of the principal component analysis were then used in further analyses to examine two sets of relationships. The first was to identify those factors that influenced students to select learning strategies, and the second examined the influence of those learning strategies on learning outcome.

Influences on the selection of learning strategies were identified by regressing the independent variables of task representation, study skills management and the type of learning task onto learning strategies. For the seven classes of learning strategy, these variables accounted for between 16.2 percent and 35.9 percent of the variance. The most important of these was task representation, that is, how the student sees the task. It is important to note that this may not always be an accurate representation of the task itself. The student's task representation reflects his or her appraisal of the learning task which then dictates how the student will respond. This is an important missing link in understanding *why* students choose particular approaches to learning. Rather than the learning strategy training that has been a feature of attempts to improve learning, these results suggest that a fruitful source of remediation might be directed at student task analysis.

The type of learning task was also expected to influence which learning strategies were selected. Four types of task were used. These included an essay (accounting), a practical project (engineering technology), and two case studies. One case involved writing a research proposal (management) and the other consisted of providing advice to a young teacher on developing and structuring her teaching notes (human resource management).

Essay Task The essay task had a strong negative correlation with *relating information to other contexts* strategies and a strong positive correlation with *background filling* strategies. These students chose to develop their background knowledge, but were not sufficiently motivated to want to relate this knowledge to other aspects of their lives. This fits with the nature of the course which was "Accounting for Non-Accountants", a compulsory paper (subject) for non-accounting students doing the Bachelor of Business Studies degree. Students would use *background filling* strategies to develop a context that would improve their understanding of the content, but were averse to the use of *relating information to other contexts* strategies which are normally associated with a deep motive and an intrinsic interest in a subject (Biggs, 1987).

Project Task The practically-based project task correlated strongly with *practice* strategies, and to a lesser degree with the more advanced *projecting/predicting* strategies. Both of these strategy groups proceduralise information and organise it in a manner that facilitates its application to a new situation, and are, therefore, highly appropriate for learning this type of task.

Case Study 1 The case study 1 task predicts *identifying key ideas* and *relating information to other contexts* strategies, both of which build comprehension. A case study task is in effect a simulated practical task, and as such, should generate some practice strategies. However, the students' reliance on comprehension strategies suggests instead that these students may have been interpreting this task as a declarative task rather than the simulated practical exercise that was intended. On the other hand, the presence of *relating information to other contexts* strategies, associated with an intrinsic interest in the subject, reflects the fact that this class was part of the students' chosen major in Human Resource Management.

The use of *grouping information* strategies for case study 1 was also interesting since no other task predicts these strategies. It seems likely that *grouping information* strategies may be used to organise and structure *key ideas*, which were favoured by these students.

Case Study 2 In contrast to case study 1, case study 2 students did use *practice* strategies. These students clearly saw the task as a simulated practical task. No other strategies were significantly predicted by this task.

The nature of the task was also found to influence the strategies students chose for learning the task. However, overall this influence was not as strong or pervasive as task representation.

A further expected influence on the choice of learning strategies was the student's study management skills. These establish the conditions under which the learning strategies are used. Less favourable conditions, such as a shortage of time due to poor planning, are likely to reduce the range of learning strategies options available to the student, or diminish the effectiveness of the strategies.

Only *planning* of the study management skills predicted learning strategy selection. *Planning* contributes to the use of *practice*, *relating information to other context*, *background filling* and *grouping information* strategies. The first three of these learning strategies have significant additional time requirements when compared to strategies, such as *identifying key ideas*. *Grouping information* strategies are organisational in nature and are probably used in conjunction with other strategies, such as *identifying key ideas*, and thus would also require time and forethought.

Influences on Learning Outcome

Having determined which variables influence the selection of learning strategies, the second step was to establish how learning strategies influenced the quality of learning outcome. Learning strategies were not expected to be the only influences on learning outcome. According to Tabanick and Fidell (1989) and Norusis (1990), leaving out significant independent variables from a regression is likely to distort the results. Prior knowledge is well established in the literature as an important determinant of learning outcome (Alexander & Judy, 1988; Chi, Feltovich, & Glaser, 1981; Chi, Glaser, & Rees, 1982) and was included for this reason. Study management skills were also included because these skills provide the conditions under which the learning process occurs. The presence or absence of these skills impinges on the effectiveness of the whole process, and thus, is likely to directly affect learning outcome. The measure of learning outcome used was the assignment mark the student received.

The regressions were run in five stages. The first was for the total sample as a check on which variables generally might influence learning outcome. However, as it was expected that the association of learning strategies with success would depend on which learning task was being performed, the regression was run again for each of the four tasks.

Learning Outcome Results

When a single regression was run for the total sample, prior knowledge emerged as the main influence on learning outcome. It was also important when each task was examined individually. Study management skills were significant predictors of learning outcome as well. Both procrastination and study materials (having clear and complete notes) showed negative correlations with assignment marks for the sample as a whole. In the case of procrastination, this result seems common sense. If students delay study, they reduce learning opportunity. The study management skill of having clear and complete study materials/notes correlated negatively with learning outcome. It seems probable that these students were overly concerned with the form and appearance of study materials to the neglect of a more substantial involvement. Planning, on the other hand, is an important ingredient for success.

No learning strategies consistently predicted outcome for all tasks. However, in the context of specific tasks some learning strategies were clearly important. The comprehension type strategies which aid understanding, including *identifying key ideas* and *relating information to other contexts* were important in predicting outcome for declarative type tasks. Declarative knowledge is generally likely to be a more superficial cognitive structure than procedural knowledge (Glaser, 1990), because procedural knowledge is declarative knowledge that has undergone further processing. It seems likely then that the student learning a declarative task will require more support for retention than the student learning a procedural knowledge task. Some evidence for this is indicated by the appearance of *remembering* strategies as a significant influence on learning outcome whenever *identifying key ideas* and *relating information to other contexts* strategies are also important. *Practice* strategies and *grouping information* strategies were negatively correlated with learning outcome for declarative tasks. This suggests that in specific contexts some behaviours are so inappropriate that their use is detrimental to performance.

For the project task, only prior knowledge and *planning* predicted successful learning. No learning strategies predicted learning outcome. This result seems consistent with the results obtained by other researchers, such as

Entwistle and Ramsden (1983) and Biggs (1970b), who found that success in Science and Technology subjects seemed to be more dependent on prior knowledge than on specific learning strategies. It is possible that highly structured procedural tasks, such as these, have the conditions of applicability (strategies) embedded in the student's knowledge structures. High prior knowledge, which includes strategies for using that knowledge, may increase the effectiveness of the strategies. Use of the same strategies, without high prior knowledge, would be less effective, leading to a relatively poor learning outcome. If this interpretation is correct, then the type of learning strategies used would not appear significant as it is the effectiveness with which the same strategies are used that makes the difference.

Conclusions

This study set out to examine two major relationships: (1) to identify factors which influence the selection of learning strategies, and (2) to determine the influence of learning strategies, in conjunction with prior knowledge and study management skills, on learning outcome.

Learning proceeds in three stages. In the first instance, the student undertakes a situation assessment in which the task context is appraised to determine an appropriate learning response. Several variables influence this appraisal process including the nature of the task, the student's representation of the task, and study management skills. Based on the student's assessment, specific strategies are used, in conjunction with prior knowledge, to learn the task.

When the student actively engages in learning, prior knowledge acts as a context for understanding and interpreting the new information (Bellezza, 1981, 1982; Montague, 1972; Wittrock, 1974, 1978). Learning strategies, which describe the nature of the interaction between the new and prior knowledge, work by integrating the new content with the existing content, resulting in new knowledge structures that are not just quantitatively increased but are more complex and sophisticated. The type of learning strategy used influences the sophistication of the hybrid knowledge structures (Biggs, 1987; Entwistle & Ramsden, 1983; Jansweijer, Elshout, & Wielinga, 1990). For example, rote rehearsal strategies will produce relatively rigid and inflexible structures, that allow only a replication of the information. Strategies that encourage the student to process the information at a more meaningful level, such as relating information to other contexts strategies, integrate prior knowledge and new information to create a new structure which is richer in content, and multi-linked to other knowledge structures beyond the immediate topic of study. It is likely that this new knowledge may be able to be used in a variety of forms and situations.

The student's study management skills, while not a direct cause of learning, help to create a learning environment that will either facilitate or hinder learning. For example, good planning ensures that students maximise their opportunities for assembling resources, reduce anxiety, deploy appropriate learning strategies and allocate time to learning activities. On the other hand, too much effort organising and re-writing study notes may lull students into a false sense of security regarding their learning effort and detract from understanding the material.

Learning outcome eventually becomes prior knowledge for the next learning task and this aids new learning. While this part of the process was not specifically tested in the study reported here, logic suggests that this prior knowledge also feeds back to influence the manner in which students interpret and understand (represent) new learning tasks. The accessibility and usability of this prior knowledge is largely determined by the quality of processing or learning previously undertaken by the student (Elshout & Veenman, 1992; Weinstein, Underwood, Wicker, & Cubberly, 1979).

The underlying assumption of the Student Approaches to Learning (SAL), as the deep/surface approach has come to be called, suggests that the quality of learning can be improved by encouraging students to use deep learning strategies (Richardson, Morgan, & Woodley, 1999; Trigwell & Prosser, 1991). This study on the role of learning strategies suggests that the interaction between learning strategies and learning outcome is more complex and that a single solution for all students and situations is unlikely. While simple answers are seductive, the more complex picture identified in this study provides greater opportunity for intervention in the learning process and allows us to target more specifically dysfunctional learning behaviour.

The first point at which students can err is in the task representation. If students are unable to develop an accurate interpretation of the learning task, they will not select the most appropriate learning strategies. Task representation showed such a strong influence on learning strategy selection that it seems clear that students would benefit from acquiring adequate skills to develop accurate mental representations of the learning task. If, as Glaser

(1988, 1990) and others have suggested in the similar area of problem representation, high prior knowledge is a significant advantage in the initial representation of a task, then providing students with either a clear task representation (from the instructor) or the skills to develop their own accurate representations may lessen the disadvantage of low prior knowledge.

Given the dominance of prior knowledge on learning outcome, the overall effect could be quite large. This information could be provided in either of two ways. First, it is suggested that students could be actively taught these skills. Wherever possible students should be given the tools to control their own learning. Equally, instructors who commence a lesson by providing students with a useful task representation (advance organisers) may enhance the learning that ensues (Cole & Chan, 1994; Pithers, 1998).

The present study clearly identified the importance of matching the appropriate learning strategies with tasks for learning outcome. All learning strategies are not equally effective for all tasks. For example, students who used practice strategies for the essay task had poor learning outcomes. While many successful students seem to use appropriate learning strategies instinctively, a large number of less successful students cannot. Thus, the second possible point of intervention could be that we could teach students to match task and strategy. While this could be done in a learning skills course, it should also be reinforced by instructors of specific subjects.

Finally, students are not necessarily good managers of their learning environment, and yet these skills impact on learning effectiveness. While a study skills course may be appropriate, instructors have a large role to play in encouraging and monitoring the development of study skills as part of the instructional process.

Summary

The constructivist view of learning is that the activities and behaviour of the student are central to the success of the learning process. This implies that students must have the appropriate skills to undertake this more active role. The study described here suggests that such skills include: task analysis (to develop an accurate task representation), matching task type to learning strategy, and good study management skills to create favourable learning conditions. Student decision making during the learning process also impacts strongly on learning outcome. To the extent that instructors can help students to make better decisions, it may be possible to improve the effectiveness of the learning process.

Reflective Activities

1. Select four or five learning objectives that you have written for a subject you normally teach, or use personal learning objectives if you are a student in the process of learning. Then, identify from the list of learning strategies (see Table 5.2) those that you would consider appropriate for the intended learning outcome.
2. Look at these objectives again and think how you could elaborate them to give students a clearer task representation. You may want to consider the following: the answer format of assessment questions (e.g., multiple choice, short answers, essays), the type of content to be learned (facts, principles, procedures, concepts or skills), and the level of learning required (memorisation, comprehension, application).
3. Look at the study management skills (Table 5.4). Assess yourself honestly on these skills. Write out two areas where you would like to improve your performance. Look for material that will help you achieve these objectives.

Suggested Readings

Biggs, J.B. (1987). *Student approaches to learning and studying*. Melbourne: Australian Council for Educational Research.

Biggs, J.B. (1993). What do inventories of students' learning processes really measure? A theoretical review and clarification. *British Journal of Educational Psychology, 63,* 3-19.

Biggs, J.B. (1994). Approaches to learning: Nature and measurement of. *The International Encyclopedia of Education, 1,* 319-322.

Birenbaum, M., & Feldman, R.A. (1998). Relationships between learning patterns and attitudes towards two assessment formats. *Educational Research, 40*(1), 90-98.

Laurillard, D. (1993). *Rethinking university teaching: A framework for the effective use of educational technology*. London: Routledge.

Reigeluth, C.M. (1999). *Instructional design: Theories and models. A new paradigm of instruction* (Vol. 2). Mahwah, NJ: Lawrence Erlbaum.

Richardson, J.T.E., Morgan, A., & Woodley, A. (1999). Approaches to studying in distance education. *Higher Education, 37,* 23-55.

References

Alexander, P.A., & Judy, J.E. (1988). The interaction of domain-specific and strategic knowledge in academic performance. *Review of Educational Research, 58*(4), 405-437.

Bellezza, F.S. (1981). Mnemonic devices: classification, characteristics and criteria. *Review of Educational Research, 51,* 247-275.

Bellezza, F.S. (1982). Updating memory using mnemonic devices. *Cognitive Psychology, 14,* 301-327.

Biggs, J.B. (1969). Coding and cognitive behaviour. *British Journal Educational Psychology. 60*(3), 287-305.

Biggs, J.B. (1970a). Faculty patterns in study behaviour. *Australian Journal of Psychology, 22,* 161-174.

Biggs, J.B. (1970b). Personality correlates of certain dimensions of study behaviour. *Australian Journal of Psychology, 22*(3), 287-297.

Biggs, J.B. (1976). Dimensions of study behaviour: Another look at ATI. *British Journal of Educational Psychology, 46,* 68-80.

Biggs, J.B. (1978). Individual and group processes in study processes. *British Journal of Educational Psychology, 48,* 266-279.

Biggs, J.B. (1985). *Study process questionnaire.* Melbourne: Australian Council for Educational Research.

Biggs, J.B. (1987). *Student Approaches to Learning and Studying.* Melbourne: Australian Council for Educational Research.

Chi, M.T.H., Feltovich, P.J., & Glaser, R. (1981). Categorization and representation of physics problems. *Cognitive Science, 5,* 121-152.

Chi, M.T.H., Glaser, R., & Rees, E. (1982). Expertise in problem solving. In R.J. Sternberg (Ed.), *Advances in the psychology of human intelligence* (Vol. 1, pp. 7-75). Hillsdale, NJ: Lawrence Erlbaum.

Cole, P.G., & Chan, L.K.S. (1994). *Teaching principles and practice* (2nd ed.). Sydney, Australia: Prentice Hall.

Cronbach, L.J. (1967). How can instruction be adapted to individual differences? In R.M. Gagné (Ed.), *Learning and individual differences* (pp. 23-39). Columbus, OH: Charles E. Merrill.

Elshout, J.J., & Veenman, M.V.J. (1992). Relation between intellectual ability and working method as predictors of learning. *Journal of Educational Research, 85*(3), 134-143.

Entwistle, D. (1982). Study skills and independent learning. *Studies in Higher Education, 7*(1), 65-73.

Entwistle, N.J. (1977). Strategies of learning and studying: Recent research findings. *British Journal of Educational Studies, 3,* 225-238.

Entwistle, N.J. (1981). *Styles of learning and teaching*. London, Wiley.

Entwistle, N.J., & Entwistle, D. (1970). The relationship between personality, study methods and academic performance. *British Journal of Educational Psychology, 40,* 132-143.

Entwistle, N.J., Hanley, M., & Hounsell, D.J. (1979). Identifying distinctive approaches to studying. *Higher Education, 8,* 129-148.

Entwistle, N., & Ramsden, P. (1983). *Understanding student learning*. Beckenham, UK: Croom Helm.

Fransson, A. (1977). On qualitative differences in learning IV: Effects of intrinsic motivation and extrinsic test anxiety on process and outcome. *British Journal of Educational Psychology, 47,* 244-257.

Gagné, R.M. (1967). *Learning and individual differences*. Columbus, OH: Charles E. Merrill.

Glaser, R. (1970). *The conditions of learning* (2nd ed.). New York: Holt, Rinehart & Winston.

Glaser, R. (1988, August 15) The reemergence of learning theory within instructional research. American Psychological Association's Distinguished Scientific Award for the Applications of Psychology Address. Atlanta, GA.

Glaser, R. (1990, August 18). Intelligence as an expression of acquired knowledge (expertise). Paper presented at a seminar, Massey University, Palmerston North, New Zealand.

Hunt, L.M. (1995a). *Approaches to learning: The selection and use of learning strategies*. Unpublished doctoral dissertation, Massey University, Palmerston North, NZ.

Hunt, L.M. (1995b). *Influences on the learning process and learning outcome: Practical implications for the trainer.* Paper presented at the First Industrial and Organisational Psychology Conference. Sydney: The Australian Psychological Society.

Jansweijer, W., Elshout, J.J., & Wielinga, B.J. (1990). On the multiplicity of learning to solve problems. In H. Mandl, E. de Corte, N. Bennet, & H.F. Friedrich (Eds.), *Learning and instruction: (Vol 2.1). Social and Cognitive Aspects of Learning and Instruction* (pp. 127-245). Oxford, UK: Pergamon Press.

Laurillard, D. (1999). A conversational framework for individual learning applied to the 'learning organisation' and the 'learning society'. *Systems Research and Behavioral Science Systems Research, 16,* 113-122.

Marton, F., & Säljö, R. (1976a). On qualitative differences in learning I: Outcome and processes. *British Journal of Educational Psychology, 46,* 4-11.

Marton, F., & Säljö, R. (1976b). On qualitative differences in learning II: Outcome as a function of the learner's conception of the task. *British Journal of Educational Psychology, 46,* 115-127.

Merrill, M.D. (1991). Constructivism and instructional design. *Educational Technology,* 45-53.

Montague, W.E. (1972). Elaborative strategies in verbal learning and memory. In G.H. Bower (Ed.), *The Psychology of Learning and Motivation* (Vol. 6). New York: Academic Press.

Norusis, M.J. (1990). *SPSS/PC+ statistics 4.0.* Chicago: SPSS Inc.

Piaget, J. (1950). *The psychology of intelligence.* London: Routledge & Kegan Paul.

Pines, A.L., & West, L.H.T. (1986). Conceptual understanding and science learning: An interpretation of research within a sources-of-knowledge framework. *Science Education, 70,* 583-604.

Pithers, R.T. (1998). *Learning through effective training.* Katoomba, Australia: Social Science Press.

Reese, H.W., & Overton, W.F. (1970). Models of development and theories of development. In L. Goulet & P.B. Bates (Eds.), *Life span developmental psychology: Theories and research* (pp. 115-145). New York: Academic Press.

Richardson, J.T.E., Morgan, A., & Woodley, A. (1999). Approaches to studying in distance education. *Higher Education, 37,* 23-55.

Svensson, L. (1977). On qualitative differences in learning: III — Study skill and learning. *British Journal of Educational Psychology, 47,* 233-243.

Tabanick, B.G., & Fidell, L.S. (1989). *Using multivariate statistics* (3rd ed.). New York: Harper and Collins College Publishers.

Trigwell, K., & Prosser, M. (1991). Relating approaches to study and quality of learning outcomes at the course level. *British Journal of Educational Psychology, 61,* 265-275.

Weinstein, C.E., Underwood, V.L., Wicker, F.W., & Cubberly, W.E. (1979). Cognitive learning strategies: Verbal and imaginal elaborations. In H.F. O'Neil, Jr. & C.D. Spielberger (Eds.), *Cognitive and Affective Learning Strategies* (pp. 45-75). New York: Academic Press.

Wittrock, M.C. (1974). Learning as a generative process. *Educational Psychology, 11,* 87-95.

Wittrock, M.C. (1978). The cognitive movement in psychology. *Educational Psychology, 13,* 15-29.

6 Factors that Affect Learning

Irene Henley

Despite all the research into how adults learn, it must be acknowledged that the teaching and learning transaction is not always a smooth process. Barriers to learning are encountered in many forms ranging from failure to participate in learning in formal settings, to participants experiencing stress about a new learning environment, or resisting new information and skill development once engaged in instructional programs. Although there are no universal remedies for these problems, an awareness of their possible causes provides greater potential for perceptive educators to overcome such barriers.

Further, despite the fact that a high level of motivation is generally one of the characteristics used to describe adult learners, especially in the aviation training context, it must also be noted that this does not always ensure that the teaching and learning transaction is straightforward and simple or that the same level of motivation will be sustained throughout the learning episode. A range of factors can, and often does, negatively impact on teaching and learning. When this occurs and the learning is compromised, it is important to be aware of the possible nature of these barriers to effective learning, especially since some barriers can be instructor-induced.

Research by Henley (1987, 1995) has shown that unfortunately many instructors in the aviation industry have not received adequate training for their role as teachers and are not necessarily aware of barriers to learning and of the effect of stress on learning. Testimonies from student pilots clearly revealed that feelings and emotions play an important role in flight training and that psychological stress is frequently instructor-induced and can be quite debilitating. Student pilots indicated that many flight instructors were unsympathetic to their feelings and failed to recognise individual differences. Further, some instructors seemed insensitive to the fact that the learning environment can at times be extremely daunting or threatening to novice pilots (Henley, 1987, 1995). According to student pilots, many instructors display stress-producing attitudes and behaviours and have faulty or unrealistic expectations of students. In contrast, some students said that their instructors, through personal insight, experience and natural teaching abilities, were able to transcend their lack of pedagogical or andragogical training and develop effective teaching strategies and humanistic ways of dealing with students. This chapter outlines some of the factors that affect learning and examines instructor attitudes and behaviours that can hamper or facilitate learning.

Barriers to Participation

There are many barriers to participation in educational and training activities for adults. The two barriers most frequently cited by adults, and to which most certainly can relate, are *lack of time* and *cost* (Darkenwald & Merriam, 1982, p. 137). Cross (1981) identified three categories of barriers to participation: situational, dispositional, and institutional.

- Situational barriers relate to factors arising from one's situation or life context, such as lack of time, cost, home or job responsibilities, lack of transportation, etc.
- Dispositional or psychological barriers are those that relate to the learner's attitudes, values and self-perception, such as being afraid of being too old to learn, lack of confidence, fear of appearing stupid or clumsy, etc.
- Institutional barriers are factors that may block potential participants from involvement in education and training in formal settings, such as inconvenient schedules, lack of part-time courses, lack of attractive or appropriate courses, inflexibility on the part of formal institutions, etc. (adapted from pp. 98-108).

Darkenwald and Merriam (1982) added informational barriers as a fourth category to Cross' three main categories because they found that the lack of readily available information about courses and on educational opportunities was also a major barrier to participation.

Cross suggests that educational and training organisations could reduce institutional barriers to participation by making university and college programs accessible to working adult students through:

- scheduling classes when and where working adults can attend;
- granting credit by examination for non-collegiate learning; and
- creating more flexible admission procedures (p. 104).

Barriers to participation in education and training can also stem from wider societal factors. Recent research into access and equity in vocational education and training shows that there are a number of identified groups in the Australian society who are under-represented in training programs. These groups include Aboriginal and Torres Strait Islands people, women, people who are from non-English speaking backgrounds, people from rural and remote areas, and people with disabilities (Smith & Keating, 1997). Similar findings would likely be revealed in many other developed countries. The implications of these findings are significant since the links between education and training and employment opportunities are clear. Simply put, people with higher levels of education are more likely to be employed. Therefore, although most teachers and trainers are consumed with teaching and learning problems and issues, barriers and breakthroughs of participants in their classrooms, it is also important to ask the question — Who is not in the classroom and what barriers may have prevented their participation?

Factors that Affect Learning

In addition to barriers to *participation*, there are numerous factors that affect *learning* once students engage in learning activities. Galbraith (1991) identifies three main risk areas for adult learners.

Risks and Challenges

One barrier to learning may be the fear of the *risks* involved (Galbraith, 1991). Risk taking in adult learning can be seen in the following ways:

- risk of commitment — where learning involves the extension of oneself into new dimensions and territories of involvement and action;
- risk of confrontation — where the role of the teacher or trainer involves assisting adult learners to challenge held beliefs, values, and demonstrated actions;
- risk of independence — not all who enter an educational activity are ready or at times willing to shed the psychological dependence they hold. This dependence can be in the form of previously held beliefs and practices or dependence on the knowledge, authority or expertise of teachers. (Galbraith, 1991, p. 5)

Fortunately, many learners soon experience the exhilaration of taking the risks and "surviving" the learning experience, so that they feel a sense of mastery over new knowledge and skills. However, for others the challenge provides an ongoing barrier to learning. Galbraith (1991) discusses the emotional complexity of these situations, where adults report feelings of threat to the ego when exploring new areas of knowledge or when approaching new sets of skills. The impact of these negative emotions clearly has the potential to produce a barrier to learning.

Negative Emotions

It is important, some even say necessary, for adults to be emotionally involved in their learning, if that learning is to be of a high quality. However, if these emotions are overwhelmingly negative, learning may be impeded. Research into the experiences of adult learners shows that the most common of these negative emotions are depression, anger, frustration, hostility and anxiety or stress (Marin, 1997). The reasons for these emotional responses may be as varied as the participants in any classroom, however, generally speaking they may have their basis in:

- the learner's response to the educational system;
- unsatisfactory relationships with individual teachers; and
- the students themselves (Marin, 1997, p. 316).

In simple terms, these emotions result in barriers to learning because they impact upon the learners' ability to process information. According to Marin, elevated negative emotions alter the way information in the learning environment is received and retained. If a participant is preoccupied with a strong emotional response, reception of information can be completely blocked.

A. Rogers (1998) also recognises that learning can sometimes be stressful or threatening to adult learners. He discusses the way participants' pre-existing knowledge may set up a barrier to learning. Often, adult learners know something about a topic, issue or skill set before they undertake further education or training on it. Their individual interests have led them to explore this area, and as such they have often devoted much time and energy pursuing their interest. Often, they may also have developed a set of strongly held ideas about their area of interest, and perhaps even a set of habitual ways of approaching the issue or carrying out the skill. Ironically, according to A.

Rogers (1998), this pre-existing interest and knowledge may lead to blocks in learning new information and skills since "adult student participants have already invested emotional capital in acquiring this knowledge and experience. They will spend much more in defending the integrity of this knowledge, so new learning changes will sometimes be strenuously resisted" (p. 207).

Participants' Perspective

When discussing factors that affect learning, it is also important to consider how adult learners experience learning. Looking at learning from the perspective of the participants can provide valuable clues for the perceptive teacher. For example, Galbraith (1991) and Henley (1987) report that many adult learners describe their learning experiences in highly emotional terms. This emotional response is often experienced very early in a course of study, with participants often doubting their capacity to continue. Many adult learners report questioning whether or not they should be there at all:

> Adults' reflections on how learning feels run the emotional gamut, from profound embarrassment at their inabilities to seem as assured and confident as they feel they ought, to deep, angry resentment at the dismissive arrogance displayed by teachers, or exhilarating relief and pleasure at being able to perform some task previously thought impenetrable. (Galbraith, 1991, pp. 41-42)

According to J. Rogers (1977), one of the major obstacles to learning faced by adult learners is the development of emotional blockages which are caused by fears about the new situation, fear of the unknown, their unwillingness to expose what they regard as their ignorance or lack of ability, or by "their anxiety that they might be making themselves look foolish, or that they might be exposing themselves to failure" (p. 33). Clearly, adult learners can sometimes be quite vulnerable in the learning context. This is especially true in the aviation training milieu where students are learning in a non-traditional, dynamic and often stressful environment. Research into student pilot stress revealed that many novice pilots experience fairly high levels of stress or anxiety during flight training, which can be a serious impediment to learning, can impair performance and interfere with decision making processes (Grandchamp, 1971; Henley, 1987; Jensen & Benel, 1977; Krahenbuhl, Marett, & Reid, 1978, 1979; Prince, Bowers, & Salas, 1994).

The claims of the literature reviewed in the next section also show that psychological stressors (emotional factors and mental workload) are as significant in their potential to impair cognitive processes, flying skills and decision making as both physical stressors (temperature, noise, "G" forces) and physiological stressors (fatigue, lack of physical fitness, missed meals). Further, psychological stressors provide even greater problems than physical and physiological stresses because they are more ubiquitous, often more insidious, and probably occur more frequently (Henley, 1986). A good number of student pilots initially feel quite anxious before flight or before new untried manoeuvres and some are afraid of appearing clumsy or stupid in front of their flight instructor or fellow students (Henley, 1987). Indeed, it is interesting to note that Melton, Hoffman, and Delafield (1969) found that stresses in private pilot flight training are equivalent in intensity to those of combat pilots and astronauts. Unfortunately, according to close to 40 percent of student pilots, the major source of psychological stress is often the instructor (Grandchamp, 1971; Haward, 1969; Henley, 1987; Krahenbuhl, Darst, Marett, Reuther, & Constable, 1980; Melton & Wicks, 1967 Termoehlen, 1987).

The Effect of Stress on Learning, Performance, and Decision Making

Stress usually results when there is a mismatch between how we *perceive* the demands placed on us (these demands can be external or self-generated) and our *perceived* ability to cope with these demands (Air Canada Human Factors Project Team, 1996; Miller & Smith, 1993; Stokes & Kite, 1999). Research has shown that during high stress periods the efficiency of an individual can be impaired severely because stress can degrade the ability of humans to process information (Alkov, Gaynor, & Borowsky, 1985; Sloan & Cooper, 1986; Smode & Meyer, 1966; Stokes, Kemper, & Kite, 1997; Stokes & Kite, 1994). Further, studies by Krahenbuhl et al. (1980), Haward (1968, 1969) and Melton and Wicks (1967) revealed that not only does a high level of stress reduce efficiency but it actually produces interfering responses which cause decrements in both learning and performance. Under stress, a person can experience reduced cognitive capabilities to such an extent that even simple tasks are adversely affected. According to Sanders and Ethell (1986), "under stress, adding two plus two and other simple mental tasks become major hurdles" (p. 9). Indeed, the incompatibility between high levels of stress and maximum intellectual functioning has long been recognised through the work of renowned stress researcher Hans Selye (1956).

Some tasks in the aviation training context are inherently stressful because they may be seen as threatening or they require the student "to attend to too many things at a time, or to perform too many operations at once" (Lazarus, Deese, & Osler, 1952, p. 298). At such times, a person's capacity for attention may become overloaded. Bond, Bryan, Rigney, and Warren (1968) explain how overloading can occur in the aviation context:

> One of the important things the student pilot learns is to shift the focus of his [sic] attention sufficiently, so that he [or she] will be able to attend to all the elements in the real-time situation. Shifting of attention can be applied both to sensory inputs from displays and to outputs of controls.
>
> Of course, there is a limit to the effectiveness of attention shifting in real-time control. It is possible to sample only a finite amount of information in a fixed time interval, or to make only a finite number of control actions. If the time interval is not within permissible time-late tolerances in the situation, the pilot will be overloaded. (p. 4-3)

One example of such a situation is when a student learns to land an aircraft. Initially, too many demands may be placed on the learner at one time. Landing an aircraft requires that the student be able to fly straight and level, perform coordinated turns, and fly a gradual descent, constantly keeping the airspeed and the attitude of the aircraft properly adjusted while compensating for crosswinds or gusts of wind. In the operational aircraft, these actions must all be performed concurrently, they cannot be broken down as separate tasks; they happen simultaneously as complex concurrent operants. Thus, the student's ability to attend to all the tasks can often be taxed. Haward (1968) points out that "without doubt, landing is the most difficult flying task the student pilot is called upon to execute, calling for greatest concentration, skill, and sensorimotor co-ordination. It is the phase when cognitive efficiency is at a premium, yet, as the data show, is the phase when stress is highest" (p. 22). The stress created by the limited time factor during the landing phase is also highlighted by Bailey and Hughes (1980) who note that:

> What makes the task unusual is that decisions and responses must be made so rapidly and flawlessly, since either a delayed response or an incorrect judgment could be fatal. It is this latter element, no doubt, that puts such stress on the pilot and which probably makes acquisition of the motor skills in the aircraft itself so labile. (p. 10)

Therefore, as stress builds up in a task, overloading often occurs and performance deteriorates or breaks down because the individual is required to handle too much information at once or must perform within critically short time periods (Smode & Meyer, 1966; Stokes, Kemper & Kite, 1997; Stokes, Kemper & Marsh 1992). Research into stress in air traffic controllers as a result of increased workload has revealed similar outcomes (Brookings, Wilson, & Swain, 1996; Costa, 1989; Farmer, Belyavin, Berry, Tattersall, & Hockey, 1991).

In addition to stress created by the time-pressed tasks and by the concomitant potential mental overloading, further stress is obviously added in flight training by the possibility of bodily harm or damage to the aircraft. Krahenbuhl, Marett and King (1977) and Henley (1987) found that power-on stalls, spin recoveries, landing and first solo flights resulted in the most pronounced stress responses (i.e., excitement, confusion and feelings of helplessness). Since these exercises are not physically demanding, the authors suggest that a large portion of the stress experienced by the students emanated from psychological sources. Students *perceive* these exercises as being potentially dangerous or threatening which in turn can adversely affect their cognitive efficiency and their flying proficiency.

Davis (1964) points out, however, that "to suppose that the most important source of fear is physical danger is far too simple a view" (p. 12). He suggests that the fear most pilots experience is more complicated than a primitive fear of death or injury, it usually contains elements which include a fear of letting someone down, or a fear of failure, as well as the fear of being unable to achieve a goal or an objective. He states that "in whatever term they were described, these elements amounted to an uncomfortable awareness of factors which endangered the successful attainment of an objective" (pp. 12-13). He explains that emotional blockages emanating from stress "are likely to arise whenever the outcome of a task in which a person is highly motivated appears to him [sic] to be in doubt — that is, at times of uncertainty, insecurity or impaired confidence" (p. 13).

Steininger (1964) concurs, stating that the stress experienced by students often relates to a fear that achievements will be graded below standards, or that one may be embarrassed due to inadequate ability with adverse consequences on one's further career, or the fear, due to various reasons, of one's instructor or the lack of confidence in one's instructor and his or her way of teaching. Stokes and Kite (1999) also found that uncertainty is an important element that can have a considerable impact on a person's level of stress and flying performance. Harss, Kastner, and Beerman (1991) further note that subjective controllability of a situation can be a key factor affecting a person's stress level.

The value of a certain amount of apprehension or stress in the aviation training environment cannot be overlooked because it raises an individual's level of awareness and vigilance, makes the student more attentive to what is happening and facilitates his or her concentration on the task or the situation (Air Canada Human Factors Project Team, 1996). Indeed, stress or emotional arousal is an essential part of life because of its energising effects, however, the incapacitating effects of overly strong emotional incitement also need to be recognised (Selye, 1956). According to Bond et al. (1968), "strong emotion makes precise movements more difficult and concentrates our field of attention to such an extent that adequate solutions to emergencies may be difficult to accomplish" (p. 9-14). Baddeley (1972) notes that such "cognitive tunneling" could severely impair performance in dangerous environments. Several aviation accidents, where the crew concentrated or fixated on a problem at the expense of monitoring all instruments and systems, seem to corroborate this, for example the Eastern Air Lines' Lockheed L-1011 Tristar Flight 401 crash in the Everglades swampland in Florida on December 29, 1972 when the crew, preoccupied by an abnormal undercarriage indication, failed to monitor the altitude of the aircraft.

The positive effects of mild stress and the negative effects of severe stress are usually related to performance in a curvilinear fashion or bell-shaped curve. At a low level of stress, performance is poor due to apathy or boredom. As stress increases to a moderate level, performance reaches an optimal point after which performance will gradually deteriorate if stress is increased and overloads a person's capability to cope with the task at hand (Air Canada Human Factors Project Team, 1996). Iampietro et al. (1972) add that performance on complex tasks that "require a high degree of alertness, mental functions, muscular coordination and time-sharing of functions is more readily susceptible to degradation under adverse conditions than is performance of simple tasks" (p. 1218).

Numerous studies have shown impairment of perceptual-motor performance under stress resulting in fixation, mental block, inattention, increased rigidity, disorganised activity, hesitation, slow learning and non-adaptive behaviour (Baddeley, 1972; Cohn, 1994; Stokes, Belger, & Zhang, 1990; Stokes & Kite, 1994). Indeed, a detrimental symptom that can be experienced during the landing phase when a pilot is stressed is tunnel vision or target fixation. Sanders and Ethell (1986) note that "under stress, peripheral vision deteriorates, leaves one with foveal vision. The former perceives motion while the latter is used for reading. Therefore, the pilot has no perception of motion and he [or she] can hit the ground without a clue of what's happening" (p. 37).

Davis (1964) agrees that the "disintegration of the sensory field" tends to occur when stress is increased. His research showed that anxiety interfered with proper scanning techniques and led to fixation and subsequently to the misinterpretation of the instruments. Steininger (1964) adds that

> If the student's mind is generally anxious and if his [sic] attention is stressed additionally by multiple tasks, a small aberration alone could release considerable confusion. This leads to blocking the normal receptivity to information, e.g. when reading or interpreting instruments or listening to radio communications. In such cases it can be seen that the student overconcentrates on a single instrument instead of making a proper cross check. Consequently, his entire perceptual organization breaks down. (p. 35)

Research by Simonov, Frolov and Ivanov (1980) further indicates that "a high stress level disorganizes the activity, complicates it with a trend towards untimely acts and reactions to the insignificant signals" (p. 46). Alkov (1974) makes the same point, stating that "overstressed individuals often engage in irrelevant activities or rigid stereotyped behavior and experience loss of discriminative skill and mental efficiency" (p. 154). Psychological stress, therefore, tends to narrow one's range of attention (perceptual tunneling) so that a person's normal "channel capacity" is materially reduced (Stokes & Kite, 1999). According to Bond et al. (1968), the way in which stress

> affects thinking is by narrowing or "channeling," attention. The frightened individual is likely to be acutely conscious of the emergency facing him [sic], but to be unable to notice things in the periphery of his [or her] attention. He [or she] may not shift attention to those things which might be important in escaping from the emergency. (p. 9-2)

Similarly, Holmes (1986) concludes that anxiety can hinder a pilot's performance because it impairs perceptual functioning; it constricts and narrows the entire field of attention:

> During stressful situations all visual, auditory and kinesthetic (seat of the pants cues) perceptual processes may be distorted and even imagined.

> Anxiety can also cause forgetfulness, such as not completing the [landing] check prior to landing. Stress and/or anxiety present serious problems under instrument conditions. The most critical is believing one's own perceptions or beliefs rather than believing the aircraft instrument indications. (p. 32)

In short, stress causes people to concentrate or focus on the difficulties or the disturbing elements rather than on the practical aspects or solutions of the situation being encountered. Similarly, emotional conflicts and personal worries insidiously push their way into the focus of our attention when we should be concentrating on the task to be performed (Bond et al., 1968; Della Rocco, Fiedler, Schroeder, & Nguyen, 1999; Sloan & Cooper, 1986; Stokes & Kite, 1999). Indeed, one of the effects of stress is that it often produces a powerful distraction. "Threatened subjects frequently report that their productive thinking is disrupted by the compelling preoccupation with the thought of the consequences of failure or danger" (Lazarus, Deese, & Osler, 1952, p. 312). Confronted with a threat (physical or to ego integrity), students narrow their attention to the threatening situation or object; all their perceptual faculties focus on the thing or condition that generated the fear or stress. Consequently, learning and performance will likely be impaired. According to Melton, McKenzie, Kelln, Hoffman, and Saldivar (1975),

> It would be expected that such perceptual tunnel vision would impede the synthesis of perceptions into meaningful wholes and to the development of insights that are essential to true learning. Thus tension would be expected to increase the number of hours required for a student to attain private pilot competency, thereby increasing the monetary cost of flight training. (p. 5)

Stress does not only affect a student's intellectual efficiency adversely, but can also create confusion which hampers his or her control movements (Stokes & Kite, 1994, 1999). Tension can cause students to overcontrol the aircraft as in overbanking, skidding, uneven or erratic control handling during the approach and landing, or "freeze" during stall or spin recoveries and, as a result, can also delay student progress (Stokes & Kite, 1999). Steininger (1964) explains that disorder of sensorimotor skill is usually a distinct symptom of anxiety which can be seen as a mix-up of control movements. He notes that "progressive overcontrol develops coarse control movements, causes tenseness, and consequently overstrains the kinaesthetic sensitivity. This does not facilitate the student's learning the correct control pressures and movements required for corrective action" (p. 40).

In addition to debilitating perceptual faculties and impaired performance, stress induces fatigue (Della Rocco, et al., 1999; Harss, Kastner, & Beerman, 1991; Sloan & Cooper, 1986; Stokes & Kite, 1994). Research by Callister, Percival, and Retzlaff (1999) into the effects of stress during Survival, Evasion, Resistance and Escape training revealed that fatigue increased and self-confidence waned under the stress of training. Bond et al. (1968) also found that "stressful missions which cause anxiety cause more subjective fatigue than longer, routine missions. Emotional arousal uses up energy at a higher than normal rate, causing subsequent feelings of extreme tiredness" (p. 10-9). Similarly, Della Rocco et al. (1999) report that pilots who experienced stress at home felt tired, distracted, and worried which in turn increased their stress at work and interfered with their flying performance.

McFarland (1953), in an investigation of the effects of fatigue and stress on flying, concludes that the timing of motor responses suffered more and more as fatigue developed, subjects became increasingly willing to accept reduced standards of accuracy and performance, they shifted from following the six primary flying instruments to making more automatic responses, and they increasingly forgot to check instruments out of their immediate range of vision (pp. 348-350). He also notes that when pilots were aware of the approaching termination of the flight there was an "end deterioration" effect, that is, a tendency for a sudden increase in errors. The indication here is that the tired or stressed pilot tends to relax or be less vigilant at the end of a flight, thus increasing the possibility of landing accidents.

In addition to perceptual impairments and fatigue caused by stress, decision-making processes and judgement also can be adversely affected (Air Canada Human Factors Project Team, 1996; Berlin, Gruber, Holmes, Jensen, Lau, Mills, & O'Kane, 1982; Hammond, 2000; Jensen, 1982; Mauro, Barshi, Pederson, & Bruininks, 2001; Orasanu, 1997; Prince, Bowers, & Salas, 1994; Stokes, Barnett, & Wickens, 1989; Stokes, Belger, & Zhang, 1990; Stokes, Kemper, & Kite, 1997; Stokes, Kemper, & Marsh, 1992). According to Bond et al. (1968), "stress has a more serious effect on central processes, such as judgment and decision-making, than on motor skill per se" (p. 10-9). The authors indicate that in times when stress is high, decision making is perhaps one of the first abilities to deteriorate (p. 7-16).

Information must be organised into meaningful patterns to permit an individual to interpret the nature of the information received and to respond accordingly. It is on the basis of such perceptions that pilots make decisions.

Thus, the decisions and subsequent responses are determined by the interpretations placed on the pattern of sensations received. Therefore, the key to correct decisions and accurate responses is, in fact, correct interpretation or perception. Bond et al. (1968) maintain that "many of the errors of perception result from the failure to properly integrate information being received through various senses" (p. 5-11). Under psychological stress or when a pilot's attention is preoccupied with strong emotions or is distracted, he or she can fail to integrate the information available and could consequently make faulty decisions and revert to wrong responses. Jensen and Benel (1977) and Berlin et al. (1982) claim that an overstressed state of mind, indeed, can increase the likelihood of poor judgement. Roscoe (1980) agrees that emotional stress can affect decision-making efficiency, stating that "the in-flight decision process is further complicated by the fact that flying is an emotional experience for many people. High levels of emotion, whether in normal or emergency flight situations, can be expected to have extreme effects on rational decisions, either adaptive or maladaptive" (p. 177).

Selye (1956) also notes that stress is both combinative and cumulative. Therefore, while some stressors may not, by themselves, be "sufficiently stressful to interfere with the pilot's ability to fly safely, together they may so interfere with the cognitive processes and operational skills of the pilot that a serious error of judgement, potentially capable of producing a fatal accident, may occur" (Haward, 1968, p. 19). Thus, any stressor can create a climate that produces a ripple effect when combined with other stressors.

In stressful situations, according to Selye (1956), the automatic defence system is activated, mobilising immediate energy so that a person is ready for emergency action, such as running away or fighting (the flight or fight reaction). These defence mechanisms put the body on alert and enable a person to react more energetically to an external threat. However, an interesting and pertinent point made by Davis (1964) is that, in situations like flying, such automatic emergency reactions could be inappropriate and thus may be an added deterrent to proper responses. Consequently, in addition to being forced to deal with a threatening or stressful situation, the pilot has to ignore or suppress automatic bodily reactions. As Davis states,

> When dangers appear imminent, behaviour tends to undergo characteristic changes. For instance, responses tend to become more readily elicited, more rapid, more forceful and more extensive. These changes are of biological value, in that they enable the subject to react

> vigorously to situations which threaten him [sic], but they are not always advantageous, for many of the danger situations met by pilots require, not vigorous activity, but restrained, deliberate and accurate responses. When these qualities are required, the emergency mechanisms may impair efficiency. This impairment may be progressive, because a vicious circle is readily instituted. (p. 13)

Further, Bond et al. (1968) caution that the effects of emotional stresses are often very subtle and may not be easily recognised.

> Reactions to the normal stresses and irritations of life vary from those of little concern to anyone, such as rationalisation, to those which seriously interfere with a person's efficiency — and safety. More often than not a stressful situation merely serves to distract attention, to make it difficult to concentrate on the task at hand. Frequently a person is impelled to take risks, to "short cut", to accept reduced standards in himself or his equipment because of his [sic] emotions. (p. 9-15)

In fact, psychological stress can induce problems that are imponderable and never fully comprehensible. Like an iceberg, their bulk lies invisible below the surface of behaviour.

It should be underlined, however, that people have different stress thresholds; some students do not experience undue anxiety about learning to fly and they are not adversely affected by the experience, or are able to cope effectively with the stresses they experience. However, there is a large number of students who initially approach the flight training situation with a high degree of trepidation which interferes with their learning, their performance and their decision-making ability. The responsibility lies with the flight instructor to reduce, or help the student control, the stress in flight training in order to facilitate learning and thus promote achievement and build the student's confidence. Unfortunately, research has shown that some flight instructors, instead of alleviating emotional stress, frequently impose additional stress on student pilots. Krahenbuhl et al. (1980) found that "one of the most salient stress producing agents in pilot training is the instructor pilot" (p. 1). Research by Henley (1987), Grandchamp (1971), Haward (1968), Melton and Wicks (1967), and Horowitz (1964) has also identified the flight instructor as one of the weakest links in the flight training process, and sometimes the element most feared by the student pilot.

The instructor: An asset or a liability? Haward (1968) and Krahenbuhl et al. (1980) conducted almost identical studies to determine the level of stress different instructors imposed on student pilots. They found that, while some instructors lowered the student's stress level, other instructors had an undesirable augmenting effect upon the stress experienced by student pilots. When examining the behaviour of the instructor who imposed additional stress, Haward (1968) notes that:

> … two clear cut factors were his manner of correcting errors in cockpit drills during the prelanding checks (which proved particularly upsetting when the student was trying to concentrate on the difficulties of the approach) and his unduly early takeover of controls prior to landing if he felt the approach was less than perfect (which proved frustrating for the student and at the same time impaired learning efficiency). (p. 22)

Haward points out that, in contrast, the instructor who reduced stress was a young instructor who displayed confidence in both the aircraft and the student. While this instructor conscientiously corrected errors "from correct flying with a quiet word, he nevertheless gave the student complete freedom to follow the exercise to the best of his [sic] existing ability, never taking over control unless there was insufficient time to give appropriate instruction for remedial action" (p. 22). Similarly, Krahenbuhl et al. (1980) found that "positive instructors" (those who relied on encouragement, acceptance, praise, and instructional cues) produced less stress on student pilots, the result being better performance, while "negative instructors" (those who relied on harsher tones, criticism, and scolding) significantly raised the stress level of students, thus inhibiting performance.

Henley's (1987) research into student pilot stress and flight instructor training also revealed that the comments, attitudes and behaviours of flight instructors greatly influence learning. The attitudes and behaviours of flight instructors identified as "negative" or stressful by student pilots included sarcasm, scolding, yanking the controls, impatience, negative criticism, intolerance of errors, lack of teaching skills, and overloading. Negative instructors also failed to recognise individual differences and appeared insensitive to student psychological stress and unresponsive to student needs. On the other hand, enthusiasm, acceptance, patience, positive rapport, trust, praise and encouragement were identified as positive traits of effective flight instructors. Interviews with experienced pilots undergoing conversion training or recurrent training identified similar barriers to learning. However, experienced pilots noted that they faced the added pressure of losing their livelihood if they were not successful in their training (Henley, 1987).

Researchers agree that when stress induced by the task is augmented by distractions produced by an instructor's behaviour or verbal disparagements of the student's performance, learning can be seriously hampered (Grandchamp, 1971; Haward, 1968; Henley, 1987; Krahenbuhl et al. 1978, 1979, 1980). When students are confronted with a situation in which they feel particularly vulnerable and where their pride is at stake, or there are threats to ego integrity, the defensive tendency to "look good" or "save face" increases and attention that should be oriented to the task is used to deal with emotions. Thus, the reaction to emotional stress is often ego-defensive, that is, the student is primarily preoccupied with the maintenance of self-esteem rather than with the task. In such a situation concentration on the task at hand is greatly reduced and flying efficiency is adversely affected "because our faculties are put to work mobilizing personal defenses" (Smith, 1982, p. 46).

Negative Attitudes and Behaviours of Instructors

Testimonies from student pilots in both Canada and Australia revealed that psychological stress is frequently instructor-induced, and can indeed hamper learning. Student pilots claimed that their flight instructors were generally competent and skilful pilots, however, the teaching skills of many were considered to be seriously lacking. All the participants in Henley's (1987) study indicated that they had encountered at least one instructor who made them "*feel clumsy or not very bright*", as one student put it. Following are a few examples of the more prevalent and commonly disturbing stress-producing or "negative" attitudes and behaviours displayed by flight instructors and the student pilots' reaction to the encounter.

Students said that on many occasions they left the flying school in a depressed state of mind with a strong disinclination to continue their training, all the while knowing that their instructor was unaware of the condition in which he or she had left them (Henley, 1987). Student pilots also stated that when they were upset, annoyed or angry with their instructor's dismissive arrogance or condescending attitude, they could not concentrate on learning how to fly but instead were preoccupied with their feelings and their emotions. As a result they made more mistakes, their performance deteriorated and some said that ultimately they lost their confidence. Frequent negative criticism and a lack of encouragement can also intensify feelings of incompetence and negative self-concepts which may precipitate the development of emotional barriers to training and hamper a student's progress.

Sarcasm Sarcastic remarks and verbal disparagement of the student's performance were identified as being most upsetting and detrimental to productive thinking and to effective learning. Students said that they could not concentrate on the task of flying, instead many lost self-confidence or became angry or upset and their performance deteriorated. For example, one student said,

> *My instructor seemed to think that humiliation techniques were an accepted way of dealing with students. He would often belittle me or make derogatory comments about my performance. What really bugged me was that he'd comment on my performance in front of others — jokingly mind you, but it still was very embarrassing and made me feel like a real klutz* (cited in Henley, 1987, p. 83)

Another student noted:

> *When he made sarcastic remarks, several times I used to have to concentrate on not crying because I would be so upset. He seemed to be able to make me feel very small or dumb. Maybe it was just my reaction but it usually happened when I was feeling pretty unconfident or more nervous than usual. Something had made me feel kind of jittery or uneasy, so an ill-placed remark would make me feel pretty upset. There were times when I would just sort of grit my teeth and say to myself, "Okay, let's just get through this until we get back on the ground and I can get out of this place".* (Personal interview, Student 4, 12 June 2000)

Sarcasm directly attacks a person's pride. It is a threat to ego integrity. Consequently, one's defensive tendency to "save face" or maintain self-esteem increases and attention that should be directed to the learning task is consumed dealing with emotions, as the following student indicated:

> *He was very sarcastic — he'd make a few sarcastic remarks about my performance; I knew he was frustrated with me. It wasn't that I couldn't hear him. I just couldn't do it [steep turn] right because I wasn't getting something. My altitude was off, so he'd make snarly comments. I felt, "forget it". The day was lost, my concentration and my drive were gone.*

> *It was a complete waste of time.* (Personal interview, Student 7, 15 June 2000)

Clearly, sarcasm and ridicule are out of place in any learning situation. Instead, as Knowles (1980) stresses, "the psychological climate should be one which causes adults to feel accepted, respected, and supported; in which there exists a spirit of mutuality between teachers and students as joint inquirers; in which there is a freedom of expression without fear of punishment or ridicule" (p. 41).

Display of impatience Instructors who scolded, took over the controls unduly early, shouted or brusquely yanked the controls definitely augmented student stress and were a hindrance to learning (Henley, 1987; Krahenbuhl et al., 1980). Students indicated that they were either frightened by such behaviour or "*felt clumsy or dumb*". Many students also indicated that they lost their self-esteem or confidence and were reluctant to show initiative with such an instructor (Henley, 1987). Understandably, performance worsened and learning was inhibited under these circumstances. One student said,

> *My instructor would scream: "Watch that!" "You're forgetting this" — tapping on the instruments, banging this and pointing at that. The absolute worst thing that I can think of is to move the control column when the student is hanging on to it. It just threw me right off. That was it, I didn't even want to touch the control column any more because you never knew when you're gonna get a rap or a pull or a push. That was very stressful and a totally negative lesson — totally wasted. The learning went right down the tube.* (cited in Henley, 1987, pp. 96-97)

Similarly, another student said,

> *My instructor believed in shock treatments, like yelling or hitting the controls. When he'd yell at me, or immediately took over as soon as I made a mistake, or hit the control column if I was too high, my mind would go blank. I couldn't think about what I was doing any more. I was more worried about trying to keep him from yelling at me than concentrating on the exercise I was doing. It made me extremely angry. I felt, I'm an adult. I don't need to be treated like a moron!* (Personal interview, Student 17, 2 May 2001)

Intolerance of errors Another stress-producing attitude identified by student pilots was an instructor's intolerance of errors. Instructors who immediately displayed their impatience when students made a mistake often made students feel incompetent. They lost confidence in their ability to learn to fly, and consequently, became apprehensive about their capability to think clearly or to cope adequately with the situation at hand. The following statements are typical of most replies by student pilots:

> *As soon as I made a mistake, I felt very inadequate because my instructor would become impatient. He seemed to be able to make me feel very dumb. He didn't scream at me or anything like that, but I could tell by the tone of his voice, I knew he thought I was slow. But, I surely didn't make mistakes on purpose! I didn't WANT to make dumb mistakes! So after that I would hesitate. I guess I lost my confidence. I doubted that I could learn to fly. It took me until my commercial before I got an instructor who helped me build by confidence again* (Personal interview, Student 7, 15 June 2000)

> *Flight training was one of the most negative types of training I ever undertook. Initially, I was very keen and really motivated. I had wanted to take flying lessons for a long time but my instructor soon made me feel really clumsy and stupid. When I made a mistake, he'd just shake his head and look disgusted. Or if I asked a question, he'd say, "Everybody knows that, it's very basic", and he'd leave me there feeling like an idiot. I got the impression that he didn't think I could learn to fly. I felt I was the most stupid student pilot there. I became very unsure of myself. I had no confidence at all in my flying ability. When I flew with him, I got more and more nervous and I would manage to do the most stupid things. The more nervous I got, the more mistakes I made; it was a vicious circle. I felt completely incompetent.* (Personal interview, Student 20, 16 May 2001)

Understandably, if feedback is excessively negative, students become discouraged and probably will lose interest in the learning activity. Instead of being overly critical, instructors should explain to students that errors are part of the normal process of learning (Cole & Chan, 1994, p. 221). Eddowes (1974) correctly noted that often in flight training,

> The student is being taught to be errorless, or perfectly accurate in his [sic] aircraft control, when in fact, perfection is not actually required and is seldom achieved in normal flight operations. Thus, the student is being misled, cognitively, with regard to his [or her] learning tasks requirements. The goal of efficient, accurate and fast information processing to detect, interpret and correct flight control errors could be achieved if the instructor taught flight path error correction on the basis of errors being normal and their correction being the desired state, rather than perfect control the goal and errors of any sort indicating a kind of incompetence not socially acceptable among real pilots. (p. 8)

Overloading Some instructors wanting to be either scrupulously thorough or friendly will overload or distract the student and may be unaware that they are interfering with effective learning. The following comments are typical responses from student pilots:

> *One thing my instructor did was he would load me up with too many ideas at one time. I would find it hard to keep up. I guess what he basically did was he used to talk all the time — chatter away about all kinds of things. He would talk all the time when I was supposed to be taking off and climbing to 2,500 feet and then he'd say: "Watch your altitude!" And I'd look at my altitude and it would be 3,000 feet and he'd asked, "How come you didn't level off at 2,500 feet?" Well, he was talking to me, what did he expect? I can't concentrate on everything. Maybe the point that he was trying to stress was that eventually I would have to deal with passengers talking to me but there were a lot of times when his talking got in the way. I appreciated that he was friendly but I don't think he quite realized how much of my concentration was required to just fly the airplane, and a lot was required. I mean, all of it was required to fly the airplane at first.* (cited in Henley, 1987, p. 109)

> *I don't think my instructor realized he made students feel dumb. And this kind of thing [constant questioning] was reinforced when we were coming in to land and he'd be yelling: "You haven't got the aircraft straight. How are you gonna straighten up the aircraft? How much rudder should you be putting on?" Asking a barrage of questions just at the time when a student needed to be quiet or concentrating and do things himself. It was terribly distracting.* (cited in Henley, 1987, p. 113)

As with the landing phase, the forced approach is an exercise where student pilots typically felt overloaded:

> *For the forced approach it was really stressed that I had to make sure to get the Mayday call out three times and do all the checks. And at first I thought to myself, all I want to do is get the plane in the field in one piece and I want to get everything set up first, then I'll start worrying about talking to people and doing all the various "cause" checks. The most important thing is to maintain control. If you get down in one piece, that's three quarters of the battle. But the way it was taught, with all the checks and Mayday call three times and checking every single little thing (laugh), by the time you've done all this, you've lost your field! They just tried to crowd in too many things at once and that way you end up losing the real purpose of the exercise which is to get down safely and do the checks IF YOU CAN.* (Personal interview, Student 24, 23 May 2001)

Lack of teaching skills One of the major problems identified by student pilots was the fact that some instructors, while being competent pilots, could not teach. They lacked teaching and interpersonal skills and thus were unable or unprepared to help students learn and at times were even an obstacle to effective learning. Students who were interviewed indicated that all too often instructors limited their instructing to the "follow me through" method of teaching in which a manoeuvre is demonstrated and the student attempts to mimic the flight instructor's movements without having a clear idea of what he or she is trying to accomplish, which cues to look for, or why a manoeuvre should be performed.

> *My instructor didn't really know how to teach. He knew how to fly, but he didn't know how to teach effectively. There was no application – the exercises weren't made real for me. He didn't explain why we did certain things, so I was learning by imitating him and by trial and error, but I didn't understand why I did certain things or why I needed to do them.* (cited in Henley, 1987, p. 107)

> *I expected that maybe the instructor would do a little more explaining so that I would understand why I was doing what I was doing. I felt that my instructor — what he did a lot was mostly demonstrate the motions but not give me the understanding behind it so I could remember and I could really understand and think for myself. I got a good model to imitate, but*

> *I don't think that's really teaching. I wanted to know what to look for, some more background explanation. Also he never really told me WHAT I did wrong, just that it was wrong. So he'd show me again and again, but a lot of times I didn't know what to look for. So when I tried to imitate, I'd fumble my way through and the instructor would say, "No, just watch it again". So he'd do it again and I'd finally get it, but I didn't really understand or feel comfortable with the exercise. So, although he was a very accomplished pilot, as a teacher it left a lot to be desired.* (Personal interview, Student 27, 30 May 2001)

Lack of interest, building flight time Several student pilots stressed that it is important that instructors show that they care about their students and about teaching and that they should not be there only to fill a hiatus between jobs or to build up flying hours in order to get hired by the airlines. As this student indicated, charter flights often take precedence over instruction so students either get shuffled around or their booking gets cancelled which makes them feel unimportant:

> *Whenever a charter flight came up, my instructor was gone in a flash. It could be in the middle of a briefing, he'd just say: "Sorry, but I've got to go on a charter. Can you come back another time?" You don't have much choice, right? He was building up his multi-engine time and I knew he'd much rather fly the twin than teach me.* (cited in Henley, 1987, p. 116)

Negative feedback Testimonies from student pilots also indicated that many instructors were not overly generous with their praise. On the other hand, these instructors seemed to be quick at pointing out something that was wrong, though they were not necessarily forthcoming with advice on how to correct what was wrong. One of the complaints most often expressed by students was: "*I never got feedback from my instructor, except when I did something wrong*" (Personal interview, Student 9, 15 June 2000). Following are two testimonies that echo the views of many participants:

> *When you get an instructor who vehemently criticises every move you make, you become full of self-doubt. You actually wonder if you really ARE capable of flying that aircraft. You have a lot of self-doubt. This instructor destroyed my confidence. There were times when I couldn't do anything right. All I heard was criticism. It was so frustrating and disheartening.* (Personal interview, Student 26, 30 May 2001)

> *It seemed as if I never did anything right. I don't remember him giving me any positive feedback. He'd say, "If I don't say anything it's because it's alright". But he wasn't talkative in the first place and even when I did something wrong, he didn't say anything. Instead, he'd point at the instruments or give me signs to turn, climb or descend and shake his head and look annoyed. I didn't really know how I was doing. I really started to doubt myself.* (cited in Henley, 1987, pp. 138-139)

Feedback should be provided to students in order to assist their learning. While feedback should inform students of the adequacy of their performance, clues should also be provided on how to improve, if the performance is deemed inadequate. If students are given a constant regimen of negative criticism they will probably lose motivation and their learning will certainly be adversely affected (Cole & Chan, 1994). Therefore, positive feedback is also needed to enhance learning and to keep students interested.

Negative criticism tends to reinforce negative self-concepts by introducing psychological barriers that inhibit the learning process. Students feel that their instructor perceives them as awkward or stupid, which in turn leads them to doubt themselves and to make more mistakes. The way instructors perceive their students has a major influence on how students will perceive themselves and how they will perform. According to the principles of the self-fulfilling prophecy, students do better when their teacher or instructor expects them to do well (Rosenthal & Jacobson, 1968). Their research revealed "how one person's expectation for another person's behavior can quite unwittingly become a more accurate prediction simply for its having been made" (p. vii). Therefore, if an instructor has a low opinion of a student's ability and expects him or her to perform poorly, chances are that these expectations will reflect negatively on the student's performance (Brophy, 1979, 1983; Butler, 1991, Cooper, 1979; Wood & Knight, 1988).

Negative or irritating attitudes and behaviours displayed by instructors have a number of implications for the training environment. Attacks on the intentions, competence, or self-regard of the learner may actually help create or reinforce the idea that the student is a poor performer. If students perceive that their instructor thinks of them as being stupid or clumsy, they sometimes attribute their failure or poor performance to a lack of ability and may withdraw from the learning activity. This withdrawal may occur not only in its most obvious physical sense, but in a variety of ways, of which the perceptive teachers should be aware. For example, students may use defence mechanisms, such as projection, rationalisation or repression (A. Rogers,

1998). Furthermore, if students persist in spite of their instructor, the training becomes unpleasant and acquiring a licence or an academic qualification becomes a matter of goal-directed perseverance (Henley, 1987). As one student said,

> *I can assure you that many times I asked myself, "why am I doing this?" "Why put up with this crap?" But when I start something, I see it through to the end. Besides, flying was a childhood dream and I wasn't about to let a lousy instructor stop me from reaching my goal.* (Personal interview, Student 4, 12 June 2000)

However, in all fairness to flight instructors it should be stressed that a concern for safety can sometimes deter from optimal forms of fault analysis or feedback. Bailey and Hughes (1980) note that "the aircraft itself is certainly a less than perfect setting for maximizing the acquisition of skills required to fly a plane. Safety requires the IP [instructor pilot] to put proper maneuvering above analyzing the instructional process and the stress involved in correcting student errors may result in less than optimal forms of feedback" (p. 9).

Finally, Hiner (2000) also identified a few additional behaviours by flight instructors that tend to increase student stress. According to Hiner, behaviours by instructors that are seen as particularly irritating to students include being frequently late or unprepared, yelling a lot, not letting the students fly, scaring the students, not listening to students, and using the students' lesson to do some flying of their own (pp. 71-73).

Positive Attitudes and Behaviours of Instructors

Fortunately, many student pilots who were interviewed also indicated that while there were "negative" flight instructors whose attitudes and behaviours were a deterrent to efficient learning, there were many flight instructors who were effective teachers and made learning to fly an enjoyable and rewarding experience (Henley, 1987). Most student pilots said that they had had "good" or "positive" instructors dedicated to flight training who helped them cope with their apprehensions or fears and made learning to fly a worthwhile and enjoyable experience. Because of their enthusiasm, empathic caring, and uncritical acceptance of students, these instructors were successful in establishing a positive rapport or relationship with their students and in creating an emotional climate that was conducive to effective learning. Their patience and tolerance of errors helped students feel more relaxed and less anxious or threatened about trying out their new skills. A perceptive understanding of students' feelings and problems, coupled with a judicious use of praise and encouragement for students' achievements, fostered the development of a positive self-concept and helped build self-confidence among students. Successful instructors were also skilful at using humour or laughter to relieve tension and to make flying lessons more pleasant and enjoyable (Henley, 1987, pp. 142-158).

Enthusiasm Student pilots identified enthusiasm as an important trait of instructors with whom they enjoyed training. An instructor's genuine interest in flight training helps students maintain their motivation and interest throughout the course.

> *I had an older instructor and I had a lot of confidence in him. He had a lot of experience and it was obvious he enjoyed instructing so you really felt pleased or privileged that you're up there with this guy and you're gonna get little tidbits of information. That was really inspiring and really relaxing. Also someone showing you an exercise and seeing their smoothness on the controls. By their demonstration, you can see their skill and that plays an important part as to how much I relax. I also like an instructor who is soft-spoken. And ENTHUSIASM — that's the most important! I like to feel that this is my lesson, not just number 14 and it could be given to any person. I like to know that this is custom-tailored for me. I also like a relaxed atmosphere in the airplane where it's not all just plain work. It should be fun and enjoyable.* (Personal interview, Student 15, 22 June 2000)

> *My second instructor was very enthusiastic. You could tell she enjoyed teaching. She made the lessons interesting and enjoyable. She made learning fun. I had a good time, I laughed, but I was still learning because even though it was a relaxed atmosphere, she wouldn't let mistakes go by uncorrected. She spent extra time with her students and she'd always be available if you had some questions or needed some help. She obviously cared for her students and liked what she was doing. I think that's very important in teaching.* (cited in Henley, 1987, pp. 143-144)

Caring or good rapport Student pilots indicated that they felt more comfortable flying with instructors who genuinely cared about them and who wanted to help them increase their knowledge and their skills. A good rapport between the student and the instructor was considered essential.

> *With this instructor I have now, we seem to work quite comfortably together. I now look forward to my lessons. I'm now leaving my lessons with a good feeling and, to me this is the way it should have been all along. Before I used to dread the lessons and leave quite frustrated because I just felt I can't do anything right. This instructor cares about his students and you can see that he enjoys instructing.* (Personal interview, Student 17, 2 May 2001)

> *My second instructor was very keen and professional-like. He was sincere and to the point. He was very honest and gave a darn about his students. He went the extra mile. He gave us extra material, extra time, he made photocopies for us. You knew he cared. He was friendly, yet he was very business-like. He could relate well with the students but he kept everything in perspective. He was very professional.* (cited in Henley, 1987, p. 147)

Acceptance and respect for the student All the students who were interviewed considered acceptance and respect for the student as important attributes of an effective instructor. Acceptance and respect appeared to be the foundation upon which is built a good rapport between the student and the instructor.

> *I was very fortunate, I think. I had two very good instructors. As far as unusual manoeuvres like stalls and spins, they always explained it on the ground first very thoroughly and then in the air they'd demonstrate it a few times before I had a go. I was a little apprehensive going into them, so I was fortunate to have them. I had total confidence in them. They also made it very clear to me that if I didn't feel good to let them know and we would do something else. They didn't make you feel like a wimp if you didn't care for stalls and spins. I felt very comfortable with them.* (cited in Henley, 1987, p. 150)

Patience Student pilots identified patience as an essential quality of an effective instructor. An instructor who can calmly correct mistakes and who can pace the lesson in order to keep students challenged yet not overwhelm them with too much material is one who was greatly appreciated by students.

> *I was fortunate in that my instructor was a pretty dedicated and patient guy. So I feel I got an above average instructor. He was very calm, a laid-back type of guy. When you made a mistake he didn't make you feel like an idiot. He'd show you how to do it correctly and stressed the fact that he was there for that. He was very patient and that's so important in instructing.* (cited in Henley, 1987, p. 152)

Challenge Students also indicated that they enjoy a challenge because "*it makes learning more interesting and enjoyable*" (Personal interview, Student 7, 15 June 2000).

> *My instructor for the private licence was very good. He'd make things a little bit of a contest. For example, we'd bet a coffee for whoever would get the closest ETA [estimated time of arrival] for check points during the cross-country. Or there was always a contest as to who could land in the shortest distance, and of course he would mess it up sometimes so I'd win and sometimes he'd do it right so he'd win. It was fun. At the same time, he knew when to put down a firm hand when I'd start to mess up pretty badly. But he made it a challenge and it was really enjoyable. He was a very good instructor.* (Personal interview, Student 19, 16 May 2001)

Encouragement Student pilots felt that instructors needed to provide encouragement in addition to constructive criticism. All the students interviewed by Henley (1987) said they wanted to be told when they had done well, not simply when they made mistakes. While students said that they wanted to be praised when they performed well, they stressed that they wanted honest feedback, otherwise they tended to lose faith in their instructor.

> *My instructor would brief me before we went flying and when I did something right, he'd sort of give me a pat on the back and say: "That was well done!" And when I did something wrong, he didn't get impatient. He'd just point out what I did and show me or tell me how to do it right. Sometimes also he'd just smile and say: "What do you think you did wrong here?" And most of the time as soon as I'd done it, I knew*

> *what I had missed. So I'd tell him and he'd say: "Right, Okay try it again". And I felt that he sort of had confidence in me. He knew that I knew how to do it. I just needed to practice more. So it was encouraging and I did much better with him than with my first instructor.* (cited in Henley, 1987, p. 155)

> *My second instructor was very positive. If I made a mistake, she wouldn't get uptight or give me hell. She'd show me how to correct it and if I did it wrong a few times, she'd just teach it again, she would try different things, different ways until I got it right. She also had a good sense of humour, so a lot of times we'd laugh and that made me feel more relaxed. She'd tell me not to worry about the mistakes I made because that's how you learn. She was very encouraging. And when I did something right, she'd pat me on the back and say: "Nice going!" It helps to get that bit of a boost. I felt good flying with her.* (Personal interview, Student 3, 12 June 2000)

Laughter On several occasions students mentioned the fact that laughter helped them to relax (Henley, 1987). The fact that they were able to laugh at their mistakes because the instructor had created an atmosphere of genuine acceptance seemed to go a long way in helping students cope with psychological stresses inherent in the learning situation. For these students, laughter did appear to be the best medicine in releasing stress.

> *Like I said before, I took up flying as recreation, not for a career and with my first instructor it wasn't enjoyable at all. He wasn't interested. But with my second instructor it was really a lot of fun. I finally got to laugh during my lessons. If I made a mistake, we'd both laugh. I didn't feel she was judging me. If I was nervous about an exercise — like the spin! — just to laugh about what happened relaxed me and I now felt, "Yeah, I can do this now". It gives you confidence.* (Personal interview, Student 3, 12 June 2000).

Strategies to Reduce Barriers to Learning and Cope with Stress

From a teacher's perspective, there are some useful strategies to reduce blocks to learning, although there are certainly no universal remedies for all individuals in all situations. However, encouraging students to express and discuss their emotions may provide a useful starting point for action. Similarly, finding out if the emotion has its source in some aspect of the teaching or learning process or stems from the training or home environment will allow the responsive teacher to decide on a possible course of action to help the participant have a more positive encounter with learning. For some participants, feeling that their fears and negative emotions have been noticed and heard may be enough to allow them to begin to deal in a more positive way with the learning situation.

As stated above, there are no set answers to dealing with all situations where barriers to learning arise. A. Rogers (1998) suggests that the application of individual judgement of each teacher in each different situation is the only realistic approach. Teachers can only really prepare themselves for these situations by having an awareness of the issues that can arise for adult learners and by being aware of some of the coping strategies suggested below. Using this awareness of the range of factors which may be impacting on the teaching and learning process will be the key to analysing each situation and deciding on a course of action to break down barriers to learning. Following are a few steps that instructors could consider in their endeavour to help reduce barriers to learning and assist students cope effectively with stress.

Obviously, one of the first steps instructors could take is to sincerely reflect on their attitudes toward learners and closely examine their behaviour with students. A genuine review of one's belief and teaching practices might reveal some attitudes and behaviours that could be detrimental to effective learning. In addition, instructors may need to acknowledge their own stress in the teaching environment. In addition to anxiety specific to the task of teaching, instructors within the aviation context often experience stress specific to the environment in which they must teach. Research has shown that instructors who teach in a dynamic and stressful milieu and who are seldom given due credit for their contribution to aviation training also experience stresses that can interfere with effective teaching (Henley, 1987, 1995; Henley & Berlioz-Nott, 1997; Henley, Nicoll, & Soros, 1994; Soros, Henley, Wyndham, & Nicoll, 1995). There is an old cliché that states that "flight training is hours and hours of boredom, interspersed by moments of stark terror". Indeed, most would agree that stress is part of a flight instructor's job. However, the contribution that flight instructors make to the aviation industry often goes unnoticed and is seldom fully appreciated, which can be an added source of stress for some instructors (Henley, Nicoll, & Soros, 1994). As Kershner (1988) correctly states, "if I had my way the really committed flight instructor would make $100,000 a year. I think he [sic] is infinitely more valuable than an airline pilot, because teaching is what it's all about" (p. 13).

Another important factor to consider is that although it is essential to correct student errors, it is also important to accentuate the positive. Students need to know when they perform well and, when they encounter problems in performing a manoeuvre or a task, the feedback provided needs to be constructive. Wilson (1990) suggests that even the statements we use to give instructions to students should reflect a positive approach. For example, instead of telling the student: "Don't let the airspeed drop below 70 knots", a more positive statement could be: "Maintain an airspeed of 70 knots".

Finally, instructors also need to be aware of stress coping strategies in order to help students develop effective ways of coping with stress. While the stress coping strategies suggested here are specific to flight training, creative instructors could adapt the majority of these strategies to other aviation training contexts.

- Take a realistic assessment of yourself — While most pilots give their aircraft a thorough preflight and meticulously plan their flight, many forget to preflight themselves. Instructors can encourage students to be aware of their limitations, their strengths and weaknesses, and assist students in sensibly assessing their readiness to undertake a flight. Pilots also need to be able to recognise situations that are likely to be stressful and be alert to the warning signs of stress in order to realistically consider those situations and signs when deciding whether to undertake a flight or to cancel it. Instructors need to make students aware that a myth about stress is that *only wimps let stress affect them*. Stress affects everyone. Although stress affects different people to differing extent, no one is immune to the effects of stress.
- Physical fitness — Exercise can reduce the effects of stress by giving the body an outlet for stress. It takes self-discipline, but it is well worth the effort.
- Mental and emotional fitness — Whenever possible, avoid stressful situations or encounters (i.e., arguments, squabbles, etc.) before undertaking a flight. Avoid flying when angry (at yourself or others) or emotionally upset. It is important to realise that one of the myths about stress is that *people can leave their problems at home*. Another myth about stress is that *mental stress does not affect physical capabilities*. As we have seen in this chapter, stress can interfere with productive

thinking, obstruct effective decision making, and impair performance.

- Create a solid social support network in your life — Family, friends and colleagues can provide needed support, however, one needs to realise that family relationships can also be a source of stress. If problems are severe, professional help should be sought.
- Be well prepared and well informed — One of the best antidotes to stress is thorough preparation. Maintaining proficiency will also help build self-confidence. Review and rehearse difficult manoeuvres. Use visualisation to rehearse procedures.
- Learn to recognise and avoid heavy pressures imposed by getting behind schedule and not meeting deadlines.
- Avoid getting into a situation where you must get to your destination at all costs — According to Nelson (1993), "get-there-itis [is] an acute disease which sometimes afflicts pilots; caused by an attack of overconfidence and time constraints; often characterized by diminished good sense and anxious mannerisms" (p. 56).
- Do not press the weather, it could be fatal — One only needs to review accident statistics to be convinced of the importance of having a healthy respect for weather. Instructors could encourage students to use computer-based programs, such as *Weatherwise* (Wiggins, O'Hare, & Hunter, 2000) to develop weather-related decision making skills (also see Chapter 12; O'Hare, 1992; Prince, Bowers, & Salas, 1994; Wiggins & O'Hare, 1995).
- Do not allow passengers to dictate when it comes to safety matters, such as the amount of baggage to carry or where and when to go.
- Avoid situations that distract you from flying the aircraft — Do not let small mistakes bother you. If you make a mistake which you detect and correct, the most sensible thing to subsequently do is to forget it and return your attention to flying. Strive for excellence, not perfection.
- When possible, reduce your workload — Use "load shedding", a form of mental discipline which means putting aside non-relevant or lower priority thoughts and concentrating on the task at hand. You can also ask experienced passengers to help look for traffic and landmarks.
- Learn to relax and make time for recreation — Condition yourself to relax and think rationally when stress appears in flight. Also make time for activities you do solely because you enjoy them. Recreation renews and recharges our energy. As the word implies, it "re-creates" us.

- Have a contingency plan or an alternate plan. Anticipation is another effective antidote to stress.
- Do not hesitate to ask for assistance. Air traffic controllers are there to assist you.
- Do not hesitate to declare an emergency (adapted from Air Canada Human Factors Project Team, 1996; Stokes & Kite, 1999; Transport Canada, 1989).

Conclusion

This chapter was designed to introduce some of the factors that affect learning. While it is accepted that a realistic training environment is important, it also must be recognised that exposure to the stress of a realistic aviation environment may initially directly affect a student's ability to learn, to perform skilfully and to make effective decisions. The perceptive practitioner will assess each situation in terms of participant characteristics and potential barriers to learning and respond appropriately to ensure that effective teaching and learning take place.

It also should be stressed that although the negative and positive attitudes and behaviours of instructors discussed in this chapter relate mainly to flight instruction, they are not special cases. There is every reason to suppose that similar attitudes and behaviours would have the same effects on student performance in all other aviation training contexts. Indeed, the stress-producing attitudes and behaviours identified by Henley (1987), Hiner (2000) Krahenbuhl et al. (1980), and Haward (1968) are not limited to flight instructors. University students interviewed by Appleby (1990) also indicated that they were irritated by teachers who displayed a condescending attitude toward their students (i.e., treating them like children), showed no enthusiasm, were unprepared for their class, were unresponsive to student needs, were overly critical of students, and did not allow questions from students. Similarly, Catsoulis (1981) found that "good college teachers" explain things well, expect students to do well and have control of their class. They also have a sense of humour, can hold their tempers, do not bore students and get along with students. "Bad teachers", on the other hand, are unfair, far too strict, will not help students, and cannot explain their material clearly. They may also be lazy, bad tempered and moody.

Reflective Activity

1. What factors may create barriers to learning? List some strategies you might use to counteract these barriers.
2. How might emotional factors impact on aviation instructional contexts? Can emotions be used in a positive way to increase learning in your context?

References

Air Canada Human Factors Project Team (1996). *Human factors for aviation: Basic handbook.* Ottawa, Canada: Transport Canada Safety and Security.

Alkov, R.A. (1974, April). Life changes and pilot error accidents. In G.A. Berry, C.V. Durham, D.C. Prathner, W.M. Moe, J.B. Stiles, & J.H. Thomas (Eds.), *Proceedings, Psychology in the Air Force, Fourth Annual Symposium* (pp. 153-155). Colorado Springs, CO: United States Air Force Academy.

Alkov, R.A., Gaynor, J.A., & Borowsky, M.S. (1985). Pilot error as a symptom of inadequate stress coping. *Aviation, Space, and Environmental Medicine, 56,* 244-247.

Appleby, D.C. (1990, Spring). Faculty and student perceptions of irritating behaviors in the college classroom. *The Journal of Staff, Program and Organizational Development, 8*(1), 41-46.

Baddeley, A.D. (1972). Selective attention and performance in dangerous environments. *British Journal of Psychology, 63,* 537-546.

Bailey, J.S., & Hughes, R.G. (1980, January). *Applied behavior analysis in flying training research.* Brooks Air Force Base, TX: Air Force Human Resources Laboratory.

Berlin, J.I., Gruber, E.V., Holmes, C.W., Jensen, P.K., Lau, J.R., Mills, J.W., & O'Kane, J.M. (1982). *Pilot judgment training and evaluation* (Vols. 1–3). Washington, DC: Federal Aviation Administration.

Bond, N.A., Bryan, G.L., Rigney, J.W., & Warren, N.D. (1968). *Aviation psychology.* Los Angeles, CA: University of Southern California.

Brookings, J.B., Wilson, G.F., & Swain, C.R. (1996). Psychophysical responses to changes in workload during simulated air traffic control. *Biological Psychology, 42,* 361-377.

Brophy, J.E. (1979). Teacher behavior and its effects. *Journal of Educational Psychology, 71*(6), 733-750.

Brophy, J.E. (1983). Research on the self-fulfilling prophecy and teacher expectations. *Journal of Educational Psychology, 75,* 631-661.

Butler, R.E. (1991, May 5). *CRM, LOFT, and Bernard Shaw: Flight crew evaluator training workshop presentation* (NASA/University of Texas Technical Manual 91-3). Austin, TX: The University of Texas at Austin.

Callister, J.D., Percival, J.L., & Retzlaff, P.D. (1999). Stress and fatigue during aircrew survival training. In R.S. Jensen, B. Cox, J.D. Callister, & R. Lavis (Eds.), *Proceedings of the Tenth International Symposium on Aviation Psychology* (pp. 843-847). Columbus, OH: Ohio State University.

Catsoulis, C. (1981). Teachers and students at Johnholme College. In S. D'Urso & R. Smith (Eds.), *Changes, issues and prospects in Australian education* (pp. 130-136). Brisbane, Australia: University of Queensland Press.

Cohn, R.L. (1994). *They called it pilot error.* New York: McGraw-Hill.

Cole, P.G., & Chan, L.K.S. (1994). *Teaching principles and practice* (2nd ed.). Sydney, Australia: Prentice Hall.

Cooper, H.M. (1979, Summer). Pygmalion grows up: A model for teacher expectation communication and performance influence. *Review of Educational Research, 49*(3), 389-410.

Costa, G. (1989). Stress and performance in air traffic controllers. *Ergonomics, 12,* 93-100.

Cross, P. (1981). *Adults as learners: Increasing participation and facilitating learning*. San Francisco: Jossey-Bass.

Darkenwald, G.G., & Merriam, S.B. (1982). *Adult education: Foundations of practice*. Sydney, Australia: Harper & Row.

Davis, D.R. (1964). Psychological mechanisms in pilot error. In A. Cassie, S.D. Fokkema, & T.B. Parry (Eds.), *Aviation psychology* (pp. 11-23). The Hague: Mouton.

Della Rocco, P.S., Fiedler, E.R., Schroeder, D.J., & Nguyen, K. (1999). US coast guard pilots: The relationship between the home and work stress and self perceived performance. In R.S. Jensen, B. Cox, J.D. Callister, & R. Lavis (Eds.), *Proceedings of the Tenth International Symposium on Aviation Psychology* (pp. 838-842). Columbus, OH: Ohio State University.

Eddowes, E.E. (1974). *A cognitive model of what is learned during flight training.* Brooks Air Force Base, TX: Air Force Human Resources Laboratory.

Farmer, E.W., Belyavin, A.J., Berry, A., Tattersall, A.J., & Hockey, G.R.J. (1991). *Stress in air traffic control II: Effects of increased workload.* Farnborough, UK: RAF Institute of Aviation Medicine.

Galbraith, M. W. (Ed.). (1991). *Facilitating adult learning: A transactional process.* Florida: Krieger Publishing.

Grandchamp, R.J. (1971). *Student pilots' attitudes toward fear concepts in flight training*. Unpublished doctoral dissertation, University of Illinois, Urbana-Champaign.

Hammond, K.R. (2000). *Judgments under stress*. New York: Oxford Press.

Harss, C., Kastner, M., & Beerman, L. (1991). The impact of personality and task characteristics on stress and strain during helicopter flight. *The International Journal of Aviation Psychology, 1*(4), 301-318.

Haward, L.R.C. (1968). Stress in flying training. *Flight Safety, 1*(3), 19-23.

Haward, L.R.C. (1969). The effect of different flying instructors upon the stress level of pupil pilots. In F. Gubser (Ed.), *Aviation Psychological Research WEAAP, Reports of the Eighth Conference for Aviation Psychology* (pp. 325-332). Zurich: Swissair.

Henley, I. (1986). Student pilot stress and flight instructor training. *Journal of Educational Administration and Foundations, 1*(1), 16-24.

Henley, I. (1987). *The association between student pilot stress and flight instructor training: A study of perceptions*. Unpublished master's thesis, University of Manitoba, Winnipeg, Canada.

Henley, I. (1995). *The quality of the development and evaluation of flight instructors in Canada and Australia*. Unpublished doctoral dissertation, University of Newcastle, Australia.

Henley, I., & Berlioz-Nott, L. (1997). Stress of flight instructors teaching students of non-native-English speaking backgrounds. In R.S. Jensen & L.A. Rakovan (Eds.), *Proceedings of the Ninth International Symposium on Aviation Psychology* (pp. 629-634). Columbus, OH: Ohio State University.

Henley, I., Nicoll, H., & Soros, S. (1994, October). Flight instructor stress. *Australian Aviation,* 36-38.

Hiner, R. (2000, March). Seven things instructors do to irritate students. *AOPA Flgiht Training,* 71-73.

Holmes, H. (1986, March). Cockpit perceptions. *Sport Aviation,* 31-36.

Horowitz, M.W. (1964). *An analysis of anxiety as it affects flight training*. A study sponsored by the Link Foundation, Suite M-10, Eastside Airline Terminal, 645 First Avenue, New York, NY.

Iampietro, P.F., Melton, C.E., Higgins, E.A., Vaughan, J.A., Hoffman, S.M., Funkhouser, G.E., & Saldivar, J.T. (1972). High temperature and performance in flight task simulator. *Aerospace Medicine, 43*(11), 1215-1218.

Jensen, R.S. (1982). Pilot judgment: Training and evaluation. *Human Factors, 24,* 61-73.

Jensen, R.S., & Benel, R.A. (1977). *Judgment evaluation and instruction in civil pilot training*. Springfield, VA: National Technical Service.

Kershner, W. (1988, May 1). Safety spotlight. *Aviation Safety,* 12-13.

Knowles, M.S. (1980). *The modern practice of adult education: From pedagogy to andragogy*. New York: Cambridge.

Krahenbuhl, G.S., Darst, P.W., Marett, J.R., Reuther, L.C., & Constable, S.H. (1980, July). *Undergraduate pilot training: Instructor pilot behavior and student stress and performance*. Tempe, AZ: Arizona State University, Department of Health and Physical Education.

Krahenbuhl, G.S., Marett, J.R., & King, N.W. (1977). Catecholamine excretion in T-37 flight training. *Aviation, Space, and Environmental Medicine, 48,* 405-408.

Krahenbuhl, G.S., Marett, J.R., & Reid, G.B. (1978, September). Task-specific simulator pretraining and in-flight stress of student pilots. *Aviation, Space, and Environmental Medicine, 49,* 107-110.

Krahenbuhl, G.S., Marett, J.R., & Reid, G.B. (1979, February). *Stress and simulation in pilot training*. Williams Base Air Force, AZ: Flying Training Division, Air Force Human Resources Laboratory.

Lazarus, R.S., Deese, J., & Osler, S.F. (1952, July). The effects of psychological stress upon performance. *Psychological Bulletin, 49*(4), 293-317.

Marin, G. (1997). A strategy for overcoming blocks to learning brought about by negative emotions. In R. Blunden (Ed.), *Teaching and learning in vocational education and training* (pp. 317-324). Katoomba, Australia: Social Science Press.

Mauro, R., Barshi, I., Pederson, S., & Bruininks, P. (2001). Affect, experience, and aeronautical decision-making. In R.S. Jensen (Ed.), *Proceedings of the Eleventh International Symposium on Aviation Psychology* (CD-ROM). Columbus, OH: Ohio State University.

McFarland, R.A. (1953). *Human factors in air transportation.* New York: McGraw-Hill.

Melton, C.E., Hoffman, M., & Delafield, R.H. (1969). *The use of a tranquilizer (chlordiazepoxide) in flight training*. Oklahoma, OK: FAA, Office of Aviation Medicine, Civil Aeromedical Institute.

Melton, C.E., McKenzie, J.M., Kelln, J.M., Hoffman, S.M., & Saldivar, J.T. (1975, January). The effect of a General Aviation Trainer on the stress of flight training. *Aviation, Space, and Environmental Medicine, 46*(1), 1-15.

Melton, C.E., & Wicks, M. (1967). *In-flight physiological monitoring of student pilots*. Oklahoma, OK: FAA, Office of Aviation Medicine, Civil Aeromedical Institute.

Miller, L.H., & Smith, A.D. (1993). *The stress solution.* New York: Pocket Books.

Nelson, J. (1993, April). Get-there-itis. *Flight Training,* 56.

O'Hare, D. (1992). The "artful" decision maker: A framework model for aeronautical decision making. *The International Journal of Aviation Psychology, 2*(3), 175-191.

Orasanu, J. (1997). Stress and naturalistic decision making: Strengthening the weak links. In R. Flin, E. Sales, M. Strub, & L. Martin (Eds.), *Decision-making under stress: Emerging themes and applications* (pp. 43-66). Brookfield, VT: Ashgate.

Prince, C., Bowers, C.A., & Salas, E. (1994). Stress and crew performance: Challenges for aeronautical decision making training. In N. Johnston, N. McDonald, & R. Fuller (Eds.), *Aviation psychology in practice* (pp. 286-305). Aldershot, UK: Avebury Technical.

Rogers, A. (1998). *Teaching adults* (2nd ed.). Buckingham: Open University Press.

Rogers, J. (1977). *Adults learning.* New York: Open University Press.

Roscoe, S.N. (1980). *Aviation psychology.* Ames, IA: Iowa State University Press.

Rosenthal, R., & Jacobson, L. (1968). *Pygmalion in the classroom*. New York: Holt, Rinehart & Winston.

Sanders, F., & Ethell, J. (1986). The stress syndrome. *International Council of Air Show News,* pp. 9, 37-39.

Selye, H. (1956). *The stress of life.* New York: McGraw-Hill.

Simonov, P.V., Frolov, M.V., & Ivanov, E.A. (1980, January). Psychophysical monitoring of operator's emotional stress in aviation and astronautics. *Aviation, Space, and Environmental Medicine, 51*(1), 46-50.

Sloan, S.J., & Cooper, C.L. (1986). *Pilots under stress.* London: Routledge & Kegan Paul.

Smith, E., & Keating, J. (1997). *Making sense of training reform and competency based training.* Wentworth Falls, Australia: Social Science Press.

Smith, R.M. (1982). *Learning how to learn: Applied theory for adults.* New York: Cambridge, The Adult Education Company.

Smode, A.F., Hall, E.R., & Meyer, D.E. (1966, November). *An assessment of research relevant to pilot training.* Wright-Patterson Air Force Base, OH: Aerospace Medical Research Laboratory.

Smode, A.F., & Meyer, D.E. (1966). *Research data and information relevant to pilot training* (Vol. 1). Alexandria, VA: Defence Documentation Center.

Soros, S., Henley, I., Wyndham, B., & Nicoll, H. (1995). Flight instructor stress. In R.S. Jensen & L.A. Rakovan (Eds.), *Proceedings of the Eight International Symposium on Aviation Psychology* (pp. 955-960). Columbus, OH: Ohio State University.

Steininger, K. (1964). Psychological factors in the training of student pilots. In A. Cassie, S.D. Fokkema, & J.B. Parry (Eds.), *Aviation Psychology* (pp. 33-44). The Hague: Mouton.

Stokes, A.F., Barnett, B.J., & Wickens, C.D. (1989). *Modeling stress and bias in pilot decision making.* Champaign, IL: University of Illinois, Aviation Research Laboratory.

Stokes, A.F., Belger, A., & Zhang, K. (1990). *Investigation of factors comprising a model of pilot decision making: Part II anxiety and cognitive strategies in expert and novice aviators.* Savoy, IL: University of Illinois Aviation Research Laboratory.

Stokes, A.F., Kemper, K., & Kite, K. (1997). Aeronautical decision making, cue recognition, and expertise under time pressure. In C.E. Zsambok & G.A. Klein (Eds.), *Naturalistic decision making* (pp. 183-196). Mahwah, NJ: Lawrence Erlbaum.

Stokes, A.F., Kemper, K., & Marsh, R. (1992). *Time-stressed flight decision making: A study of expert and novice aviators.* Savoy, IL: University of Illinois Aviation Research Laboratory.

Stokes, A.F., & Kite, K. (1994). *Flight stress: Stress, fatigue, and performance in aviation.* Aldershot, UK: Ashgate.

Stokes, A.F., & Kite, K. (1999). Grace under fire: The nature of stress and coping in general aviation. In D. O'Hare (Ed.), *Human performance in general aviation* (pp. 47-85). Aldershot, UK: Ashgate.

Termoehlen, J. (1987, April). *The role of the flight instructor: An important psychological factor in flight training.* Paper presented at the Fourth International Symposium on Aviation Psychology. Columbus, Ohio.

Transport Canada. (1989). *Pilot decision-making: Manual for private pilot training.* Ottawa, Canada: Author.

Wiggins, M.W., & O'Hare, D. (1995). Expertise in aeronautical weather-related decision-making: A cross-sectional analysis of general aviation pilots. *Journal of Experimental Psychology: Applied, 1,* 305-320.

Wiggins, M.W., O'Hare, D., & Hunter, D. (2000). *Weatherwise* [Computer-Based Training Program]. Washington, DC: Federal Aviation Administration.

Wilson, K. (1990, March/April). Accentuate the positive. *Canadian Flight,* 22-23.

Wood, E., & Knight, J. (1988). 'I feel sorry for supply teachers…': An ethnographic study. *UNICORN, 15*(1), 36-43.

Part 2
Teaching Strategies

7 Planning and Preparation for Effective Instruction

Jayne Bye and Ann Reich

Introduction

Course or program planning is a key activity for trainers and educators and it lays the basis for effective educational practice. It can be understood as a process of decision-making stages, ranging from the overall course structure to the planning of an individual session. Education and training theorists have generated substantial useful knowledge about these processes and some have endeavoured to describe the processes involved by creating planning models.

This chapter provides an overview of the key models of program development and also considers the constraints on planning activity and the importance of analysing the social, institutional, public policy and industry contexts of the planning situation. It discusses the four key stages of program planning processes, that is, needs analysis, program design, program implementation and program evaluation, and provides detailed outlines to guide a new trainer in this process.

The chapter emphasises the creative process of planning where trainers, even in the most constrained training context, can have important input. While the chapter does not attempt to provide a formula for planning, it provides useful planning guidelines for teachers and trainers in the aviation context.

Understanding the Terms

A number of different terms are used in the literature in this field to describe an organised pattern of study, such as *course*, *program* and *curriculum*. Similarly, a range of terms, such as *program development*, *course planning*, *curriculum design* and *instructional design* are used to describe the overall planning of the activities that make up a course or program. These terms are sometimes used interchangeably while at other times they are used by different authors who have a very specific meaning in mind. For example, Jarvis (1995) notes that theorists in adult education use the term program planning rather than curriculum planning due to the fact that the word curriculum has long been associated with initial schooling. For this reason, this chapter will utilise the terms *program planning* and *program development* to describe the activities in which educators engage when trying to conceptualise, organise and implement an educational encounter. However, some definitions used by theorists in the area of course or program planning are included here to assist in gaining an understanding of this terminology.

Curriculum

Curriculum is often seen as a body of knowledge, the content of education to which the students need to be exposed. But curriculum is much wider than a list of subjects to be studied; it is not only what you say but how you say it! Curriculum is all the planned experiences to which the learner may be exposed in order to achieve the learning goals (Rogers, 1998, p. 176).

Course

In the United Kingdom, a course is the total study undertaken by a student in a year, or study taken leading to a particular qualification. In the United States of America, it is used to mean a series of class sessions during one term or semester.

Program

The use of the term *program* also requires some explanation because this term too is subject to varied use. Brookfield (1986) provides a broad working definition to assist in formulating an understanding:

> The word program has a wide range of applications. It can denote a single educational or training event, a formal course, a collection or set of courses, an individual learning project, a workshop, a colloquium, a conference, or a public education campaign. Educational programs may be difficult to define but there are a number of features that are common to them all. People have come together for a purpose and, for some at

> least, that purpose is to learn. What these people are doing has an order or sequence. And what they are doing has a time limit, in that the program has a discernible beginning and end. (p. 204)

To maximise clarity, it is also useful to differentiate between *planning* and *preparation*. According to Cole and Chan (1994),

> *planning* involves the selection of goals and the design of appropriate instructional procedures. It is concerned with the formulation of instructional ideas and is essentially schematic and conceptual. Preparation, on the other hand, evolves from planning and is more practical and programmatic. *Preparation* is concerned with the applied aspects of the planning process and entails the organisation of lessons and the provision of teaching aids, worksheets, laboratory tools, concrete materials and related equipment. The planning process normally precedes the preparation, but very often both aspects have to be considered concurrently. (p. 84)

Models of Program Planning

Planning education and training activity is a complex process involving many steps and stages. The complexity of this process has engaged educational theorists and their attempts to come to grips with this area have resulted in the conceptualisation of theoretical models which seek to capture and describe these processes. The main differences between the models are:

- the values or philosophy of the author; and
- the purpose for which the model was developed.

These two points are illustrated below in the overview of program planning models. There has been such a wealth of models developed that it is easy to become overwhelmed by the choices. Indeed, Boone (1985) notes the difficulties educators face when attempting to use the differing programming models as a guide for practice due to the fact that "a consistent application of any model is made difficult because of the wide variety of contexts within which adult education is practiced" (p. 20). Therefore, when attempting to choose a programming model, trainers and program designers face the problem of which planning model to use.

However, one key suggestion for beginning trainers is to realise that these models have mostly been designed with particular contexts in mind. As Boone (1985) indicates, since adult education and training occurs in such a variety of contexts, no one model is universally applicable. Therefore, the specific nature of the training context will have implications with regards to which programming model might be usefully employed. As Brookfield (1986) also notes, program planning is an area in which many trainers and educators have described a disjuncture between the formal models and their experience of practice. For example, the models do not take into account the lack of resources, tight time scales, or the impact of the cultural issues of a particular organisation.

Overview of the Models

The following overview does not provide a comprehensive coverage of program planning models but rather it offers some insights into the influential theorists in this field. For the purposes of clarity, the models have been placed into the following categories:

- institutional models
- human resource development models
- experiential models
- social action models

These categories are not exhaustive and are indeed very broad. However, they serve as a basic organising principle through which a range of models might be understood. In vocational education and training contexts, particularly in highly regulated fields such as the aviation industry, with many prescribed outcomes of earning, an institutional model or a human resource development model may be the most appropriate. However, a short description of the other models is included to ensure educators and trainers are aware of the diversity of the field and may use them in specific contexts for which they are best suited.

Institutional Models

The term institutional model is used by Brookfield (1986) to group together a set of programming models which he sees as including the following common stages:

- identify needs;
- define objectives;
- identify learning experiences to meet the predefined objectives;
- organise learning experiences; and
- evaluate the program in terms of how well the objectives were met.

These models are generally acknowledged as the most influential and widely used models in adult education and training. Within this category of models, numerous key theorists have made major contributions to thinking about program planning. The models of two theorists, Ralph Tyler and Malcolm Knowles, are discussed because of their significance in this field.

Tyler — objectives model In 1949, Tyler released his book entitled *Basic Principles of Curriculum and Instruction*. In this book, he sets out a programming process organised around four main questions:

1. What educational purposes should the organisation seek to attain?
2. How can learning experiences be selected that are likely to be useful in attaining these objectives?
3. How can the selected learning experiences be organised for effective instruction?
4. How can the effectiveness of these learning experiences be evaluated? (cited in Newman, 1995, pp. 61-62).

Tyler (1949) elaborates on the significance of each of these questions as the basis for the discussion of his model which can be summarised as:

- In setting educational objectives, programmers should take into consideration the learners, society, philosophical considerations within the institution conducting the program, the intellectual climate of society and experts in the subject.
- Learners need to be given opportunities to practice new skills since learners learn from what *they do*.
- The content and its presentation must hold meaning for the learner. While ways of presenting material may seem logical to the teacher, this may not be the case for the learners. Continuity, sequence and integration are the key factors in helping learners make meaning of new ideas.
- Evaluation of the learning can be measured in changed behaviours. If learners are tested at the beginning of a program of instruction and then

again at the end, the program objectives have been met if there is a change in behaviour in line with the training objectives (Newman, 1995, pp. 61-62).

The key underlying assumptions of this model are that all program outcomes can be stated in terms of predefined objectives and that all learning can be measured in terms of changed behaviours. In this way, it can be seen that this model draws heavily on the ideas of behavioural psychology (as discussed in Chapter 1). These features of the model have made it very attractive to programmers over the years. Set objectives are thought to offer trainers and learners alike a very clear direction for learning. Furthermore, a focus on measurable outcomes provides all stakeholders with clear feedback and the basis of evaluation.

Although still a very popular approach to program planning and the basis of much program planning particularly in institutional settings, the model can be seen as restrictive. As Newman (1995) notes, "once the objectives are set, and with the evaluation relying on meeting those objectives, the educators and learners will be less likely to explore, take risks, go off on tangents, pursue their own interest and engage in kinds of learning that might take them into unexplored territory" (p. 63). In fact, Tyler himself has since revised his ideas and urged more flexibility to overcome concerns of the impact of too closely specifying objectives in advance (Brookfield, 1986).

Knowles — andragogical model Malcolm Knowles' (1970) well-known concept of andragogy (see Chapter 2 for an explanation) leads to an approach to program planning that puts greater emphasis on the needs of the learner. The key points of an andragogical approach to program planning are:

- adults move towards self-direction;
- adults use experience as a learning resource;
- adults are ready to learn in accordance with socio-developmental tasks; and
- adults desire immediate application of learning.

There are clear implications for program planning in this concept of the adult learner. Firstly, if adults desire self-direction, then learner self-diagnosis of needs is important in order for the learner to feel ownership of the learning goals. Similarly, learner participation in the planning and the learning processes is crucial if experience is to be utilised as a resource and the learning is to be meaningful. Finally, assessment planning and evaluation strategies must also be based around the learner's goals since these are the bases upon which the outcomes have been set.

Despite the widespread influence of the concept of andragogy since its promotion by Knowles in the 1970s, many theorists have questioned the ability of this concept to cover all adult education and training contingencies. For example, Jarvis (1995) enters into the debate around the work of Knowles and the programming implications of his concept of andragogy suggesting that

> individuals having little experience in a subject might want a more pedagogic approach whilst those having a lot of experience in an area might prefer an andragogic approach. This leads on to the possibility of conceptualising the initial vocational education curriculum in pedagogic terms and the continuing vocational education programme in andragogic terms. (p. 198)

Clearly Jarvis' points are worth consideration. If the participants have no experience of the content and skills they wish to acquire, then a more traditional teacher-led program may be appropriate, at least in the short term. However, the needs of the learners still need to be kept in mind.

Human Resource Development Models

Human Resource Development models emerged in the 1980s in response to the development of the field of human resource development in organisations and the shift to organisations providing more training and development activities "in-house". Many of the models used are based on a Tylerian model although the training for management often adopts Knowles' models (Newman, 1995). However, a major difference has been the shift from a focus on individual or organisational need to a focus on the employees gaining skills and performing tasks to achieve organisational goals. Therefore, there is an emphasis on the analysis of the structure and functions of the organisation and the tasks and roles of employees (Newman, 1995). A key application of this model today is Competency-Based Training (CBT). The features of CBT which align it to the human resource development models include:

- organisational analysis;
- competency analysis;

- design of programs to match the competencies; and
- assessment based on the pre-defined competencies.

In view of the importance of CBT as the model for much workplace and vocational training in Australia and in other countries, the key aspects of CBT are outlined below in order to understand the implications for program planning under this recent predominant national system.

What is Competency-Based Training? Defining CBT is a more difficult task than it may initially appear. CBT is a framework for training and assessment that has been interpreted in a number of ways in a variety of countries. Also, in the current Australian context, the implementation of CBT has undergone changes since its emergence in the late 1980s. The following definition of CBT is often cited.

> [CBT is] a way of approaching (vocational) training that places primary emphasis on what a person can do as a result (the outcome), and as such represents a shift away from an emphasis on the process involved in training (the inputs). It is concerned with training to industry specific standards rather than an individual's achievement relative to others in the group. (Australian Chamber of Commerce and Industry, 1992 cited in Smith & Keating, 1997, pp. 101-102)

As Smith and Keating (1997) point out, the key points in this definition of CBT are:

- the focus of the training is on the outcome of the training;
- the outcome is measured against specified standards not against other students; and
- the standards relate to industry.

Competency standards At the centre of this framework is the concept of competency standards and despite the fact that there have been a number of changes to the way CBT is implemented in Australia (as will be discussed below), the centrality of competency standards has been maintained. According to Smith and Keating (1997), "industry competency standards describe the competencies which a worker in that industry should possess. The standards consist of units of competency, each of which in turn is divided into elements of competency with performance criteria" (p.105).

Training Packages (Syllabus) The latest change to the implementation of the CBT system to have impact on program planning in Australia was the introduction of Training Packages announced at the end of 1996. With the introduction of Training Packages, the CBT system abandoned the notion of accredited curriculum and moved toward packaging competency standards, national qualifications and assessment guidelines together into comprehensive kits intended to guide training providers in the design and assessment of competency standards. It also included a move towards assessment of competency standards being a key factor, signalling less concern with how or where the competencies were learned. The following example shows one application of CBT to the aviation industry in Australia.

> The Australian National Competency Standards for Private and Commercial Pilots lists 27 Units of Competency which must be achieved before a Private Pilot Licence (PPL) qualification is granted. Each unit is also described in more detail in terms of the elements that comprise it. Clearly, some units of competency can be described in greater detail than others. For example, Unit 12 — Navigate Aeroplane (PPL) is made up of 9 elements, while Unit 10 — Control Aeroplane Solely by Reference to Full Instrument Panel (PPL) has only one element.
>
> The Training Package (Syllabus) also contains a lengthy section devoted to describing the evidence needed to confirm competence in a trainee. This section of the package is known as the assessment guide and contains valuable information for program planners and trainers concerning the criteria by which trainees should be assessed. Range of Variables statements further describe the conditions under which these tasks should be demonstrated.
>
> The crucial point for trainers to understand about this document and these descriptions of Units and Elements of Competence, Performance Criteria and Evidence is that it is not a training document. These descriptors do not indicate anything about *how* a trainee pilot acquires these skills. It is up to the planner of training to make decisions about how the training will occur, based on information in the Training Package (Syllabus).
>
> By examining the units and elements of competence, in conjunction with the evidence guide, trainers devise learning outcomes to guide training. Specifically, a trainer must:

1. consider which knowledge, skills and attitudes trainees need to be able to perform the tasks, and write them as learning outcomes;
2. devise assessment criteria for testing outcomes of the training which describe the task in measurable and observable stages;
3. design activities to develop the necessary skills; and
4. organize opportunities and resources necessary to practice the skills

(adapted from Australian National Training Authority, 1999).

Experiential Models

Despite the prevalence of institutional models of program planning, many theorists look to what might be seen as a less restrictive and pre-defined approach to learning. Experience-based learning (EBL) is one example of this approach. EBL is founded on the following set of assumptions about learning:

- experience is the foundation of, and the stimulus for, learning;
- learners actively construct their own experience;
- learning is a holistic process;
- learning is socially and culturally constructed; and
- learning is influenced by the socio-emotional context in which it occurs (Boud, Cohen, & Walker, 1993 cited in Foley, 1995, p. 208).

In contrast to Tyler's model, this type of program draws on humanistic rather than behaviourist psychology (see Chapter 1). This influence assumes that all human beings have a need for "growth" or "self-actualisation" and stresses the importance of the needs of the individual. It also emphasises the need to consider the whole person in the training context — the emotions, body and mind.

Experienced-based learning is also the basis of many of the contemporary approaches to planning learning experiences particularly in workplaces, such as approaches using critical reflection, mentoring, action learning, and the basis of management training especially approaches using outdoor experiences for team building.

The implications for program planners using this framework are many. Firstly, there is little or no emphasis on pre-defined objectives. If individuals construct their own learning on the basis of experience, there is no way this learning can be exactly anticipated in written objectives. Certainly, the educational experience can have holistic aims, such as "personal growth", however, this will mean different things to different participants. Furthermore, assessment, as traditionally conceived, does not occur. Participants and facilitators evaluate the learning in terms of their perceived "growth" or insights from the experience.

Social Action Models

The most influential social action theorist in the area of education and program planning is Paolo Freire. In 1970, Freire released a very significant text called *Pedagogy of the Oppressed.* In this work, Freire outlines his concept of pedagogy which grew out of his work on literacy programs for the poor and oppressed in Brazil. For Freire, literacy was a way of engaging learners in education that was centrally concerned with freedom from oppression, aiming for learners to experience a humanising and humanistic liberation. To achieve this aim, participants in Freire's programs begin to *question, analyse* and eventually *act* upon their world (Newman, 1995, p. 73). Freire's ideas, grounded in the empowering process of "conscientization", are widely applied in a range of educational contexts, including community education, literacy and activist education programs.

Comparisons Between the Models

Clearly, there are many differences between the models briefly described above. Boone (1985) suggests that programming models can be analysed in terms of their variation in the following features: context, scope, underlying philosophy, perspective on programming, primary situational applicability, and main theme (p. 33). However, Boone also notes the following essential similarities between most models of program planning:

1. problem/need definition
2. setting of objectives, goals and means
3. some formal and informal learning activity
4. either an explicit or an implicit evaluation (p. 34).

Constraints on Program Planning — The Importance of Context

The program planning process can be a creative, exciting and sometimes daunting prospect as the range of activities outlined in the stages above might suggest. However, the reality of most training environments means that there will always be some kind of constraints impinging on the planning process and this is an important realisation for beginning course planners. Brookfield (1986) critiques much of the literature on course planning for its lack of attention to this reality. He notes that "the practitioners in these [program planning] manuals seem to operate in contexts where adequate resources, congenial colleagues, supportive superiors, and a sympathetic political or institutional climate always exist" (p. 206). Of course, this is rarely the case in most training environments. Cranton (1992) provides a realistic analysis of most contexts when she states that:

> A wide variety of institutional or organisational constraints can influence learning situations. Constraints within the physical environment, such as room size, furniture, lighting, room temperature, and availability of appropriate resources or equipment, can plague the educator who is attempting to create a comfortable working atmosphere. Administrative constraints, such as requirements for examinations, required grading distributions, scheduling, multisectional courses, deadlines, and registration procedures, can severely inhibit [planning] … Curriculum constraints, such as required textbooks or course outlines, preset evaluations, and the use of prepackaged materials, can restrict self-directed learning. Financial constraints which lead to extremely large groups, restrictions on materials or photocopying, outdated equipment, sharing of equipment, or restrictions on the use of field sites can limit the effectiveness of the educator's work. (pp. 199-200)

Limitations specific to planning vocational education and training — as discussed above — come from the current framework of Competency-Based Training. This framework provides an outline of the required knowledge, skills and attitudes that must be taught to ensure competent workers. If trainers are teaching in an accredited course that leads to a nationally recognised qualification, there is little room for creativity in setting at least the general goals of training. In this case, Cranton (1992) suggests that

> Given legitimate constraints, the educator's next step is to determine what remaining aspects of the learning experience are flexible.... For example, most often, the methods by which learners reach prescribed objectives need not be constrained. The educator can work with learners to determine appropriate methods for specific groups or individuals. The steps that can be taken depend greatly on the context, but the adult educator can introduce considerable learner self-direction in almost any situation. (pp. 203-204)

A key part of any program planning process is therefore to identify aspects of the context in which the program will occur. As Jarvis (1988) suggests, the program planner needs to take account of:

- The *social context* — who are the potential learners; what are their socio-cultural backgrounds and previous educational experiences?
- The *organisational context* — what are the particular requirements, constraints, opportunities in the organisational context?
- *Public policy and industry context* — what are the current public policies that will impact on the program, such as national competency standards processes? What are the particular industry policies, regulatory frameworks and policies (e.g., from the International Civil Aviation Organization, and in Australia, from bodies such as the Civil Aviation Safety Authority) that will impact on the program?

Steps in the Planning Process

Figure 7.1 represents the key steps that are often described as the program planning process. Based on the institutional model, the four key steps provide a guide to the process. It is diagrammatically set within the three context areas described above: social, organizational, and public policy and industry contexts. These contexts will have a major impact on the planning process and the nature of the program developed and each of these contexts needs to be considered in general as well as in relation to each stage.

The simplicity of the diagram may be somewhat deceiving given the amount of thought, time, resources and sometimes anxiety that go into the process. However, it does capture the main tasks program planners must consider and plan for as well as the cyclical nature of the activities. The first three stages of the planning cycle will be described below. Evaluation will not be covered here since it is examined in detail in Chapter 15.

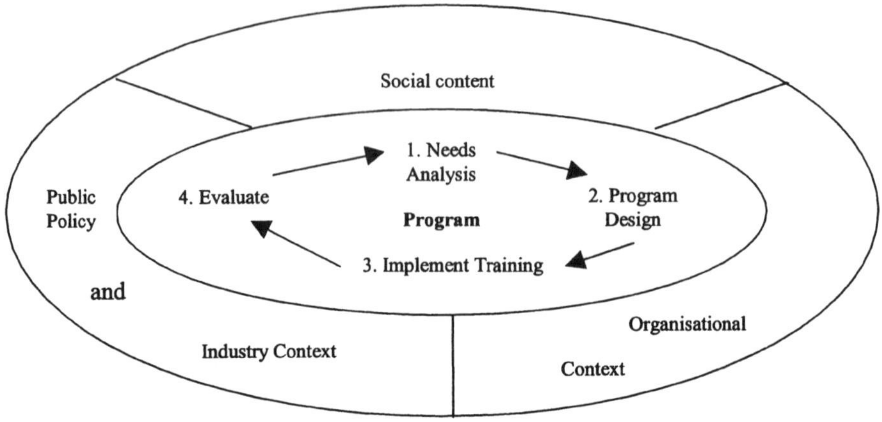

Figure 7.1 Planning in context

Needs Analysis

The goal of a needs analysis is to identify the gap between the current skills and knowledge of an individual or group and the skills and knowledge needed by an individual or a group for a specific purpose. Information on needs can be collected in many ways. For example, if the training context is not subject to prescribed training goals, needs analysis can be largely an informal process. It might be possible to discuss the issue with other experienced trainers to find out what they have done with groups at similar levels.

Alternatively, making contact with the participants prior to the beginning of the course may provide useful information about their needs. For example, approaching potential students at the enrolment session or holding an informal meeting prior to the course and asking them about their expectations of the course may provide sound information concerning their needs (Daines, Daines, & Graham, 1995, p. 18).

This approach is clearly aimed at meeting the needs of the learners. However, often this is not the only consideration. Brookfield (1986) divides needs into two categories: the felt needs of the learners (assessed easily using any of the above methods) and prescribed needs which are defined by educators. Needs can also be expressed by other stakeholders in the training context. For example, industry representatives and employers may define training needs for workers and trainees in their area. Licensing requirements and regulatory bodies, such as the International Civil Aviation Organization (ICAO) and in Australia, the Civil Aviation Safety Authority (CASA), may also prescribe the skill requirements of workers in a specific context, therefore, defining the training needs of the employees or the trainees.

In the current Competency-Based Training framework used in vocational education and training in Australia and in other parts of the world, competency standards describe the knowledge, skills and attitudes needed by competent workers in many aspects of the aviation context. These descriptions function as training guidelines because they define the elements of the work that job-holders need to accomplish if they are to be successful (McLeod, 2001, p. 63), and therefore inform trainers of what the industry requires of competent workers. However, the learning needs of particular participants in a course will still need to be assessed.

Course Design

Many decisions come under the heading of course design. These include:

- writing goals, objectives and learning outcomes;
- setting assessment criteria;
- sequencing objectives and learning outcomes;
- deciding on and organising content; and
- selecting teaching and learning strategies.

Writing Goals, Objectives and Learning Outcomes

Rogers (1998) notes that a lack of understanding of the terms *goals* and *objectives* prevails in education, therefore, he defines these terms and clarifies the nuances of their meaning. He defines the goals of a program as "the intended outcomes of the teaching-learning activities" (p. 117). He then outlines the various levels of goals into:

1. overall goals (aims)
2. specific goals

3. immediate goals (objectives/learning outcomes).

The *aim*, as indicated by Rogers (1998), is usually a statement of the general learning outcome expected from having taken part in the learning activity (goals toward which to work). The aim therefore describes the more general (overall) goals. However, each aim will need to be defined further by writing a list of *specific* types of behaviour that can be used as evidence that the general aim or goal has been achieved. In most aviation instructor guides or manuals, trainers will generally encounter only the terms *aim* and *instructional objectives*, while the recent private and commercial pilot syllabi in Australia now refer to *learning outcomes*, therefore, this will be the term used here.

Pithers (1998) explains that learning outcomes usually indicate some behaviour about the learner that will change as a result of the training. In practice, a common stem used in writing a learning outcome is: "At the end of this lesson (instructional unit or training session), the trainees should be able to…." For example, "At the end of this lesson, students will be able to explain the forces acting on an aircraft during a climb" or "At the end of this training session, the student will be able to maintain level flight within + or – one hundred feet". These statements about learner performance are generally expressed as statements beginning with a verb (usually an action verb), therefore focusing on the skill that the learner will be able to perform. For example, typical learning outcomes in the aviation context might state that:

At the completion of this training session, participants will be able to:

identify the 3 axes about which an aircraft moves;
explain how lift is generated;
maintain a constant altitude and heading at nominated airspeeds; and
detect and correct attitude, heading and altitude errors;

Or

outline the correct spin recovery procedure;
recognise and recover safely from an incipient spin with a minimum loss of altitude; and
appreciate the importance of the early recognition of the symptoms of a spin.

Pithers (1998) also notes that in writing learning outcomes, "sometimes the learning outcome decided upon is something which is difficult to express in behavioural terms; its attainment very difficult to measure and assess" (p. 83).

Typical examples in the aviation context are "airmanship" issues or human factors, although Transport Canada (1996) maintains that most of human factors is behavioural (see pp. 5-12). Kemp (1985) remarks that "when we turn to the affective domain — attitudes, feelings, and appreciations — we find difficulty in specifying objectives in clearly observable and measurable terms. Most often we indicate these objectives only indirectly by inferring from what can be observed" (p. 85). Some see this aspect of learning outcomes as a shortcoming of this framework and argue that "CBT is a throwback to behaviourist notions of education, which are generally seen as inferior" (Smith & Keating, 1997, p. 114).

Regardless of one's opinion on this issue, this point does alert planners to the fact that not all learning is the same. As program planning becomes more focused upon making decisions about specific learning events, it is important to be able to break down the learning outcomes into clear statements about the nature of the learning. This is usually achieved with reference to what is known as the three domains of learning: cognitive, psychomotor and affective. These three domains are briefly described below.

1. Cognitive (thinking) or knowledge objectives are written as learning outcomes covering the information requirements of the program.
2. Psychomotor (doing) or skills objectives are written as learning outcomes stating the practical skills to be acquired.
3. Affective (feeling) or attitude objectives are written as learning outcomes which cover attitudes or values to be developed.

These three domains are thought to cover and describe the full range of learning that can take place in the training environment. Furthermore, each of these three domains is broken down into a hierarchy — levels which rank the objectives from simple to complex. Table 7.1 shows this hierarchy.

Clearly, it is useful for trainers to understand these domains and their hierarchies when writing learning outcomes for training sessions. Attention to all three domains will ensure that learning is holistic and not merely technical. This approach may provide part of the answer to the criticisms levelled at CBT touched upon above. Similarly, understanding the notion of hierarchy will allow trainers to set outcomes at the level appropriate to the learners' level and experience, thus allowing them to apply some valuable adult learning principles.

Table 7.1 Taxonomy of domains and category of learning

	Cognitive	Affective	Psychomotor
Simplest	Knowledge	Receiving	Perception
	Comprehension	Responding	Set
	Application	Valuing	Guided response
	Analysis	Organisation	Mechanism
	Synthesis	Value complex	Complex overt response
	Evaluation		Adaptation
Most Complex			Origination

Source: Lehrer, 1993, p. 277.

Setting Assessment Criteria

Setting the assessment criteria is the next planning step after stating the learning outcomes. Each learning outcome is broken down into a set of more detailed steps and descriptors which serve the purpose of providing the measurement against which competency is judged. Pithers (1998) notes that it helps if assessment criteria "describe behaviour which is observable and hence, measurable" (p. 85). As with learning outcomes, statements of assessment criteria begin with a verb and describe precisely the actions the students will perform in showing that they have achieved the learning outcome (see Chapter 15). Pithers (1998) provides the following set of action words useful for assessment criteria:

Table 7.2 Action words for assessment criteria

Knowledge	**Skills**	**Attitude**
Define	Assemble	Judge
Reorganise	Perform	Listen
Recall	Operate	Accept
Explain	Activate	Reflect
Write	Adjust	Influence
State	Change	Reject
List	Repair	
Evaluate	Comment	
Contrast	Compute	

Source: Pithers, 1998, p. 86.

Sequencing Objectives and Learning Outcomes

How do teachers decide what to teach first to aviation students? Is there a particular sequence of ideas and information that learners need to be taken through? Once again, the ideas of Tyler (1949), who warns against designing programs or courses according to the apparent logic of a subject, can be drawn upon. He draws a distinction between logical organisation, which he describes as "the relationship of curriculum elements as viewed by an expert in the field", and psychological organisation, which he describes as "the relationship as it may appear to the learner" (p. 97). This is an important distinction to consider when session planning. Trainers often make assumptions about there being only one way to organise content and skills in a course. This results from being in a position of having the overall or "big picture" in mind and, therefore, seeing how it all fits together. The learners, being new to the area, may not share this framework and so objectives must be carefully sequenced to allow the framework to develop.

Cantor (1992) provides some suggestions and guidelines on how to sequence objectives that focus on the learners' point of view. He outlines three sequencing methods and notes that each method has its uses depending on the nature of the task being learned and resource availability:

- *Job performance sequence* — where skills are learned in the order they would be performed on the job;
- *Simple to complex cognitive sequence* — where instruction progresses from the known to the unknown (this is used extensively in flight training);
- *Logical performance sequence* — where a balance between job performance and the psychological order best understood by the learners is maintained (adapted from pp. 109-110).

Deciding On and Organising Content

The content to be covered in a particular course is often set out in a syllabus or training manual. Individual teachers often do not have much control over deciding the parameters of this course content. In some areas of training, such as flight training, air traffic controller training and aircraft maintenance training, regulatory bodies and licensing provisions will strictly prescribe what trainees need to know before they are judged competent. However, decisions must still be made at the level of session planning regarding *which content* and *how much* of it can be covered in any particular class session.

Rogers (1998) warns that teachers have a tendency to overload learners with content and that this can happen for a variety of reasons. One often quoted reason is the imperative of covering the whole course within the allotted course timeframe. This is a realistic concern for most trainers. Time — both the trainer's and the learners' — is a resource and as such has great value. The adage that "time is money" is unfortunately true in the training environment and, therefore, most training courses are expensive to run and often costly to undertake. For this reason, course contact time is generally decided according to budget constraints as well as educational need. This can lead to trainers putting pressure on themselves and on the learners to get through content and often leads to the imposition of quite formal methods of instruction such as lectures, at the expense of other teaching methods.

The logic behind these types of choices needs to be examined closely. If it is true that adults learn best in a participatory learning environment, what will be the effect of this type of teaching? As Rogers (1998) reminds us, "if covering the whole of the material means going so fast that the learner cannot learn deeply enough and with understanding, then the class is not effective. What is the point of completing the syllabus if the learner does not really learn?" (p. 181).

Rogers (1998) claims that this approach to content coverage is harmful and unnecessary. It has as its basic assumption that learning can only take place if the teacher is present to "dispense" knowledge and skills. Clearly, there are alternative views about how adults learn. If the assumption about training and teaching is that the teacher's role is to *facilitate* learning (rather than simply imparting knowledge and "getting through the content"), and that in facilitating learning we help learners develop the skills to continue to learn independently, then these pressures of time and content can be approached differently with a range of creative teaching and learning strategies.

Questions about how the content will be organised are also often best decided by the teacher in relation to considerations such as the availability of training resources and the immediate needs of a particular group of learners. Cantor (1992) suggests that deciding how to organise the content in a session is a key component of session planning. He presents a range of ways to approach this task where content may be organised:

- *Relationally* — related ideas and activities are presented together;
- *Positionally* — content is presented in terms of how the ideas relate to each other (e.g., presenting the pros and cons of a particular idea or approach to using a skill); or
- *Structurally* — using an established framework such as who, what, when, where and why or cause and effect (p. 113).

Selecting Teaching and Learning Strategies

Choosing teaching strategies can be a challenging task for new trainers. Knox (1986) concedes that the majority of experienced trainers "use only a few methods with which they happen to be familiar" (p. 77). While this may lead to effective use of a limited range of strategies due to trainer familiarity and confidence, it is important to remember that no single method suits every occasion. In fact, as Daines, et al. (1995) suggest, "a combination of methods is likely to be more effective than any one used singly and … a variety will help maintain people's interest and motivation" (p. 33). The reasons behind this guideline for practice is that not all methods will appeal to all learners. In Chapter 4, the concepts of preferred learning styles and the range of learner motivation and characteristics were discussed. Through the utilisation of a variety of teaching strategies, teachers and trainers can put this knowledge into practice.

Matching teaching strategies to content, objectives and participants The literature on teaching methods suggests that there is a range of ways of categorising and then choosing teaching methods. For example, Daines et al. (1995) categorise strategies in a way that overcomes what they consider to be the "overly simplistic categories of 'learner-centred' and 'teacher-centred' which have been in vogue recently" (p. 33). They note that even when these terms are "recognised as being the ends of a continuum, [they] often carry with them an implicit assumption that the former is in some way philosophically and psychologically superior to the latter" (p. 33). This is an important point for teaching and learning in the aviation context, since it is true that much valuable learning will be closely directed and sometimes led by trainers. Daines et al. go on to suggest that there are three main ways of categorising teaching strategies based on the type of activity involved:

- *presentation* — where the tutor predominantly transmits ideas, information or skills, for example by lecturing and demonstrating, often using a chalkboard, flip chart, etc.;

- *interaction* — where knowledge and experience are shared between the teacher and the learners and/or amongst the learners themselves, for example by discussion, questioning, role play and brainstorming; and
- *search* — where the learners explore and discover for themselves, either on their own or in small collaborative groups, for example by individual practice, problem solving tasks and case studies. (p. 33)

One way of approaching learning is to categorise it in terms of domains of learning. In choosing teaching strategies, this may be one of the considerations trainers entertain. It has been suggested that different teaching strategies are appropriate for each of the three domains of learning: cognitive, psychomotor and affective. For example, Daines et al. (1995) propose the following:

- Since the affective domain is concerned with attitudes, methods such as discussion, case study, role play and simulation are most appropriate in facilitating this type of learning.
- Since the psychomotor domain is concerned with skills, the methods such as demonstration, individual practice, and coaching are most appropriate in facilitating this type of learning (see also Chapter 9).
- Since the cognitive domain is concerned with knowledge, methods such as lectures, small group work and problem solving tasks are most appropriate in facilitating this type of learning. (p. 34)

However, the reality of most teaching and learning situations is that each of the three domains will be involved in the learning, therefore, an appropriate mix of methods is generally the key to effective teaching.

Ultimately, regardless of how individual trainers choose to conceptualise categories of teaching strategies, Knox (1986) suggests that the three main "considerations in planning learning activities are program objectives, learner characteristics, and your perspective. In each instance, pertinent information can help you plan more responsive and effective learning activities" (p. 79).

Implementing Training

There are a number of steps in the implementation stage of program planning. These include session planning, organising resources and designing assessment events. Given the increasing focus on the assessment of competence in vocational training, assessment issues will be dealt with in a separate chapter (see Chapter 15).

Session Planning

Due to the fact that the education and training of adults occurs in such a variety of contexts serving a multitude of purposes, there clearly can be no strict rules concerning the planning of training sessions. Furthermore, within one training context there will be different types of training sessions ranging from knowledge-based, theoretically-focused sessions to skills-based, work simulation sessions. Each type of session has its own learning outcomes and activities. Each will require different things from the teacher and learners. However, there are some key principles of planning which seem applicable in a range of settings and are worthwhile for novice and experienced trainers alike to consider.

The structure of a training session After surveying the literature in the area of planning training sessions, Pithers (1998) notes a consensus among many theorists regarding the basic structure of a training session. This consensus seems to be that the "general framework for each training session consists of an introduction, a body, in which there is a step-by-step exposure to the material to be learned and finally, a conclusion" (p. 101).

This concept seems simple enough to follow, however, within each of these stages, there are decisions to be made by the trainer. The following outline suggests some guidelines and ideas about how to approach these three key structures.

Session introduction There are different types of introductions that trainers may utilise when attempting to capture the attention of trainees and motivate them to learn. The introduction should be brief but may include the following techniques:

- *Link or review* When using this approach, previously learned knowledge or skills can be reviewed at the beginning of a session. The advantage of this technique is that it reveals to learners the relationship between the knowledge, skills and attitudes learned across a series of sessions and allows the trainer to gauge the level of learner retention. This is an easy type of introduction and is simple to plan and implement.

- *Current topic* This type of introduction focuses on a current, topical issue. The trainer may, for example, refer to a current issue or event that provides a link with the external environment or workplace, for example, a recent accident or contentious issue. It indicates relevance that the training room content is in contact with outside issues and provides a link between something familiar to learners and the new content to be introduced in the session.
- *Anecdotal* Here a common personal but relevant experience is used to illustrate a concept or principle. In this approach, an analogy is applied and linked to the subject matter under consideration. The approach uses a story or narrative, real or imaginary, to make a relevant training point clearer.
- *Background and historical introduction* This approach provides a brief overview of the historical development of some piece of knowledge or skill or the development of some attitudes. This type of introduction is useful when learners need to establish a depth of understanding about a subject or issue but may not be the preference for learners who need to see the direct relevance of the topic if they are to become motivated to learn (adapted from Pithers, 1998, pp. 103-104).

Body of the session This stage of the session, as its title suggests, forms the bulk of the session. During this stage, the training methods selected by the trainer are utilised. Generally speaking, these methods should reflect sound principles of adult learning and, as such, involve opportunities for active trainee practice and learning through, for instance, hands-on-practice, exercises, note taking, questions, answer and follow-up exercises, case studies, role-play or small group work and discussion groups (Pithers, 1998, pp. 103-104). The details of some of these techniques and their uses are discussed in the following chapters.

Session conclusion Like the introduction, the conclusion can be brief and is generally used to:

- review or summarise the content (only the major points);
- overview the coherent, whole picture that has been developed;
- assess trainees' questions and give feedback;
- explain how to apply the learning to a new setting (e.g., in the workplace); and

- provide links to what is to follow in the next session (Pithers, 1998, p.107).

Writing a Session Plan

A written session plan functions in a number of ways. Most importantly, "it is the physical evidence of the completion of the planning phase of training" (Pithers, 1998, p. 111) and draws together all the important elements of effective teaching and learning: clearly stated outcomes and assessment criteria, organised and relevant content, and appropriate teaching strategies. In a very practical sense, it can provide a script for the trainer to follow and, therefore, is highly effective for novice trainers as well as for experienced trainers teaching new material. In short, a session plan acts as a guide and keeps the instructor on track during the presentation; ensures the teacher meets the instructional objectives and as such adds to professionalism, ensures that important teaching points are not overlooked, and ensures more rapid progress (instead of rambling on) and more satisfaction. However, Blunden (1997) provides a most useful piece of advice about written session plans when he states that we should

> Always remember that the plan is there to serve you — don't stick slavishly to it if you feel it is not working — be prepared to be flexible and change to suit your learners. This may mean adjusting the time allocation and activities to come. Very occasionally you may even have to abandon the whole plan and *go with the group*! (p. 41)

Flexibility is of utmost importance when teaching in a dynamic environment, such as aviation, because at times situations may present themselves where you can take advantage of a spontaneous occurrence to make a point more vividly. For example, if en route to the training area the flight instructor noticed that cumulonimbus (CB) clouds were forming, it would be a good idea to take advantage of this phenomenon to describe and explain the conditions under which CBs develop and point out their hazard.

There are no strict rules governing what should go into a session plan but there are useful and well-tried guidelines. Many experienced trainers would agree that a session plan should include:

- the objective(s) or learning outcome(s);
- the time breakdown for segments of the session;

- an introduction which should be motivating and show relevance;
- the development of the topic, which provides the order of delivery, including student activity where appropriate (sufficient notes to jog your memory);
- a conclusion, reviewing appropriate key points (preferably through questioning) and connecting to subsequent training, if appropriate;
- assessment of learning (questions and answers); and
- resources and equipment needed (adapted from Blunden, 1997, p. 40).

Organising Resources

The use of teaching and learning resources is taken for granted by most training participants. As a participant we even come to be able to judge when resources are used ineffectively. For example, "death by overhead" is a common complaint made by participants forced to endure endless amounts of information presented using this often over-utilised teaching aid. As a trainer, there are a number of key questions to be asked about this important aspect of session planning:

- What is a teaching resource?
- What are the characteristics of a good resource?
- What things should a teacher think about when planning to use resources in the classroom?
- Why do teachers use resources?

What is a resource? The word "resource" is used to describe anything in the classroom that helps teaching and learning to take place. Resources are often divided into two groups — human resources and material resources.

Human resources The most important human resources are of course the participants and the teacher. As discussed in previous chapters on how adults learn, the experiences of the participants can be seen as a resource upon which to draw in the classroom. Using this experience and starting with what the participants know is seen as a useful way of actively engaging the learners in the learning. Classroom methods that put students and their activities at the centre of the learning are known to help students learn by giving them an active role in the classroom, allowing them to show their skills, interests and personalities. The theories behind these aspects of adult learning were discussed in some detail in Chapters 1 and 2.

Similarly, most trainers have a wealth of experience to share with participants. A student-centred teaching and learning environment does not mean that the expertise of the trainer is not a crucial ingredient in effective learning. Indeed, most participants would feel less than satisfied with a teaching and learning experience that did not involve significant teacher input. It is only when this type of input is the only valued aspect of the learning that the participatory atmosphere is lost.

Another type of human resource can be found in the wider context of the industry. Guest presenters chosen for their expertise in a particular area of industry practice can provide stimulating input. Additionally, visits to actual workplaces can provide access to industry experts. Participants can gain insights into current industry practices and issues and take the opportunity to network with future employers or their representatives.

Material resources Material resources are anything the teacher or students bring into the classroom to aid teaching and learning. Pictures, books, videos, overhead transparencies, maps, model aircraft, instruments, engine parts, slides and computers are all examples of material resources. If used effectively, these resources can make a lesson more interesting and varied.

Knox (1986) provides the following list of suggestions of the variety of purposes that material resources might serve:

- to help learners become more independent in future learning projects;
- to facilitate interaction among participants around a common theme and guidelines;
- to present a clear explanation of a complex set of relationships.
- to provide an overview of an unfamiliar topic, which may also serve as a checklist for subsequent review;
- to facilitate and encourage application with handouts, for example.
- to help learners visualise;
- to allow individualisation and self-pacing, especially for learners who need remedial help or want to progress faster in special directions; and
- to present memorable examples from outside the learning setting (p. 109).

Clearly, resources must be carefully chosen. They should only be used if they are suitable for the learning tasks planned for the lesson and the needs of the learners.

Choosing the right resource The points set out below offer some criteria for choosing or making resources.

- *The learners* It is important to think about the learning group when choosing resources and activities. For example, think about the age, ability, and experience of the learners. Is the resource suitable for their stage of learning? Will it help them to extend their knowledge and their skills? Does the activity provide an interesting learning experience?
- *The purpose of the lesson* Resources must fit in with the aim of the lesson. Beware of finding one favourite resource, for example a video, and using it too much or trying to build the lesson around it. There may be better ways to achieve the learning outcomes of the lesson. Instructional aids are exactly what the term implies; they are resources to assist or aid teaching and learning.
- *Be selective and know the resources* Become familiar with resources before using them in the classroom. In this way it will be possible to make sure that none of them contain sexist language, cultural bias or anything else that might offend learners. Material that is out of date, contains mistakes, or is difficult to use will also make the learning less effective. A poor resource can be worse than no resource at all!
- *Preparation and practice* It is always a good idea to make sure resources are easy to access and use. Remember to book equipment if necessary and also to make sure it is functional. There is nothing more frustrating for a teacher than to have an interesting resource-based lesson planned, but no equipment to deliver the lesson.

Cultural Inclusion

All training contexts reflect the wider community. Therefore, it is important that the activities and resources used in the training context do not discriminate against gender, race, culture, religion, sexual preference, lifestyle or physical ability. Some key questions to ask about the resources you utilise in your training sessions are:

- Does the material (print, graphics, video content, etc.) represent people from the wider community?
- Do the resources and activities reinforce stereotypes?
- Can participants see their lifestyles and experiences reflected in classroom activities and language?

In reality, trainers are often restricted by the availability of inclusive resources. Sometimes a key text may have been in use for many years and written in less enlightened times when sexist language, for example, had not been highlighted as an issue. This can make the use of such materials difficult in a modern training context. However, if trainers are aware of these issues and their potential to negatively impact on the teaching and learning experience, they are better equipped to handle problems when they arise. For example, some resources may be introduced with a brief disclaimer, others can be subjected to the critical analysis of learners as part of their use. In these ways, training materials can be used to enrich and reinforce the learning experience of *all* learners.

Conclusion

This chapter has provided an overview of key aspects of the process for planning and preparing for instruction. It has discussed some of the key models of program planning and then illustrated a useful way to approach program planning. This includes firstly considering the social, organisational and public policy and industry contexts and then considering four stages of program planning: needs analysis, program design, implementing training, and evaluating training. This chapter has provided guidance for beginner trainers and educators on each of these stages and the issues to consider in implementation.

Program planning, as Newman (1995) suggests, is a creative process. There are no set rules that each trainer can easily follow that will work in every situation. The discussion above has hopefully provided a helpful guide to stimulate thinking and action. The hard part — implementation — has to happen for each trainer, taking into account the specific context in which they are training.

References

Australian National Training Authority (1999). *Australian national competency standards for private and commercial pilots*. Canberra: Airservices Australia.

Blunden, R. (Ed.). (1997). *Teaching and learning in vocational education and training*. Katoomba, Australia: Social Science Press.

Boone, E. (1985). *Developing programs in adult education.* Englewood Cliffs, NJ: Prentice Hall.

Brookfield, S. (1986). *Understanding and facilitating adult learning: A comprehensive analysis of principles and effective practices.* San Francisco: Jossey-Bass.

Cantor, J.A. (1992). *Delivering instruction to adult learners.* Toronto, Canada: Wall & Emerson.

Cole, P.G., & Chan, L.K.S. (1994). *Teaching principles and practice* (2nd ed.). Sydney, Australia: Prentice Hall.

Cranton, P. (1992). *Working with adult learners.* Toronto, Canada: Wall & Emerson.

Daines, J., Daines, C., & Graham, B. (1995). *Adult learning adult teaching* Chippenham, UK: Anthony Rowe.

Foley, G. (Ed.). (1995). *Understanding adult education and training.* Sydney: Allen and Unwin.

Freire, P. (1972). *Pedagogy of the oppressed.* Harmondsworth, UK: Penguin.

Jarvis, P. (1988). *Adult and continuing education: Theory and practice.* London: Kogan Page.

Jarvis, P. (1995). *Adult and continuing education: Theory and practice* (2nd ed.). London: Routledge.

Kemp, J.E. (1985). *The instructional design process.* New York: Harper & Row.

Knowles, M.S. (1970). *The modern practice of adult education: Andragogy versus pedagogy.* New York: Association Press.

Knox, A.B. (1986). *Helping adults learn.* San Francisco: Jossey Bass.

Lehrer, H.H. (1993). Instructional design and curriculum development in aviation. In R. Telfer (Ed.). *Aviation instruction and training* (pp. 271-290). Aldershot, UK: Ashgate.

McLeod, N. (2001). *Training design in aviation.* Aldershot, UK: Ashgate.

Newman, M. (1995). Program development in adult education and training. In G. Foley (Ed.), *Understanding adult education and training* (pp. 54-74). Sydney, Australia: Allen and Unwin.

Pithers, R.T. (1998). *Improving learning through effective training.* Katoomba, Australia: Social Science Press.

Rogers, A.R. (1998). *Teaching adults.* Buckingham, UK: Open University Press.

Smith, E., & Keating, J. (1997). *Making sense of training reform and Competency Based Training.* Wentworth Falls, Australia: Social Science Press.

Transport Canada (1996). *Human Factors for aviation: Instructor's guide.* Ottawa, Canada: Author.

Tyler, R.W. (1949). *Basic principles of curriculum and instruction.* Chicago: University of Chicago Press.

8 Facilitating Critical Thinking and Reflective Practice

Irene Henley and Jayne Bye

Introduction

The acquisition of complex skills specific to aviation is obviously a key aim of aviation instruction (strategies for this type of instruction are discussed in Chapters 9-14). However, many educators of adults see the benefits of providing adult learners with a larger framework for thinking about their training and work practice so that they will benefit more fully from their experiences. For many, the skills of *critical thinking* and *reflective practice* provide this framework.

In this chapter, the essential role of critical thinking and reflective practice is addressed and strategies to promote the development of these skills are proposed. Although these two key concepts are separate, it seems useful to look at them together in this chapter since they are very closely related and are both frameworks for thinking beyond what are generally seen as the immediate concerns of most teaching and learning. Further, when combined, these two concepts lead to a powerful learning tool known as *critical reflection*, which will also be discussed in this chapter.

The concept of critical thinking will be examined from different viewpoints since it is generally agreed that the term is used in different ways. From definitions provided, it will be seen that some uses of the term refer to processes that can be learned to aid creative problem solving and decision making. In this way, critical thinking is seen as a useful skill to be applied in a range of settings. In other uses of the term, critical thinking is seen as a tool for examining and questioning the more global assumptions under which we operate. This rather more politicised application of the term sees it imbued with the power to fundamentally alter our perceptions about our actions in the world of work and beyond.

Reflective practice is also a term that is used in a number of ways. Descriptions of reflective activity labelled as technical, practical and critical will be examined with special reference to how beginning teachers and trainers can develop their reflective capacities and those of their students to achieve a range of goals.

The potential advantages for use of both critical thinking and reflective practice in the aviation context will also be specifically discussed. Suggestions are made as to how instructors can help students become reflective practitioners through reciprocal *reflection-in-action* combined with *reflection-on-action* and the use of reflective journals in order to promote critical thinking and self-assessment, and to encourage greater self-regulation or autonomy in students. However, it is important to note that these skills should not be taught in isolation or as separate entities. Students should be taught these skills and strategies for use in conjunction with factual and conceptual knowledge and with other practical skills.

Traditional Aviation Instruction

Traditional instruction in the aviation context usually follows an apprenticeship or coaching model and, in so doing, aspires to provide students with insight, skills, attitudes, procedures and techniques derived from a pre-existing body of knowledge and from the practical experience gained by instructors as practitioners in the aviation industry. However, the conventional approach based on telling, demonstrating, and imitating often inhibits the critical analysis, decision making, and self-directed growth of students, and thereby fails to promote their full professional development.

Professionals in the aviation industry agree that students must acquire the skills necessary to make reasoned decisions in unfamiliar situations, deal with complex, ill-defined problems efficiently, maintain situational awareness in a dynamic environment, adapt to change, and work productively as team members (Diehl, 1991; Hackman, 1993; Johnston, 1993; Smith-Jentsch, Zeisig, Cannon-Bowers, & Salas, 1997). These skills are essential because problems of real-world practice seldom present themselves as well-formed structures. Instead, they tend to present themselves as murky, indeterminate situations (Michelson, 1996; Schön, 1987). Therefore, to operate safely in a complex and ever-changing environment, such as aviation, professionals need to become *reflective practitioners* who can define problems, make decisions consciously and rationally and continuously monitor and evaluate their situation. They need to be professionals who "actively reflect" on their performance and on the result of their actions. Indeed, Reason (1997) has identified *ongoing reflection* as an essential component of a "positive" aviation safety culture. However, critical analysis, effective decision making, reflective judgement, and self-evaluation are usually the product of extensive practical experience (Boud, Keogh, & Walker, 1985; Brookfield, 1993b; Kolb, 1984).

Consequently, our task as aviation educators is to find ways of assisting trainees so that they will benefit more fully from each of their experiences throughout their training. Further, it should be stressed that experience is not just what happens to us; more importantly, it is what we *think about* what happens to us (Machado, 1993). Brookfield (1996b) also adds that "simply having experiences does not imply that they are reflected on, understood or analysed critically.... Experiences are constructed *by* us as much as they happen *to* us" (pp. 35-36). Therefore, experience itself is not enough. As Criticos (1993) notes, the experience of itself has no real value, "what [is] valuable [is] the intellectual growth that follows the process of reflecting on the experience. Effective learning does not follow from a positive experience but from effective reflection" (p. 162). Consequently, an important goal for all aviation educators, be they training captains, flight instructors, trainers for air traffic controllers, maintenance engineers, or flight attendants, should be to prepare aviation professionals who are able to think critically and who will be reflective practitioners responsible for their own continual learning.

The aims of aviation education must include the development of critical reflective thinking if we want to help accelerate the development of the experiential base amongst trainees and facilitate the growth of the reflective practitioner. Our experience indicates that an effective way of developing competencies such as self-awareness, self-evaluation, and independent, critical thinking can be through reflection-in-action, that is, thinking about what we are doing while doing it (Schön, 1987), combined with reflection-on-action during debriefing and/or through the use of reflective journals (Anderson & Henley, 1997, 1998; Henley & Anderson, 1999; Henley, Anderson, & Schick, 1999; Henley, Anderson, & Wiggins, 1999; Henley, Turney, & Schick, 2001).

Definitions and Characteristics of Critical Thinking

Brookfield (1987, 1995), perhaps the most prominent adult educator writing about critical thinking, notes that the term critical thinking has been used in a number of ways in education and training. Generally, the term has been used to describe the development of logical reasoning abilities, the process of reflective judgement, the ability to surface assumptions and beliefs held by oneself and others, and it is also associated with the creation, use, and testing of meaning (Brookfield, 1994, 1995, 1996a, 1996b; Merriam & Caffarella, 1999; Norris & Ennis, 1990). Brookfield (1987) goes on to suggest his own definition of the concept; one that has been widely quoted in the adult education and training literature and also one that might provide a useful focus for thinking about this skill in the aviation context. He writes that

> When we become critical thinkers we develop an awareness of the assumptions under which we, and others, think and act. We learn to pay attention to the context in which our actions and ideas are generated. We become skeptical of quick fix solutions, of single answers to problems, and of claims to universal truth. We also become open to alternative ways of looking at and behaving in, the world. (p. ix)

Norris and Ennis (1990) also note that critical thinking includes an element of judgement and decision making. Critical thinkers evaluate statements (what we believe) and actions (what we do). Further, critical thinking is consciously directed, that is, it has a purpose. Norris and Ennis stress that including "a focus on decisions about *action* is intended to highlight the practical role that critical thinking can play in our lives" (p. 4). In a more recent addition to the vast literature on this concept, Romain (1997) describes the process as one that involves "becoming aware of our emotions and reflecting on them, identifying our values, assessing information and the authorities who provide it, analysing and clarifying language, imagining solutions to problems, evaluating alternative solutions, and assessing and producing arguments" (p. 1).

Several theorists have also listed characteristics and capacities that indicate critical thinking, such as

- being well-informed and seeking credible information or reasons in reaching conclusions;
- being able to judge the credibility of sources of information and of observations;
- displaying analytical and argumentative capacities;
- identifying contradictions in arguments;
- expressing curiosity, flexibility, open-mindedness, skepticism and honesty;

- withholding judgement when the evidence and reasons are insufficient;
- seriously considering other points of view;
- showing sensitivity to the feelings, level of knowledge, and degree of sophistication of others;
- taking a position and/or changing a position when the evidence and reasons are dependable and sufficient to do so; and
- demonstrating the capability to distinguish bias from reason and fact from opinion, and the ability to evaluate inferences (adapted from Brookfield, 1987, p. 11, and Norris & Ennis, 1990, p. 12).

It can be argued that all these aspects of critical thinking could be usefully applied in the aviation work environment and as such, critical thinking should have an important place in any type of training for the aviation context. This idea is supported by Hager (1992) in a discussion about why vocational education and training should teach critical thinking. He presents a convincing range of reasons for its importance, two of which include:

- people will be better equipped to compete effectively for educational opportunities, jobs, recognition, and rewards in our society; and
- good thinking is increasingly needed to perform effectively in the workplace (p. 150).

For these reasons alone, trainers in the aviation context might consider encouraging and indeed facilitating the development of the skills of critical thinking in their trainees.

Brookfield (1987) proposes a model of critical thinking that comprises five "commonly experienced phases" (p. 25):

1. A trigger event — some unexpected happening or situation prompts a sense of inner discomfort and confusion;
2. Appraisal — the second phase includes an examination of the situation, reflecting or "brooding" about our discomfort, and finding others who may have experienced or are experiencing a similar problem or situation;
3. Exploration — during this phase, we examine or "explore" new and different ways of explaining or adjusting to the unforeseen happening or situation that has led to our discomfort or confusion;

4. Developing alternative perspectives — in the fourth phase, we try new ways of thinking about the problem or situation and new ways of behaving and gradually we gain confidence in the new perspective;
5. Integration — in the final phase, we are capable of integrating new ways of thinking and/or behaving "into the fabric of our lives" (adapted from p. 27).

Similarly, Garrison (1991) suggests that critical thinking consists of five stages that are almost identical to many aeronautical decision making models (e.g., Jensen, 1995, Jensen, Adrion, & Maresh, 1986; Klein, Orasanu, Calderwood, & Zsambok, 1993; Kochan, Jensen, Chubb, & Hunter, 1997; O'Hare, 1992; Wiggins & O'Hare, 1993). These steps are: identify the problem, define the problem, explore ways of dealing with it, apply one of the strategies to the problem, and integrate the new perspective.

In more recent writings, Brookfield (1993a, 1994, 1996a, 1996b) has expanded his notion of critical thinking to include a questioning and replacement of commonly held assumptions and beliefs about ourself, others and also about social and political structures. Finally, in explaining how critical thinking and experience are related, Michelson (1996) writes that experience is "immediate but messy, [and this] messiness can be transcended only through … the application of sustained and self-conscious rational thought that, by encouraging distance and objectivity, will allow us to identify our prior assumptions, use our minds to critique them for validity and serviceability, and reconstruct them to make them more accurate, inclusive and empowering" (p. 439). Therefore, it is clear that critical thinking is informed by reflection; the two processes are closely intertwined.

Strategies to Encourage Critical Thinking

Regardless of the finer points of definitions proposed in the literature, most key theorists in the area agree that the value of critical thinking is in its application or, as Romain (1997) states, "you can't understand critical thinking just by learning a definition; you understand critical thinking by practicing it" (p. 1). Therefore, as educators, an important first step in facilitating critical thinking should be to examine how we personally engage in critical thinking. The following questions, proposed by Brookfield (1987), can be used to reflect on your own learning style and practice, since as

Brookfield stresses, "stock taking and reflective evaluation are a central part of critically informed action" (p. 79):

- How do you develop and maintain your motivation for various learning adventures?
- To what extent are extrinsic (social contact, job advancement) and intrinsic (innate fascination with learning, being tantalised by problem solving) motives interrelated?
- How do you integrate new ideas and insights generated by critical thinking into your existing analytical and interpretative frameworks?
- What general approaches do you take toward exploring new areas of knowledge?
- What learning strategies do you use? Do you follow your instinct and intuitions, or use trial-and-error methods, or problem solving strategies, or careful planning of short-, intermediate-, and long-term goals, or do you experiment with different approaches?
- What resources do you use? Are you field-dependent (strongly dependent on cues and directions provided by others, such as experts, friends or other learners, needing social supports to maintain your motivation) or field-independent (designing your own individual learning plans, preferring to work without the company of others) in your learning activities?
- To what extent do previous experiences act as hindrances or stimuli to your development of critical thinking?
- In what ways do you feel most comfortable entering the new and potentially frightening intellectual terrain represented by critical thinking?
- How do you chart your progress? Do you use objective measures to gauge success (i.e., standardised chart or comparison with experts), or do you use subjective measures (i.e., experiencing a feeling of self-confidence)?
- What are some of the difficulties you have encountered? and
- What was most enjoyable about your learning activities (adapted from pp. 83- 85).

These questions can also be adapted to facilitate the development of critical thinking in students. According to Brookfield (1987), finding answers to such questions can help individuals "become aware of idiosyncratic tendencies, preferences, biases, habits, blockages, and aptitudes in their personal learning styles" (p. 83). By assisting students in becoming more aware of their habitual learning styles, instructors can help them in exercising greater control in their learning and in making informed choices when selecting learning strategies (see Chapters 3 and 4 for more information on learning styles and learning strategies).

One of our roles as teachers should be to help students identify and challenge assumptions, and explore alternative ways of thinking and acting (Freire, 1973; Knowles, 1990; Kolb, 1984). During problem solving activities, we need to encourage lateral thinking and flexibility and genuinely accept diversity and divergence, so that students become aware that risk taking, spontaneity, openness, and critical analysis are truly valued. A healthy scepticism of final answers or quick solutions should also be supported (Brookfield, 1987, 1995). Brookfield (1987) also suggests the following strategies for facilitating critical thinking:

- *affirm critical thinkers' self-worth* — we need to ensure that searching and challenging questions do not threaten the fundamental integrity of individuals. As Brookfield remarks, "there is an uneasy tightrope to be walked in developing critical thinking in others; we must balance between respecting their integrity, so that they do not resist our efforts, and ensuring that sufficiently hard and challenging questions are asked to prompt them to scrutinize habitual assumptions" (p. 73);
- *listen attentively to critical thinkers* — active listening means being able to take on others' perspectives and understand things from their point of view. It also means being attentive to both verbal comments and nonverbal communication and being sensitive to group dynamics;
- *show that you support critical thinkers' efforts* — initially people may lack confidence, therefore it is essential to patiently listen to them and "allow people to *risk* failure without feeling that in doing so they have actually failed" (p. 75);
- *reflect and mirror critical thinkers' ideas and actions* — it can be extremely difficult and challenging to view our own thoughts and actions as an external observer. However, as Daloz (1986) stresses, "one of the more important aspects of the special mirror that mentors hold up to their students is its capacity to extend the students' self-awareness. To see oneself in new ways, from a range of different vantage points, is the chief way we distill what we are learning from challenges and supports of our world" (p. 234). Further, realising how others perceive our views and actions is an important first step in being able to unravel and understand

the complexities of conflicts and destructive misunderstanding (crucial elements of aviation resource management);

- *motivate people to think critically* — although it is crucial to encourage students to think critically, educators must also create an awareness of the range of potential risks involved in critically analysing and challenging established values and practices. Again, it must be stressed that there is a fine balance between critical analysis that fosters a sense of empowerment and leads to transformation, and critical thinking that can have disastrous consequences and become humiliating or devastating for the individual(s) involved;
- *regularly evaluate progress* — because critical thinking is a dynamic and ongoing process, it is essential to provide time for reflective evaluation or stock taking and to assist students in reviewing where their critical analysis has taken them. As will be seen in the section dealing with reflective practice, it is sometimes difficult to convince both instructors and students of the value of reflective evaluation. Yet, it is important to understand that "stock taking and reflective evaluation are a central part of critically informed action" (p. 79);
- *help critical thinkers create networks* — social networks serve to motivate their members and provide support when needed. Because identifying and challenging one's unquestioned assumptions, beliefs and actions, and exploring various alternatives involve elements of threat and risk taking, "the peer support provided by a group of others also trying to do this is a powerful psychological ballast to critical thinking efforts" (p. 79);
- *be critical teachers* — Paolo Freire (1970, 1973, 1994), a pre-eminent figure in experiential learning and global popular education, stresses that one of the crucial roles of effective educators is to raise the critical consciousness of students. According to Brookfield (1995), "critically reflective teaching happens when we identify and scrutinize the assumptions that undergird how we work. The most effective way to become aware of these assumptions is to view our practice from different perspectives. Seeing how we think and work through different lenses is the core process of reflective practice" (pp. xii-xiii). Critical teachers are essential catalysts in helping students focus on taken-for-granted assumptions and behavioural "givens", and in challenging uncritically accepted ideas, perspectives, and actions;
- *make people aware of how they learn critical thinking* — as was stressed in Chapter 4, helping students develop an awareness of their personal

styles and patterns of learning is a crucial element in raising one's consciousness and in coming to know one's critical thinking habits. This can be done by asking probing and challenging questions, such as those listed on page 209; and

- *model critical thinking* — role models can exert significant influence in shaping student attitudes and behaviour. Teachers who question their own assumptions, acknowledge ethical dilemmas in their own practice, refer to inconvenient theories, facts, and philosophies that they may have deliberately overlooked, and demonstrate an openness to alternative viewpoints encourage students to do likewise (Brookfield, 1995). Some of the characteristics of good modelers include consistency, openness, communicativeness, honesty, and accessibility. By displaying such attributes, teachers help students gain a clearer picture of someone who is a critical thinker (adapted from Brookfield, 1987, pp. 71-88).

While most agree on the potential for personal growth through critical thinking, it should be noted that several theorists concede that some transformation in perspective and behaviour can, at times, be effected without conscious critical thinking and reflection (Mezirow, 1998; Taylor, 1994).

Definitions of Reflective Practice

Foley (1995) notes that in the past two decades, there has been a growing interest in the concept of reflection in both education and training. A number of influential theorists have discussed and described the process of reflection as a tool for self-development for a range of professional practitioners, including educators and trainers of adults. Yet, reflection is a term that has long been associated with learning. In fact, discussion of its importance to the learning process dates back to the work of Dewey in the 1930s. According to Dewey (1933),

> *Reflective* thinking, in distinction from other operations to which we apply the name of thought, involves (1) a state of doubt, hesitation, perplexity, mental difficulty, in which thinking originates, and (2) an act of searching, hunting, inquiring, to find material that will resolve the doubt, settle and dispose of perplexity. (cited in Loughran, 1996, p. 4)

Since the 1930s, many influential theorists have added to this understanding of the process of reflection, while others have sought to differentiate and add emphasis to different aspects of the process. For example, Boud et al. (1988), in a response to what they consider to be the limitations of Dewey's conception of the process as being "rational and controlled" (p. 21), emphasise the affective aspects of learning through reflection in their model of the process. They indicate that "reflection is a form of response of the learner to experience" (p. 18) and it involves the "whole person". Their model of reflection in the learning process is based mainly on Kolb's (1976, 1984) learning cycle and includes two main components: the experience and the reflective activity based upon that experience. This reflective activity involves taking stock or making sense of what happened (returning to the experience), attending to both negative and positive feelings experienced during the learning process or the situation that was encountered, and finally, re-evaluating the experience in order to appropriate new knowledge and skills to previous conceptual frameworks and, most importantly, to guide future action and to realise the value of their own experiences (these stages in the reflective process will be examined further later on in this chapter).

Perhaps one reason why theorists are unable to come to a consensus on a definition of the reflective process is because of its complexity. As Rogers (1998) states, reflection "involves making a judgement on experience, assessing it in the light of some other standard which is drawn from other experience, either one's own or other people's experience. It means trying to determine how to explain the experience, to assess in what ways the experience could have been different" (p. 107).

Once again, despite the fact that definitions of the process are clearly not set, the perceived value of reflective practice for teachers and trainers continues to grow. This is reflected in the fact that guidelines for developing reflective skills are seen as central to much teacher education. As Loughran (1996) points out,

> one aspect of teacher education that continually receives attention in both curriculum and research is the way teachers think about their practice. Since at least the time of Dewey, such thinking about practice has been termed reflection and in teacher education courses there has been a focus on developing reflective practitioners.... One reason for this is the perceived common-sense link between reflection and learning, hence the value of its use in teaching and teacher education. (p. 3)

Merriam and Caffarella (1999), drawing on the works of Cervero (1988), Osterman and Kottkamp (1993), Peters (1991) and Schön (1983, 1987), state that three major assumptions underpin the process of reflective practice:

- *Assumption One* — individuals involved in reflective practice are committed to both problem finding and problem solving, however, real-world problems are usually murky and ill-defined. Therefore, we need to be open to discovering new possibilities or different ways of looking at the problems we encounter;
- *Assumption Two* — because reflective practice means making judgements about what actions will be taken and these actions usually involve seeking change in ourselves, other people, or in systems, there is an ethical dimension to reflective practice; and
- *Assumption Three* — reflective practice results in some form of action; it is purposeful. In short, it should result in the most thoughtful and useful solutions to practice problems (adapted from Merriam & Caffarella, 1999, p. 233).

Different Types of Reflection

In addition to the noted complexity of the reflective process itself, theorists also discuss a range of types of reflection. In an illuminating survey of the literature, James (1997) draws upon the work of several theorists in bringing together the following overview of the major distinctions between three different types of reflection: technical, practical, and critical. Briefly, James writes that

- *technical reflection* is concerned with "know-how" and "know-what" rather than "why" questions and seeks to promote efficient and effective action through strategies and skills which "work";
- *practical reflection* is concerned with asking "what ought I to do?" and involves interpreting the meaning of the situation, in order to act rationally and morally *with* people, rather than *on* them in the manipulative fashion of a more instrumental approach; and
- *critical reflection* is concerned with the development of self-regulation or autonomy and responsibility through self-reflection and exploring those cultural ideologies or beliefs, values, and assumptions which are limiting to ourselves and others. Such reflection may lead to changes in

> attitudes, values and practices, possibly resulting in the development of strategic actions to modify oneself and/or one's social world. For example, critical reflection might occur on the origins, purposes, and consequences of teaching practices, challenging taken-for-granted assumptions with reference to democratic values and a concern for justice and social equity (adapted from p. 93).

Critical Reflection

In the first part of this chapter, two important concepts, critical thinking and reflection, have been introduced and briefly discussed. As mentioned before, these skills can and do stand alone as useful strategies to increase the quality of student learning and teacher effectiveness. However, combined, these concepts lead to a powerful learning tool known as *critical reflection*. In the discussion above, three types of reflection were described. Yet, it is generally conceded that it is the last of these three, critical reflection which leads to the most powerful learning for both teachers and learners (Kolb, 1984).

Cranton (1992) describes the stages in the critical reflection process in the following way:

> The reflection process begins with a new awareness of the assumptions that have just been challenged, ranging from a simple, "Oh, I never thought of it that way", to a crisis in the individual's life. Awareness of the assumptions may be followed by an examination of them — or making them explicit, thinking about them, and reflecting on them.
>
> If the process is fostered by an educator (or if the learner possesses strong self-analytical skills), the next step will be to examine first the sources and then the consequences of the assumptions.... This kind of reflection leads naturally to considering whether or not the assumptions are valid. The questioning of previously accepted assumptions comprises the process of critical reflection, and when an individual questions his or her own assumptions, it is critical self-reflection. (pp. 148-149)

Brookfield (1995) agrees, underlining that "recognizing the discrepancy between what is and what should be is often the beginning of the critical journey" (p. 29). Brookfield (1990) also suggests that the process of critical reflection comprises three interrelated phases:

- identifying assumptions that underlie our thoughts and actions;
- scrutinizing the accuracy and validity of these in terms of how they connect to, or are discrepant with, our experience of reality (frequently through comparing our experience with others in similar contexts); and
- reconstituting these assumptions to make them more inclusive and integrative (p. 177).

Boud et al. (1988) also have identified important stages in the critical reflective process: *returning to experience*, *attending to feelings*, and *re-evaluating experience* (p. 26).

Returning to Experience

The first stage in the reflective process, *returning to experience*, is the recollection of salient aspects of the experience, or the recounting of what happened. This phase includes a narration or a recapturing the experience in as much detail and richness as possible. Oftentimes, however, this phase remains strictly factual and is devoid of any reference to personal concerns, attitudes or emotions and does not reveal the student's understanding of the situation or what was learned from the experience (Boud et al., 1988). Nevertheless, this stage is useful since it serves to "contextualise" and clarify what happened, and it provides students with a better understanding of what took place because as the experience is reviewed chronologically, details begin to emerge which may have been ignored at the time of the experience. Boud et al. stress the importance of "replaying the experience in the mind's eye, to observe the event as it has happened and to notice exactly what occurred and one's reactions to it in all its elements" (p. 27).

The authors also suggest that it may be helpful to commit the description to paper or to describe it to others because the narration helps students focus on what actually happened, clarify personal perceptions, and extract from it what is important. Writing a description of the event or reporting it to others also provides a check on whether the description is free from judgements because judgements "tend to cloud our recollections and may blind us to some of the features which we may need to reassess" (p. 28). An added benefit of writing the description, as Machado (1993) notes, is that the "descriptive entry provides fodder for post-flight reflection" (p. 38).

Yet, reflection is more than simply "thinking about" a host of events. As mentioned previously, reflection has a purpose; it should be pursued with intent. Reflection involves the logical and analytical thinking needed to resolve a problem in a rational manner. Consequently, to encourage students to develop their reflective skills to a higher level, the instructor needs to prompt students to go beyond the description (*what*) to consider *how, why, so what,* and *what if?*

The task of the instructor is to assist and encourage students to become active receivers, that is, to seek meaning from previous experience and incorporate the new information into existing knowledge (Brookfield, 1987, 1993b; Kolb, 1984). By asking prompting questions that promote reflection, students begin to realise the importance of context, that is, students come to the realisation that theory is not isolated from practice. However, as mentioned above, during this initial reflective process both instructors and students need to refrain from making judgements. Yet, what can emerge in the descriptive process or through prompting questions "is the observation of judgements and interpretations which took place at the time of the experience itself or shortly thereafter" (Boud et al., 1988, p. 28). This should then lead to the observation of feelings experienced during the event.

Attending to Feelings

Developing an awareness of emotions, or *attending to feelings*, is the second stage in becoming reflective. Students need to consider the affective dimension of their experiences. Reflection includes the need to emphasise the positive feelings and remove impediments or obstructing feelings in order to examine the experience more objectively and more thoroughly without blinders. On occasion our emotions can override our lucidity to such an extent that we react irrationally or with blurred perceptions. Negative feelings may have obstructed an appropriate response to the event resulting in a less effective or inflexible response. On the other hand, our emotions can be positive ones, thus increasing our confidence and giving us the drive which leads to optimal performance or they can help us focus more sharply and provide the impetus to persist during challenging or emergency situations.

In re-examining the event, students can realise how they were feeling and re-examine what prompted those feelings. They should also be encouraged to examine how their feelings may have affected their action or their response to the situation. This stage helps to clarify the student's personal perception of the experience. The student is allowed to stand back from the experience and review it from a different perspective or in a wider context. "Learners who do not observe this affective dimension of their experience may undermine the value of their reflections by restricting them to one aspect of their response to the world around them, thus placing artificial barriers on their response to experience" (Boud et al., 1988, pp. 28-29). Therefore, the instructor needs to encourage students to address the affective domain and to integrate elements of description, observation, analysis, interpretation and introspection as part of the reflective process. The student's experiences, concerns, and emotional responses should all contribute to the reflection.

Instructors should also encourage students to accentuate the positive, not only the negative aspects of their experience. Positive feelings can be enhanced by focusing on what was satisfying and stimulating about the experience or by reviewing situations in which students were successful and in which they felt good about themselves and their performance. Boud et al. (1998) stress that "unless we believe in ourselves and our own capabilities we can constrain ourselves to such an extent that we deny ourselves learning opportunities and fail to extract what is available to us in any given situation" (p. 29).

At times, when negative feelings are strong, students may repress or distort them, however, such a reaction "does not remove the potentially debilitating effects that they have on our mental processes" (Boud et al., 1988, p. 30). Strong emotional reactions need to be addressed and resolved, however, this needs to be handled extremely carefully and sensitively. Strong negative feelings need to be discharged in a way that will remove their undesirable influence and will facilitate future learning and that can only be done in a supportive, non-threatening environment. Again, Boud et al. suggest that some people may feel more at ease expressing such feelings through forms of writing: "Through such approaches learners can be freed from the mental bonds which were acting as constraints on them and they may be able to respond freely, flexibly and creatively once again" (p. 30). The main aim of attending to feelings is to draw attention to the role of emotions in the learning process and to be aware of the importance of the affective domain in critical reflective practice.

Re-Evaluating Experience

The third stage, *re-evaluating experience* (what does it mean? what should I do next time?), is that of consolidation and application. According to Boud et al. (1988), this final stage includes four distinguishable aspects that can contribute to reflection and enhance its outcome: association, integration, validation and appropriation. Association refers to the integration of new data to that which is already known; integration refers to the seeking of relationships among the data; validation is used to determine the authenticity of the ideas and feelings which were experienced; and appropriation means making the new knowledge one's own (p. 30).

Instructors can help students achieve this final stage in the reflective process through supportive questioning. They can act as a sounding board to help students clarify their understanding, their interpretation and their intentions. Re-examining the experience and thinking through the implications are essential to synthesising new knowledge and skills with previous conceptual frameworks in order to guide future action (Boud et al., 1988). In short, instructors should encourage students to describe the event and then reflect ("*look back*") on what they have done, and finally they should be urged to "*look ahead*" to what they should do the next time. In addition, students should be encouraged to acknowledge the reasons for their actions and consider how they will approach similar tasks in the future. Reflection, in essence, is proactive; students are encouraged to reflect not so much to dwell on the past, but to guide future action. Indeed, the main purpose of the reflective effort is to improve practice. In the final stage of reflection, students should be able to describe and analyse a learning situation, gather and evaluate information regarding the event, generate several alternative solutions, look at their potential implications and finally, integrate all of the information into a tempered conclusion or solution for the problem identified and appropriate the new knowledge for future application.

Finally, it is important to reiterate that critical reflection is often a difficult learning process in which to engage because to reflect critically we must go beyond simply thinking about the experience; we must stand outside ourselves and honestly "examine the underlying beliefs and assumptions that affect how we make sense of the experience" (Merriam & Caffarella, 1999, p. 328). And, critically questioning one's assumptions can be challenging, and sometimes threatening to a teacher's or a learner's sense of self. Further, according to Brookfield (1995), very few of us can critically reflect on our practice alone, because "our most influential assumptions are too close to us to be seen clearly by an act of will" (p. 29). Most of us need others who will reflect back to us an accurate picture of how our most deeply held values and beliefs influence the way we are and the way we perform our work. However, prompting students to review *what* they did and *why* they did it can, at times, be quite uncomfortable because it involves risks to ego integrity. Therefore, critical reflection, when used as a teaching and learning tool must be facilitated with care and in a supportive environment (Brookfield, 1994, 1995, 1996b; Knowles, 1990).

Becoming a Critically Reflective Teacher

Reflection is clearly an important part of the learning process but it is also an essential part of professional practice. Yet, the various conceptualisations of reflective practice that have emerged over the years have resulted in the fact that there is no simple formula to follow in becoming a critically reflective practitioner. How practitioners develop the skills of critical reflection will depend upon how they perceive their teaching role, and how they understand the goals of their practice and the purposes for which they direct reflection. However, several theorists and researchers offer worthwhile suggestions that may prove useful to practitioners interested in developing their critical reflective skills.

According to Brookfield (1995), the critically reflective process happens when teachers identify and examine the assumptions that underpin their practice through "four distinct, though interconnecting lenses" (p. xiii). Each of the lenses proposed by Brookfield can illuminate a different part of our teaching practice and alert us to distorted or incomplete aspects of our assumptions and behaviours that may need further examination which could eventually lead to changes in our way of thinking or in our behaviour:

1. The lens provided by autobiographical reflection — examining our autobiographies as learners and as teachers is an important source of insight into our practice. Firstly, our experiences as learners are likely to have profound and long-lasting influences on how we teach. "We may think we're teaching according to a widely accepted curricular or pedagogic model, only to find, on reflection, that the foundations of our practice have been laid in our autobiographies as learners. In the face of crises or ambiguities, we fall back instinctively on memories from our times as learners to guide us in our responses" (p. 31). Under close scrutiny, we may realise that teaching preferences (e.g., favouring group work over lectures) that seem logical or instinctual can often be traced back to learning situations in which we felt valued and inspired or demeaned and demotivated as learners.

   Secondly, analysing our autobiographies as teachers allows us to become aware of taken-for-granted assumptions and instinctive reasonings that frame how we work and permits us to start testing their accuracy and validity. Some of the tools we can use to reflect on our practice include concept mapping, teaching logs, videotaping, reflective journals, and self-administered inventories to identify our philosophical and methodological orientations (Anderson & Henley, 1997; Conti, 1998; Deshler, 1990; Henley, Anderson, & Schick, 1999; Henley, Turney, & Schick, 2001; Zinn, 1998);

2. The lens represented by our students' eyes — seeing ourselves through our students' eyes can be extremely illuminating and at times quite surprising, and it can help us teach more responsively. However, we need to be aware that understandably students may be reluctant to honestly comment on a teacher's performance, even under the cloak of anonymity. Therefore, it is essential to earn the trust of students in order for them to believe that we truly value critical reflection. This can be done in part by consistently modelling critical reflection through critical scrutiny of our own ideas, views, and actions;
3. The lens provided by our colleagues' perceptions and experiences — participating in critical conversations with our colleagues about our practice can unravel some of our unchallenged assumptions and behaviours. In our quest for critical clarity regarding our practice, our peers can serve as mirrors reflecting back to us images of our actions. Further, "as they describe their own experiences dealing with the same crises and dilemmas we face, we are able to check, reframe, and broaden our own theories of practice" (Brookfield, 1995, p. 35). Critical discussions with our peers can also be quite reassuring and help increase our self-confidence because we realise that what we thought were unique problems and idiosyncratic failures are shared by others in the teaching profession; and
4. The lens of literature — theoretical literature on our practice can provide multiple perspectives on familiar situations and help us understand that what we may have interpreted as signs of personal failings as teachers are actually inevitable consequences of the teaching process. Britzman (1991) notes that often beginning teachers accept the myth that everything depends on the teacher and when things go amiss, they blame themselves rather than reflect upon the complexity of pedagogical encounters. Reading a theoretical analysis can offer an alternative interpretative framework for situations we come across. For example, critical theoretical literature may help us realise that students' disinterest may be a "consequence of a system that forces students to study disconnected chunks of knowledge at a pace prescribed by curriculum councils and licensure bodies" (Brookfield, 1995, p. 37).

Educational theory and research can help teachers understand the consequences of their practice and prevent them from mistakenly blaming themselves for situations that are created by broader political processes. "This means that the emotional energy they would have spent on criticizing themselves for their supposed lack of animation, compassion, or creativity is channelled into working for change" (p. 37). Thus, the study of theoretical literature can be seen as a psychological survival necessity. However, getting practitioners to read theoretical literature can be difficult because many seem to think that educational theorists and researchers do not understand the reality of the actual teaching environment. However, there is an increasing body of literature emerging that is grounded in practitioners' concerns and experiences which could be of significant value to most practitioners (Bell, 1995; Bullough, Knowles, & Crow, 1992; Day, Calderhead, & Denicolo, 1993; Dollase, 1992; Grossman, 1990; Henley, 1987, 1995; Henry, Huntley, McKamey, & Harper, 1995; Horton & Freire, 1990; Huberman, 1994; Jalongo & Isenberg, 1995; Newman, 1990, 1991; Shor, 1992; Thomas, 1995). Although most of this literature does not deal specifically with aviation education, many of the experiences and critical reflections that are described are nevertheless applicable to the aviation training context (adapted from Brookfield, 1995, pp. 29-39).

Critical reflection on our practice through the lens proposed by Brookfield can assist us in our teaching in several ways:

- it helps us take informed actions;
- it helps us develop a rationale for practice, that is, it provides a foundational reference point that can serve as a guide to how we should act in unpredictable situations;
- it helps us avoid self-laceration;
- it grounds us emotionally;
- it enlivens our classroom because a teacher who models critical reflective thinking acts as a powerful catalyst for critical reflective thinking in students and as a result, the classes are more challenging, interesting, and stimulating for students (Osterman, 1990); and
- it increases democratic trust because when students are taken seriously and treated as adults, they learn that independence of thought is truly valued (adapted from Brookfield, 1995, pp. 22-26).

The Application of Reflective Practice in the Aviation Context

Recent research involving experienced airline captains suggests that one of the distinguishing factors between pilots who are successful in completing captaincy training and those who are unsuccessful may be self-regulation which includes metacognition through critical reflection (Beaumont, 1999, 2002). Briefly, self-regulation is defined as the "automatic and deliberate processes by which people control and direct their cognitions, emotions and behaviours in the pursuit of their goals" (Barone, Maddux, & Snyder, 1997, p. 303. Also see Chapter 2). Metacognition refers to the self-monitoring and conscious use of learning strategies (Borkowski, Carr, & Pressley, 1987; Borkowski, Carr, Rellinger, & Pressley, 1990. Also see Chapters 2 & 3).

Bandura (1997) argues that "the development of self-appraisal skills ... relies on growth of self-reflective metacognitive skills to evaluate the adequacy of one's self-assessment" (p. 115). Indeed, central to the process of effective self-evaluation is the concept of critical reflection. As seen earlier, this is a concept widely discussed in the literature of effective training practice and requires further elaboration here. It is generally agreed that there are two basic reflective processes central to reflective practice: *reflection-in-action* and *reflection-on-action* (Brockbank & McGill, 1998; Merriam & Caffarella, 1999; Schön, 1983). The theorist most closely associated with describing these two key processes is Donald A. Schön, perhaps one of the most well-known theorists in this area. In his influential work, *The reflective practitioner — How professionals think in action,* first published in 1983, Schön describes the powerful insights to be gained about practical and professional knowledge through understanding reflection-in-action and reflection-on-action (also see Schön, 1998). Indeed, Ferry and Ross-Gordon (1998), in their investigation of the links between experience and practice, found that educators who were reflective in their practice used both reflection-in-action and reflection-on-action to build their expertise.

Reflection-in-Action

Essentially, training in the aviation domain is premised on learning by doing. *Reflection-in-action* is a type of reflection about practice that occurs while the practitioner is working. For example, it is the process that takes place during training when an instructor analyses the student's performance, tries to discern what the student understands and offers instructions, feedback, advice and correctives. It is the practical basis of decision making practitioners use when drawing on their own informal theories about practice, that is, the knowledge that they have gained by carrying out their practice as distinct from knowledge that might be gained from reading about formal theories of practice. Schön (1998) describes reflection-in-action as recognising that "sometimes we think about what we are doing. Phrases like 'thinking on your feet', 'keeping your wits about you', and 'learning by doing' suggest not only that we can think about doing but that we can think about doing something while doing it" (p. 54). Similarly, Boud and Walker (1992) believe that "we experience as we reflect and we reflect as we experience" (p. 167). Boud and Walker (1990) suggest that this involves two interactions between the learner and the milieu: first, "*noticing*, by which a person becomes aware of the milieu, or particular things within it, and uses this for the focus of reflection" (p. 68), and second, "*intervening*, in which the person takes an initiative in the event" (p. 68). Boud and Walker (1992) add that "noticing the outcomes of the intervention and planning further intervention completes the cycle of reflection-in-action" (p. 168).

This level of self-awareness in relation to actions can lead to increased understanding of how effective practice occurs. In this way, practitioners are said to be able not only to learn more about effective practice but also monitor and increase their repertoire of responses to the varied situations a work context can create (Foley, 1995). In short, reflective practitioners are conducting continuous self-evaluations which feedback immediately into their practice. Schön (1998) notes that competent and experienced professionals use reflection-in-action as a regular part of their practice, although they may not often verbalise that they are involved in this process.

Initially, students learning practical skills are asked to replicate manoeuvres or tasks without fully understanding what is required to perform each task correctly. They often do not recognise all the cues needed to replicate the manoeuvre or the task demonstrated by the instructor. Therefore, a verbal description by the instructor of how a manoeuvre or a task is being performed and perceived is needed to provide the cues to essential features of the demonstration. However, once students have gained some experience, in order to maximise learning, instructors should also encourage their students to join them in the descriptive and reflective process. Instructors need to gradually step back and allow the student to develop the skills needed to reflect-in-action and to think and analyse what they are doing while performing a task. This will allow them to develop their skills of critical analysis and self-assessment (see Chapter 9 for further information on this process).

These are essential skills because most situations in aviation are unique and frequently fall outside the procedures learned. Consequently, there is often a need to improvise and to test the strategies being used. This requires ongoing reflection (Reason, 1997). Furthermore, the instructor's or the student's reflection on one's own or on the other's performance can often yield a description that highlights inaccurate cues or reveals subtle differences in perceptions or misunderstandings that need to be clarified. Unfortunately, reciprocal reflection-in-action is frequently neglected in training within the aviation domain. Too often, instructors tell students what they should be seeing and doing instead of asking the students to describe their perceptions and reflect on their actions. Yet, reflection-in-action by students is essential to ensure that they have an accurate picture of each of the manoeuvres or tasks to be performed.

Reflection-on-Action

Reflection-on-action is an activity that can occur after an event. Foley (1995) describes reflection-on-action as "a retrospective process, in which practitioners return to the experience, attend to their own and others' feelings about the experience, and re-evaluate the experience" (p. 47). This re-evaluation can lead to a questioning of what occurred and of the assumptions or preconceptions that may have guided decision making, and sometimes, can result in a new perspective on a practice, issue or situation. If this type of reflection is carried out in a systematic way, it becomes a powerful self-evaluation tool. Reflective journal writing (described below) is an effective way to record this process and can be used as a tool for teachers to improve their practice as well as a learning tool for students (Henley, Anderson, & Schick, 1999).

Most aviation educators have long recognised that one of the most important ways to enhance learning is to strengthen the link between the practical learning experience and the reflective activity that follows it. In flight training, for example, this link is normally planned to be achieved during the debriefing period at the end of each flight where the instructor and the student take the time to step back and assess the student's performance and reflect upon the flight in a critical way (Henley, Turney & Schick, 2001). The debriefing period provides an opportunity for both the instructor and the student to engage in critical reflection, whereby experiences are used for experience-based learning (Pearson & Smith, 1988, p. 70).

Unfortunately, too frequently because of time constraints, the debriefing phase is curtailed. Yet, the opportunity to engage in reflection is a crucial element of vocational training, including flight training. As Machado (1993) appropriately notes, "a flight unexamined is not worth logging" (p. 38). Therefore, to ensure reflection takes place after a learning event, students should be encouraged to actively participate in the debriefing and they should also be encouraged to write a reflective journal because journal writing is a way to structure self-monitoring, critical analysis and self-evaluation (Boud et al., 1988; Walker, 1988; Zeichner & Liston, 1987).

In addition, educational research suggests that learning can be augmented and can be more profound when students are encouraged to reflect on the learning event and exercise their critical judgement about the content and the processes of learning. When students are engaged in the learning process, they become more acutely aware of their preconceived ideas, biases, unquestioned assumptions and behaviours (Land & Hannafin, 1996). The benefits of this "consciousness raising" are that students learn to chart their development, identify barriers, patterns and strategies, challenge their own beliefs and avoid repeating the same mistakes (Freire, 1973, 1994). Reflection can also lead toward greater confidence and assertiveness or even to a change in attitude, perspective or priorities.

Reflective Journals

As mentioned above, writing reflective journals is another way to foster reflection-on-action and critical self-analysis because it allows students to reach a higher level of correlation between theory and practice; it allows them to make sense of their experiences through a critical examination of their perceptions of the event, their emotional response, and their cognitive processes (Brookfield, 1990; Stockhausen & Creedy, 1994; Walker, 1988). Through reflection on prior learning experiences, students can derive insights that will assist them in coming to terms with their approach to learning; this in turn, can guide future actions. Writing a reflective journal encourages students to be more aware of what they do, how they do it, and why they do it (Boud et al., 1988; Schön, 1983, 1998). Journal writing also allows them to identify useful problem solving strategies, as well as recognise their own strengths and weaknesses with regard to their understanding of content knowledge, procedures and practical skill development and application. The first step in becoming a reflective practitioner needs to incorporate self-awareness or metacognition.

Another aim of reflective journal writing is to develop an orientation toward open-mindedness and a willingness to accept responsibility for self-directed learning as well as to foster a keen sense of observation, critical thinking and reasoned analysis (Brookfield, 1995; Sparks-Langer, Simmons, Pasch, Colton, & Starko, 1990; Walker, 1988). In addition, reflective writing allows students to acknowledge more freely emotions and feelings that otherwise may form barriers to learning. Further, when reflective journals are incorporated into debriefing sessions, they can offer the instructor a unique insight into student learning as well as provide the impetus for deeper critical reflections. According to Stockhausen and Creedy (1994), journal writing promotes reflection-on-action because "it helps students capture the experience for later analysis, debate, reframing, further action or resurfacing theories in use" (p. 76).

However, it should be stressed that reflective writing does not just happen. Reflective writing is an active and effortful activity. It is a skill that needs to be developed, nurtured and refined. Although some students will eventually, through trial and error, develop insights on their own without the use of a reflective journal, the learning process can be aided and time to learn decreased through the use of reflective journals (Anderson & Henley, 1997, 1998; Henley & Anderson, 1999; Henley, Anderson, & Schick, 1999; Henley, Anderson, & Wiggins, 1999; Henley, Turney, & Schick, 2001; Stockhausen & Creedy, 1994; Walker, 1988). This is due, in part, to the skills and mental processes involved in writing that require students to contemplate and anchor relationships between perceptions as they occur (Boud et al., 1988; Loughran, 1996).

Initially, to work well reflective writing needs to involve a dialogue between the instructor and the student, as well as with other students (Brockbank & McGill, 1998; Walker, 1988). However, for reflective dialogue to take place a good rapport between the instructor and the students has to be established. Further, the students need to be the centre and the focus of the reflective dialogue, with the instructor adopting the role of facilitator (Brockbank & McGill, 1998; Smith, 1993).

There is no set formula or format to write reflective journals. However, it might be helpful to encourage students to organise their ideas into sections. When reviewing a training episode, an experience or a critical incident, students could be encouraged to answer questions or complete statements such as:

- What happened? – What did I do? – How did I do it? – Why did I do it?
- What puzzled me?
- How did I feel? – Why?
- I was frightened when …
- I got angry when …
- What did I enjoy or dislike?
- The most exciting experience …
- What did I learn from the discussion, reading, exercise?
- What should I have done differently? What specific areas do I need to improve?
- Which aspects were successful?
- What am I most or least confident about; and
- How does this relate to previous knowledge and experience and to future learning activities?

Facilitating the Use of Reflective Journals

Instructors can help students develop reflective skills while using reflective journals by adopting or adapting some of the following:

- give students some introductory advice on how to approach reflective journal writing;
- explain *why* reflection is an essential skill for professionals in the aviation industry. According to Walker (1988), reflection is an essential counterpart to experience;
- explain the relationship between reflective writing and learning (i.e., it allows the writer to tap valuable inner resources, to reflect calmly upon knowledge or insights that come from within);
- initially, provide students with specific assignments. Be precise about what you want. Ask them to describe what they did, how they felt, how they reacted, and what they might do differently next time;
- stress that a reflective journal is a personal document and ensure students are fully aware that there is no right or wrong way to keep the journal. It is an extension of each individual;
- encourage students to be spontaneous, to write freely, to say what they truly feel, to let it flow uncensored, in whatever order it comes. Once they have done their initial writing, they can reflect further on what they have written;

- students should feel free to try different methods or approaches so they can discover their own personal style of writing. Visual learners could be encouraged to use diagrams, concept maps, pictures or other material to illustrate their thoughts or feelings. Students should use the journal as a workbook where they can underline, circle, use different colours to highlight important aspects or to draw attention to significant ideas, concepts, or feelings;
- encourage students to be frank and honest when they describe their thoughts, feelings, worries, achievements. They also need to focus on things that are important to them, and encourage them to also include the "fun" or exceptional moments that they experience. Make the reflective activity interesting and enjoyable as well as challenging;
- encourage students to persevere even if they find reflective writing difficult at first. The effort will provide rich returns after a while. As Peters (1991) notes, "reflective practice is not always pleasant, but it is almost always rewarding" (p. 95);
- as students progress through their training, allow them to choose the learning event upon which to reflect. This encourages them to participate more fully in the learning process and to become more self-directed;
- although students may want to keep their journals private and confidential, they also should be made aware of the importance of reflective dialogue. Openness to share with others and to listen to others is an important aspect of reflective writing. During the reflective dialogue, students are confronted with the need not only to closely observe the learning of others, but also to take account of their own learning. Collaboration with others also offers the possibility of a further check on self-deception and allows students to analyse their own uncertainties and dissatisfactions with others who may share them. As Walker (1988) notes, "talking about one's ideas, thoughts and reflections will bring feedback that can help deepen them" (p. 56);
- during each briefing or debriefing, review students notes and writings. Discuss and address any questions they may have raised. Correct misconceptions they might have formed about a concept, a technique or a manoeuvre. However, it is important to avoid judgement and criticism. Use their reflections for further discussions on how a particular event they experienced might apply to other situations;
- insist on the importance of making journal writing part of their routine and encourage them to set aside a regular time each week to review previous entries, to reflect on their reflections. The continuing reflection

on what was written previously (reflection on reflection) is important in enhancing reflective skills (Brockbank & McGill, 1998). Some learners leave a blank page in their journal after each reflection for when they go back to reflect on their prior reflections; and

- use the students' reflections to help them see the progress they are making, not just in terms of their achievements and the improvements in their understanding and performance skills, but in terms of their enhanced critical reflective skills as well. Reflective journals provide a record of significant learning experiences that have taken place throughout the learning journey (adapted from Henley, Anderson, & Schick, 1999, pp. 57-58 and Walker, 1988, pp. 54-56).

The Importance of Constructive Feedback

Since writing a reflective journal is a skill, it is also imperative that instructors provide timely feedback to students to help them develop their reflective thinking. However, it must be stressed that the feedback needs to be non-threatening, non-judgemental, constructive, and supportive. It is also important to include questions that will allude to, or prompt students to make connections between theory and practice.

With appropriate feedback and prompting questions, students can move from simple description and unsubstantiated judgements to a greater frankness and open-mindedness, and thereby acquire the ability to link theory to practice and even demonstrate a willingness to search for alternatives. As students develop their ability to reflect, they can be challenged to question their understanding and their perceptions, thus becoming more critical. Through carefully scaffolded guidance from the instructor, the student can gradually progress to higher levels of critical reflection, eventually achieving self-directed and independent learning.

Barriers to Reflective Practice

There are several barriers that instructors may encounter in their attempt to encourage students to reflect critically on their learning activities and on their performance. As Hunt indicates in Chapter 5 of this book, students are generally unused to reflect on their learning experiences, therefore, many may lack reflective skills. Further, James (1999) warns that the process of critical reflection may not be easy for many learners and sometimes evokes "considerable stress, together with resistance and hostility towards the purveyors of new ideas. Treasured beliefs and deeply held values are rarely surrendered easily, instead, more usually, involving considerable emotional turmoil" (p. 329).

In addition, some students may not see the value of reflection, and more specifically, the worth of reflective journals. Young male students, especially, may be reluctant to engage in what they possibly view as simply keeping a "diary", which is not an activity in which "real men" normally engage (Henley, Anderson, & Schick, 1999). Therefore, instructors need to explain very clearly the aims and the stages of the reflective process, why reflection is an important skill for aviation professionals, and the purpose of using a reflective journal. In some instances, this may be problematic since many instructors may not have engaged in reflective practice before and may not have used reflective journals during their own training. Therefore, they may not be convinced of the value of the reflective process for practitioners. This could entail additional training for current instructors and it means that the trainers of instructors should be encouraged to use reflection-in-action and reflection-on-action as well as reflective journals during instructor training.

Another major obstacle is the additional time involved for both the student and the instructor. Students need to be willing to spend time to write in their journal. This can be a major stumbling block for students who may not be reflective learners by nature and for adult learners who lead active lives. Furthermore, the reflective process initially requires that instructors spend additional time discussing journal entries with their students. This can be especially problematic in flight training since many flight instructors are not paid for the time they spend on the ground with students. As a result, this could prove to be a major barrier in encouraging instructors to adopt the use of reflective journals with their students.

Some instructors have also expressed the fear that their students may unknowingly get them in trouble because their students' account or perception of the event could be incorrect or may have been misunderstood (Henley, Anderson, & Schick, 1999). However, if instructors understand that journals can be used to clarify such misunderstanding or distorted perceptions on the part of students, they may feel less threatened.

Some students may not want to admit their failures or their apprehension to their instructor because they feel that their instructor may be part of their problem. As mentioned previously, reflection is not easy since it involves having to admit the truth when we see it. It means being able to acknowledge mistakes and weaknesses. This can be very daunting to some students. Therefore, it is the responsibility of the instructor to create a positive learning atmosphere where students feel free to express their feelings and their apprehensions. However, as students mature and begin to define and accept their behaviour and their motives, they will find the process less intimidating and will begin to view new possibilities and accept new challenges. They will also become increasingly aware of the complexity of operating in the aviation environment, yet they will feel more prepared and start to gain more confidence in their practice.

Conclusion

This chapter has provided a number of definitions of critical thinking and reflective practice, but more importantly, it has also suggested a number of strategies that could be used by instructors to help students become critical thinkers and reflective practitioners. However, all the strategies are contextually variable. None can be easily standardised because they are all substantially affected by context. What should be evident is that encouraging critical thinking and reflective practice is an activity as complex as the processes of critical thinking and reflective practice themselves. Encouraging the development of such skills requires perceptive and confident instructors who can withstand criticism and disagreement and who are willing to question their own interpretations and points of view. As Brookfield (1987) notes, instructors "need to be skilled in a number of systematic instructional methods, but sceptical enough of the value of these to be able to abandon them when it seems appropriate" (p. 88).

This chapter has also suggested that reflective journals can provide students with a way of organising their ideas and of relating knowledge and experience. Reflection-in-action and reflection-on-action enable students to analyse and evaluate their own practice, adopting an analytical approach towards their learning. Reflection also makes students face themselves and accept the consequences of their actions. When students reflect, they become researchers in the practical context. They do not separate thinking from doing, and thus, reflection becomes the core of practice. It must be stressed that instructors play a crucial role in preparing students for the demands of professional practice and their feedback is of paramount importance in assisting students to become reflective practitioners. Through insightful critique, instructors can respond to students' accounts, providing alternative responses and interpretations to students' writing that challenge their views and beliefs, and encourage them to question and analyse their perspective or preconceptions. Finally, an essential attribute of critically reflective teachers is sensitivity to the broad range of responses elicited by the facilitation of the processes of critical thinking and reflective practice.

References

Anderson, P.M., & Henley, I. (1997). *Developing self-awareness through the use of reflective journals*. In R.S. Jensen & L. Rakovan (Eds.), *Proceedings of the Ninth International Symposium on Aviation Psychology* (pp. 1166-1171). Columbus, OH: The Ohio State University.

Anderson, P. & Henley, I. (1998). Using reflective journals to develop self-awareness and self-evaluation skills in ab-initio flight training. In D. Findahl (Ed.), *Elevating our profession: Flight Instructor Conference Proceedings* (pp. 13-22). Daytona Beach, FL: Embry-Riddle Aeronautical University.

Bandura, A. (1997). *Self-efficacy: The exercise of control*. New York: W.H. Freeman.

Barone, D.F., Maddux, J.E., & Snyder, C.R. (1997). *Social cognitive psychology: History and current domains*. New York: Plenum Press.

Beaumont, G. (1999). A construct of pilot performance. In R. Jensen, B. Cox, J.D. Callister, & R. Lavis (Eds.), *The Tenth International Symposium on Aviation Psychology* (pp. 1197-1202). Columbus, OH: Ohio State University.

Beaumont, G. (2002). *An investigation of the management of flight aspects of airline captain performance*. Unpublished doctoral dissertation, University of Western Sydney, Australia.

Bell, J. (1995). *Teachers talk about teaching: Coping with change in turbulent times*. Bristol, PA: Open University Press.

Borkowski, J.G., Carr, M., & Pressley, M. (1987). Spontaneous strategy use: Perspectives from metacognition theory. *Intelligence, 11*, 61-75.

Borkowski, J.G., Carr, M., Rellinger, E., & Pressley, M. (1990). Self-regulated cognition: Interdependence of metacognition, attributions, and self-esteem. In B.F. Jones & L. Idols (Eds.), *Dimensions of thinking and cognitive instruction* (pp. 53-92). Hillsdale, NJ: Lawrence Erlbaum Associates.

Boud, D., Keogh, R., & Walker, D. (1988). Promoting reflection in learning: A model. In D. Boud, R. Keogh, & D. Walker (Eds.), *Reflection: Turning experience into learning* (pp. 18-40). London: Kogan Page Ltd.

Boud, D., & Walker, D. (1990). Making the most of experience. *Studies in Continuing Education, 12*(2), 61-80.

Boud, D., & Walker, D. (1992). In the midst of experience: Developing a model to aid learners and facilitators. In J. Mulligan & C. Giffin (Eds.), *Empowerment through experiential learning: Explorations of good practice* (pp. 163-169). London: Kogan Page.

Britzman, D.P. (1991). *Practice makes practice: A critical study of learning to teach*. Albany, NY: State University of New York Press.

Brockbank, A., & McGill, I. (1998). *Facilitating reflective learning in higher education*. Buckingham, UK: The Society of Research into Higher Education & Open University Press.

Brookfield, S. (1987). *Developing critical thinkers*. San Fransisco: Jossey-Bass.

Brookfield, S. (1990). *The skillful teacher*. San Francisco: Jossey-Bass.

Brookfield, S. (1990). Using critical incidents to explore learners' assumptions. In J. Mezirow & Associates (Eds.), *Fostering critical reflection in adulthood: A guide to transformative and emancipatory learning* (pp. 177-193). San Francisco: Jossey-Bass.

Brookfield, S. (1993a). Self-directed learning, political clarity, and the critical practice of adult education. *Adult Education Quarterly, 43*(4), 227-242.

Brookfield, S. (1993b). Through the lens of learning: How the visceral experience of learning reframes teaching. In D. Boud, R. Cohen, & D. Walker (Eds.), *Using experience for learning* (pp. 21-32). Buckingham, UK: Society for Research into Higher Education and Open University Press.

Brookfield, S. (1994). Tales from the dark side: A phenomenography of adult critical reflection. *International Journal of Lifelong Education, 13*(3), 203-216.

Brookfield, S. (1995). *Becoming a critically reflective teacher.* San Fransisco: Jossey-Bass.

Brookfield, S. (1996a). Breaking the code: Engaging practitioners in critical analysis of adult educational literature. In R. Edwards, A Hanson, & R. Raggatt (Eds.), *Boundaries of adult learning* (pp. 57-81). New York: Routledge.

Brookfield, S. (1996b). Helping people learn what they do: Breaking dependence on experts. In D. Boud & N. Miller (Eds.), *Working with experience: Animating learning* (pp. 27-40). New York: Routledge.

Bullough, R.V., Knowles, J.G., & Crow, N.A. (1993). *Emerging as a teacher*. New York: Routledge.

Cervero, R.M. (1988). *Effective continuing education for professionals*. San Francisco: Jossey-Bass.

Conti, G.J. (1998). Identifying your teaching style. In M.W. Galbraith (Ed.), *Adult learning methods: A guide to effective instruction* (2nd ed., pp. 73-90). Malabar, FL: Krieger.

Cranton, P. (1992). *Working with adult learners*. Middleton: Wall & Emerson.

Criticos, C. (1993). Experiential learning and social transformation for a post-apartheid learning future. In D. Boud, R. Cohen, & D. Walker (Eds.), *Using experience for learning* (pp. 157-168). Buckingham, UK: Society for Research into Higher Education and Open University Press.

Daloz, L.A. (1986). *Effective teaching and mentoring: Realizing the transformational power of adult learning experiences*. San Francisco: Jossey-Bass.

Day, C., Calderhead, J., & Denicolo, P. (Eds.). (1993). *Insight into teacher thinking: Understanding professional development*. Bristol, PA: Falmer Press.

Deshler, D. (1990). Conceptual mapping: Drawing charts of the mind. In J. Mezirow & Associates (Eds.), *Fostering critical reflection in adulthood: A guide to transformative and emancipatory learning* (pp. 336-353). San Francisco: Jossey-Bass.

Diehl, A.E. (1991, Fall). Cockpit decision making. *FAA Aviation Safety Journal, 1*(4), 14-16.

Dollase, R.H. (1992). *Voices of beginning teachers*. New York: Teachers College Press.

Ferry, N., & Ross-Gordon, J. (1998). An inquiry into Schön's epistemology of practice: Exploring links between experience and reflective practice. *Adult Education Quarterly, 48*(2), 98-112.

Freire, P. (1970). *Pedagogy of the oppressed*. New York: Seabury Press.

Freire, P. (1973). *Education for critical consciousness*. New York: Seabury Press.

Freire, P. (1994). *Pedagogy of hope: Reliving pedagogy of the oppressed*. New York: Continuum.

Foley, G. (1995). Teaching adults. In G. Foley (Ed.), *Understanding adult education and training* (pp. 31-53). Sydney: Allen and Unwin.

Garrison, D.R. (1991). Critical thinking and adult education: A conceptual model for developing critical thinking in adult learners. *International Journal of Lifelong Education, 10*(4), 287-303.

Grossman, P.L. (1990). *The making of a teacher: Teacher knowledge and teacher education*. New York: Teachers College Press.

Hackman, R.J. (1993). Teams, leaders, and organizations: New directions for crew-oriented flight training. In E.L. Wiener, B.G. Kanki, & R.L. Helmreich (Eds.), *Cockpit resource management* (pp. 1-45). San Diego, CA: Academic Press.

Hager, P. (1992). Teaching for critical thinking. In A. Gonczi (Ed.), *Developing a competent workforce: Adult learning strategies for vocational educators and trainers* (pp. 149-166). Leabrook, SA, Australia: National Centre for Vocational Education Research.

Hanson, A. (1996). The search for a separate theory of adult learning: Does anyone really need andragogy? In R. Edwards, A. Hanson & P. Raggatt, (Eds.), *Boundaries of adult learning* (pp. 99-108). London: Routledge.

Henley, I. (1987). *The association between student pilot stress and flight instructor training: A study of perceptions*. Unpublished master's thesis, University of Manitoba, Winnipeg, Manitoba, Canada.

Henley, I. (1995). *The quality of the development and evaluation of flight instructors in Canada and Australia*. Unpublished doctoral dissertation, University of Newcastle, Australia.

Henley, I., Anderson, P. (1999, June). Using reflective journals to develop self-awareness, self-regulation, and self-evaluation skills in *ab initio* flight training. *Asia-Pacific Air Safety, 22*, pp. 10-11.

Henley, I., Anderson, P., & Schick, D. (1999, April 9-10). The role of the flight instructor in teaching students to become reflective practitioners. In D.L. Burke (Ed.), *Innovations in flight instruction: Proceedings of the 2nd Annual Flight*

Instructor Conference (pp. 55-62). Daytona Beach, FL: Embry-Riddle Aeronautical University.

Henley, I., Anderson, P., & Wiggins, M. (1999). Integrating human factors into general aviation: Issues and teaching strategies. In D. O'Hare (Ed.), *Human factors in general aviation* (pp. 89-117). Aldershot, UK: Avebury Aviation.

Henley, I., Turney, M.A., & Schick, D. (2001). Practical strategies for aviation educators: Reflective practice and student-centered debriefing. In R.S. Jensen (Ed.), *Proceedings of the 11th International Symposium on Aviation Psychology* [CD-ROM]. Columbus, OH: Ohio State University.

Henry, E., Huntley, J., McKamey, C., & Harper, L.T. (1995). *To be a teacher: Voices from the classroom*. Thousand Oaks, CA: Corwin Press.

Horton, M., & Freire, P. (1990). *We make the road by walking: Conversations on education and social change*. Philadelphia: Temple University Press.

Huberman, M. (1994). *The lives of teachers*. New York: Teachers College Press.

Jalongo, M.R., & Isenberg, J.P. (1995). *Teachers' stories: From personal narrative to professional insight*. San Francisco: Jossey-Bass.

James, P. (1997). Models of vocational development. In R. Blunden (Ed.), *Teaching and learning in vocational education and training* (pp. 83-105). Katoomba: Social Science Press.

James, P. (1999). Stress and adult learners. In J. A. Athanasou (Ed.), *Adult educational psychology* (pp. 321-349). Katoomba: Social Science Press.

Jarvis, P. (1995). *Adult and continuing education — Theory and practice*. London: Routledge.

Jensen, R.S. (1995). *Pilot judgment and crew resource management*. Aldershot, UK: Avebury Aviation.

Jensen, R.S., Adrion, J., & Maresh, J. (1986). *A preliminary investigation of the application of the DECIDE model to aeronautical decision making* (Report AOPA 86-3). Columbus, OH: The Ohio State University.

Johnston, N. (1993). Integrating human factors training into ab initio airline pilot curricula. *ICAO Journal, 48*(7), 14-17.

Klein, G.A., Orasanu, J., Calderwood, R., & Zsambok, C.E. (Eds.). (1993). *Decision making in action: Models and methods*. Norwood, NJ: Ablex.

Knowles, M. (1990). *The adult learner: A neglected species* (4th ed.). Houston, TX: Gulf.

Kochan, J.A., Jensen, R.S., Chubb, G.P., & Hunter, D.R. (1997). *A new approach to aeronautical decision making: The expertise method* (NTIS DOT/FAA/AM 97/6). Washington, DC: Federal Aviation Administration.

Kolb, D. (1976). Management in the learning process. *California Management Review, 18*(3), 21-31.

Kolb, D. (1984). *Experiential learning: Experience as the source of learning and development*. Englewood Cliffs, NJ: Prentice Hall.

Land, S.M. & Hannafin, M.J. (1996). Student-centered learning environments: Foundations, assumptions, and implications. ERIC document: ED 397810.

Loughran, J.J. (1996). *Developing reflective practice: Learning about teaching and learning through modelling.* Washington, DC: Falmer Press.

Machado, R. (1993, February). Experience: A flight unexamined is not worth logging. *Flight Training*, pp. 38-40.

Merriam, S.B. & Caffarella, R. (1999). *Learning in adulthood: A comprehensive guide.* San Fransisco: Jossey-Bass.

Mezirow, J. (1998). On critical reflection. *Adult Education Quarterly, 48*(3), 185-198.

Michelson, E. (1996, November-December). Usual suspects: Experience, reflection and the (en)gendering of knowledge. *International Journal of Lifelong Education, 15*(6), 438-454.

Newman, J.M. (Ed.). (1990). *Finding our way: Teachers exploring their assumptions*. Portsmouth, NH: Heinemann Educational Books.

Newman, J.M. (1991). *Interwoven conversations: Learning and teaching through critical reflection.* Toronto, Canada: Ontario Institute for Studies in Education Press.

Norris, S.P., & Ennis, R.H. (1990). *Evaluating critical thinking.* Cheltenham, Australia: Hawker Brownlow Education.

O'Hare, D. (1992). The "artful" decision maker: A framework model for aeronautical decision making. *The International Journal of Aviation Psychology, 2*(3), 175-191.

Osterman, K.F. (Ed.). (1990). Reflective practice in education [Special issue]. *Education and Urban Society, 22*(2).

Osterman, K.F., & Kottkamp, R.B. (1993). *Reflective practice for educators: Improving schooling through professional development.* Newbury Park, CA: Corwin Press.

Pearson, M., & Smith, D. (1988). Debriefing in experience-based learning. In D. Boud, R. Keogh, & D. Walker (Eds.), *Reflection: Turning experience into learning* (pp. 69-84). London: Kogan Page.

Peters, J.M. (1991). Strategies for reflective practice. In R.G. Brockett (Ed.), *Professional development for educators of adults* (pp. 89-96). San Francisco: Jossey-Bass.

Reason, J. (1997). *Managing the risks of organisational accidents.* Aldershot, UK: Ashgate.

Rogers, A.R. (1998). *Teaching adults*. Buckingham: Open University Press.

Romain, D. (1997). *Thinking things through: Critical thinking for decisions you can live with.* Mountain View, CA: Mayfield Publishing.

Schön, D. (1983). *The reflective practitioner: How professionals think in action.* New York: Basic Books.

Schön, D. (1987). *Educating the reflective practitioner: Toward a new design for teaching and learning in the professions*. San Francisco: Jossey-Bass Publishers.

Schön, D. (1998). *The reflective practitioner — How professionals think in action.* London: Ashgate Publishing Limited.

Shor, I. (1992). *Empowering education: Critical teaching for social change.* Chicago: University of Chicago Press.

Smith, G. (1993). Self-analysis of LOFT as a strategy for learning CRM in undergraduate flight training. In R.S. Jensen & D. Neumeister (Eds.), *Proceedings of the Seventh International Symposium on Aviation Psychology* (pp. 533-537). Columbus, OH: The Ohio State University.

Smith-Jentsch, K.A., Zeisig, R.L., Cannon-Bowers, J.A., & Salas, E. (1997). Defining and training tower cab teamwork. In R.S. Jensen & L. Rakovan (Eds.), *Proceedings of the Ninth International Symposium on Aviation Psychology* (pp. 201-206). Columbus, OH: The Ohio State University.

Sparks-Langer, G.M., Simmons, J.M., Pasch, M., Colton, A., & Starko, A. (1990). Reflective pedagogical thinking: How can we promote it and measure it? *Journal of Teacher Education, 41*(4), 23-32.

Stockhausen, L., & Creedy, D. (1994). Journal writing: Untapped potential for reflection and consolidation. In S.E. Chen, R.M. Cowdroy, A.J. Kingsland, & M.J. Ostwald (Eds.). *Reflections on problem based learning* (pp. 73-86). Sydney, Australia: Australian Problem Based Learning Network.

Taylor, E.W. (1994). Intercultural competency: A transformative learning process. *Adult Education Quarterly, 44*(3), 154-174.

Thomas, D. (Ed.). (1995). *Teachers' stories*. Bristol, PA: Open University Press.

Walker, D. (1988). Writing and reflection. In D. Boud, R. Keogh, & D. Walker (Eds.), *Reflection: Turning experience into learning* (pp. 52-68). London: Kogan Page.

Wiggins, M., & O'Hare, D. (1993). A skill-based approach to aeronautical decision-making. In R. Telfer (Ed.), *Aviation Instruction and Training* (pp. 430-475). Aldershot, UK: Gower.

Zeichner, K.M., & Liston, D.P. (1987, February). Teaching student teachers to reflect. *Harvard Educational Review, 57*(1), 23-48.

Zinn, L.M. (1998). *Identifying your philosophical orientation.* In M.W. Galbraith (Ed.), *Adult learning methods: A guide to effective instruction* (2nd ed., pp. 37-72). Malabar, FL: Krieger.

9 Teaching Complex Psychomotor Skills

Irene Henley, Mark W. Wiggins, Jayne Bye and Mary Ann Turney

Introduction

Effective teaching strategies are of utmost importance in achieving expected learning goals. Interesting and well-presented lessons challenge and motivate students, thereby increasing learning outcomes. There are many teaching techniques and strategies available to teachers. However, the selection of the appropriate teaching strategy will be vital to producing effective teaching episodes. It has been suggested that different teaching strategies are appropriate for each of the three domains of learning: cognitive, psychomotor and affective. The vocational nature of the aviation environment means that quite often the competencies required by aviation trainees will call for training in highly specialised skills (the psychomotor domain). According to Pithers (1998), "demonstration is usually the means of teaching a skill. This allows for initial observational learning of the skilled processes and operations in sequence" (p. 177). Accordingly, the key focus of this chapter will be on teaching for skill acquisition and on the demonstration approach, since much of the teaching and learning that takes place in the aviation context will depend on this particular approach. In the first part of the chapter, the concept of skill acquisition will be examined as will the theorisation of the stages of this process. The second part of this chapter will expand on the three basic stages of skill acquisition and will examine the demonstration approach as well as provide some practical guidelines for skill acquisition within the aviation context.

While sound planning and preparation provide the foundations for teaching and learning, carrying out these plans requires another complex and valuable set of skills. However, examining the skills that combine to produce effective lessons is a complex process. We all know when we are in the company of a skilled teacher (and when we are not) but breaking down the attributes of a successful presentation or lesson can be difficult. As Foley

(1995) notes, "when we watch [effective] teachers we can see, and admire, their grasp of technique. But these people are much more than a bundle of skills" (p. 31). Undoubtedly, there is a "wisdom of practice" in addition to the set of skills that make up an effective teacher. Some of the literature in education refers to the "art and craft" of teaching or to the teacher as an "expressive artist" and a "master craftsman" (Brandt, 1992; Dunkin, 1987; Eraut, 1987; Leinhardt, 1990; Telfer, 1993). However, it should be noted at the outset that the "something extra" that makes effective teachers greater than the sum of their skills probably cannot be taught in this chapter, but perhaps it can be learned through practical experience founded on a sound understanding of teaching principles, and on critical reflection upon this experience. This chapter can really only address the "bundle of skills", leaving the rest to practice and experience. Having said this, it should also be noted that having this set of skills is no small consolation, because such skills are the starting point for all effective teaching, including teaching complex psychomotor skills.

Like most workplace skills, the "bundle of skills" needed for effective teaching can be broken down and studied. Essential skills, such as explanation, demonstration, observing and coaching are outlined due to their importance to learner acquisition of workplace skills. Key principles of these approaches are described in the second part of the chapter and practical suggestions are made to enhance their implementation. Finally, the skills of providing feedback and correctives are examined and their centrality to skill acquisition is discussed.

The Acquisition of Complex Skills

Before providing a description of the guidelines for teaching complex psychomotor skills, it is probably a good idea to examine the concept of skill acquisition and the theoretical approaches that underpin skill acquisition. This will allow for the "contextualisation" of the demonstration approach as the fundamental teaching strategy in this area by answering two key questions:

1. What is "skill"?
2. How are skills learned?

What is "Skill"?

Cronbach (1954) indicates that a skill is easy to describe but hard to define. Nevertheless, he defines skilled performance as "a very complex process involving differentiation of cues and continual correction of errors" (p. 356). Cratty (1964) adds that, "the term skill denotes that some learning has taken place and that a smoothing or an integration of behavior has resulted" (p. 25). The skills that make up a skilled activity can be broken down into three types: motor, perceptual and cognitive. When workplace skills, such as those required to fly an aircraft, repair a jet engine or control air traffic for example, are carried out, they usually combine each of these types of skill and are, therefore, referred to in combination as "psychomotor skills" (Kroehnert, 2000, p. 54). Pithers (1998) makes reference to current research in the literature when he analyses this concept of skill and the various components of its definition:

> A skill may be thought of as a learned coordinated pattern of behaviour directed to an object or sensory display under appropriate conditions (Cornford, 1996; Magill, 1993). The word 'learned' suggests that skills acquisition occurs gradually via practice and is strengthened by positive feedback. 'Coordinated pattern' indicates the movements are not random but sequential, each movement influencing the formation and direction of others. The notion of 'directed' is that the skilled behaviour serves some purpose, whilst 'under appropriate conditions' indicates that some discriminative signal(s), or environmental context, acts as the initiator of the piece of learned behaviour. (p. 172)

This definition can be explored through the examination of a range of typical activities carried out in aviation occupations. However, one example should suffice. During an approach for landing, a pilot will reduce the power setting and adjust the aircraft attitude to achieve the desired descent profile for the runway in use (sequential and coordinated patterns). If there is a strong headwind or a crosswind, the power setting, the aircraft configuration, and the control input will also need to be adjusted to meet the prevailing conditions (environmental context providing discriminative signals and feedback on the appropriateness of control input) in order to achieve a successful landing (skilled behaviour that serves a purpose).

How Skills Are Learned — Theoretical Approaches to Skill Acquisition

Understanding the process of skill acquisition is important for both trainers and trainees. If teachers and trainers understand this process, they can much more effectively facilitate this process through the training they provide. If trainees understand this process, they will approach their learning with greater insight into how they will become competent.

Of the considerable number of models of skill acquisition, Dreyfus and Dreyfus (1986) and Anderson (1983) provide two of the most detailed in terms of the cognitive mechanisms involved in the progression from novice to expert. While the two models differ in a number of ways, including the number of stages of skill development, it can be argued that the cognitive processes involved in both share more similarities than differences.

According to both Dreyfus and Dreyfus (1986) and Anderson (1987), there are at least three general cognitive changes that appear to occur during skill acquisition or during the transition from novice to expert. The first of these, known as the cognitive or novice stage, is based upon the observation by a number of researchers that novices tend to adopt task-independent cognitive strategies, such as general problem solving strategies, during the interpretation of complex tasks (Anderson, 1987; Lesgold, 1988; Trappl, 1985). These types of strategies contrast with those adopted by experts in which the interpretation of problem-solving tasks is governed, to a large extent, by task-specific experience (Klein, 1989). Accordingly, both Dreyfus and Dreyfus (1986) and Anderson (1987) propose that, at some point during the process of skill development, task-independent strategies become redundant and learners begin to utilise the procedures derived from their own task-related experience.

Anderson (1987) extends the notion of procedural development and suggests that task-related experience is encapsulated within a series of task-related productions or rules incorporating both statements of condition (IF this happens ...) and statements of action (THEN I need to do this). In short, productions specify the steps needed to solve a problem. During the initial stages of skill development, the nature of these productions or rules is such that learners often make reference to general strategies associated with previous problem resolution. Such strategies might include the application of trial-and-error in the case of an extremely inexperienced practitioner, or the use of similar — comparable situations by those individuals with relatively more experience. However, in many situations, it is conceivable that a combination of strategies will be adopted, depending upon the task requirements (see Chapter 5 for a discussion of the role of learning strategies in the learning process).

Concurrent with the evolution of task-dependent problem-solving strategies, Dreyfus and Dreyfus (1986) propose a second principle of skill development, in which a transition occurs in terms of an individual's understanding of the relationship between the components that comprise a problem-oriented environment. In particular, the transition from novice to expert can be characterised by a fundamental shift from a superficial understanding of the inter-relationship between components to one that emphasises a level of interdependence, based upon functional properties (i.e., mental models). Evidence to support this assertion can be drawn from Chi, Glaser, and Rees (1982), who noted a qualitative difference between the principles employed by experts and novices when they sorted physics problems. Where experts sorted problems according to major physics principles, novices sorted problems on the basis of surface features such as the nature of the objects that were under consideration. This observation is consistent with broader empirical evidence in which novices appear to possess both a superficial and a relatively limited understanding of the inter-relationship between problem-oriented components in comparison to their expert counterparts (Adelson, 1984; Gott, Lajoie, & Lesgold, 1991; Terranova, Snyder, Seamster, & Treitler, 1989).

In apparent contrast to Dreyfus and Dreyfus (1986), Anderson's (1987) model of skill development does not explicitly incorporate a shift from a lesser to a greater level of understanding in terms of the functional nature of problem components. Instead, the cognitive mechanism of *generalisation* is advocated, where existing productions or rules (If—Then) can be applied successfully to a variety of relatively unfamiliar situations. This mechanism implies that, through experience, a learner develops an understanding of those types of situations in which a task-specific production is more or less likely to be successful. However, it might be argued that this proposition is consistent with the perspective outlined by Dreyfus and Dreyfus (1986), since there is an implication that expertise is based upon an understanding of the functional inter-relationships that may exist between various problem-oriented situations.

Dreyfus and Dreyfus' (1986) final principle of skill development is based upon an observation that, in comparison to novice chess players, experts appear to recognise and respond to situations in the absence of any apparent conscious processing (automaticity). An important part of this principle of automatic processing appears to be reflected in the capacity, amongst experts, to immediately focus upon the intrinsic or essential features of a task, almost to the exclusion of extraneous factors (Gordon, 1986; Klein, 1989). This capacity to identify the appropriate task-related cues from the milieu of the problem environment is presumed to facilitate the relatively rapid response to a problem that is evident amongst experts: A notion that is supported by Anderson (1990).

Anderson (1990) also notes the paradox that appears to occur in terms of experts' ability to retrieve task-related information from long-term memory. From a rational perspective of the function of human memory, it might be assumed that as the number of items in memory increases beyond a threshold, the response time needed to locate information would also increase. Clearly, this is not the case in respect of expert performance, since the retrieval of task-related information is often significantly faster than that of novices (Chase & Ericsson, 1982).

As part of an explanation of this paradox, Anderson (1990) refers to three theoretical principles through which task-related information might be retrieved from long-term memory. Firstly, information is presumed to possess an inherent level of activation that determines the sequence and, therefore, the rate at which information is acquired and processed in memory. Features such as previous experience or enhanced motivation may increase the pertinence of particular types of information during a given situation (Anderson, 1983; McDonald, 1991). Therefore, this information may trigger an inherent level of activation within memory that is greater than information that is not perceived as relevant.

Secondly, though no less important, is the principle of node strength. A node can be defined as a point which links inter-related information within long-term memory (Anderson, 1990). The principle of node strength determines the probability that the process of activation will continue beyond a node within long-term memory. Thus, a greater level of node strength is generally associated with an increase in the probability that activation will occur within associated nodes. The result is a more efficient pattern of information activation than might otherwise be the case.

The pattern of activation between nodes is determined by Anderson's (1990) final principle of link strength, in which increases in the strength of an association are connected with an increase in the probability of activation. The aim of this process is to facilitate a spontaneous match between the cues associated with a situation, and the associated production (IF—Then) that will determine what action, if any, needs to be taken. However, the success of this pattern recognition is dependent upon the perception of appropriate task-related cues. Moreover, the response time is dependent upon the strength of a certain pattern of activation in comparison to others that occur within long-term memory.

According to Anderson (1983, 1987), the strengthening of a pattern of activation occurs at the final stage of skill development (autonomous stage), since the process is dependent upon the accumulation of extensive, task-related experience. Therefore, Anderson (1983) provides a cognitive mechanism that may explain the superior recognition performance amongst experts that was observed by Dreyfus and Dreyfus (1986) and others (Charness, 1979; Ericsson & Polsen, 1988; Patel & Groen, 1991; Voss, Vesonder, & Spilich, 1980).

The general principles of skill development espoused by Dreyfus and Dreyfus (1986) and Anderson (1983, 1987) provide a description of the general cognitive processes that are presumed to occur during the transition from apprentice to master. However, differences do exist between the two models in terms of the number of stages that comprise the development of expertise. Whereas Dreyfus and Dreyfus (1986) encapsulate skill development within five distinct stages, Anderson (1987) and others (Ackerman & Kyllonen, 1991; Cornford, 1999; Fitts & Posner, 1967; Rasmussen, 1983) propose three stage models (see Figure 9.1).

Dreyfus & Dreyfus (1986)	Novice →	Advanced Beginner →	Competent →	Proficient →	Expert
Anderson (1983)	Interpretive →		Knowledge Compilation →		Tuning
Rasmussen (1983)	Knowledge →		Rule →		Skill
Fitts & Posner (1967)	Cognitive →		Associative →		Autonomous

Figure 9.1 A comparison between the stages of expertise proposed by various skill acquisition theorists

According to Dreyfus and Dreyfus (1986), the five stages of skill development include: Novice, Advanced Beginner, Competent, Proficient, and Expert. The Novice stage can be characterised by the application of problem-solving strategies independent of context. Moreover, the attentional resources required to perform during this stage often precludes the successful completion of secondary tasks that may involve the processing of concurrent verbal or visual information (e.g., the student or novice can maintain straight and level flight but cannot talk on the radio at the same time). These observations are consistent with both Fitts and Posner's (1967) cognitive stage of skill development and the later incarnation of the model as Anderson's (1983) interpretive stage.

Dreyfus and Dreyfus (1986) attempt to provide a cognitive link between novice performance and competency using the intermediate or advanced beginner stage. They suggest that with the accumulation of task-related experience, the learner begins to develop a limited repertoire of links between the cues associated with situations and their related consequences. This repertoire increases performance beyond that of the novice stage, although there is often a requirement to resort to general problem-solving strategies evident during the novice stage. On the basis of this observation, Anderson (1983, 1987) would argue that the advanced beginner stage simply represents a quantitative increase in novice performance, rather than a distinct, qualitative change in the cognitive process, which would be necessary to constitute a different stage in the process of skill development.

Dreyfus and Dreyfus (1986) suggest that a qualitative change in performance occurs as part of the competency stage of skill development. The competency stage involves a re-organisation of knowledge into cognitive structures (schemata or mental models) based upon the cues associated with each situation. During problem solving, the competent performer will classify the type of problem involved and will subsequently use the cues to formulate an effective strategy for problem resolution. This process of association between specific conditions and actions is consistent with Anderson's (1983, 1987) compilation stage of skill development in which learners develop task-specific productions using the mechanisms of proceduralisation and composition.

Proceduralisation is the elaboration of the process in which conditions and actions are linked to form a production (If—Then). This is a significant advance beyond the cognitive strategies employed during Anderson's (1987) interpretive stage, as the productions adopted during problem solving have become task specific. However, the process of composition is dependent upon proceduralisation having occurred, since this process involves combining productions to further increase the efficiency with which they are both triggered and executed.

The penultimate stage associated with Dreyfus and Dreyfus' (1986) five-stage model of skill development is referred to as proficiency and involves a distinction between automated and non-automated information processing skills. While the recognition of a particular situation as familiar or unfamiliar is considered an automated process at this level, the proficient performer must continue to rely upon the conscious deliberation of appropriate solutions which was evident during the competency stage. However, since the performance of the task is not fully automated, Anderson (1987) would perhaps argue that the progression beyond the knowledge compilation stage is insufficient to warrant the inclusion of an additional stage of skill development.

Both Dreyfus and Dreyfus (1986) and Anderson (1983, 1987) contend that automated behaviour is the primary feature that distinguishes performance during the final stage of skill development from performance during previous stages. Referred to as *expertise* by Dreyfus and Dreyfus and *automation* by Anderson (1983, 1987), this stage is characterised by the spontaneous recognition of situations as familiar, combined with the unconscious application of task-appropriate strategies. Performance at this stage can be characterised as both maximally effective and efficient, and the underlying cognitive processes can be difficult for the expert to verbalise. Support for the latter proposition can be drawn from Ghilsen (1952), who notes a number of anecdotal examples where experts appear to undertake detailed problem-solving strategies in the absence of conscious processing. In one example, the expert mathematician Jacques Hadamard describes a situation in which "solution long searched for appeared to me at once without the slightest instance of reflection" (cited in Ghilsen, 1952, p.15). Hoffman (1987), Hoffman, Shadbolt, Burton, and Klein (1995), Rouse and Morris (1986), and Cooke (1990) also highlight the difficulties associated with the verbalisation of reasoning amongst experts. This explains why many flight instructors, for example, find it difficult to break down all the steps and describe the cues needed to perform a successful landing.

Anderson (1983) suggests that the difficulty associated with the verbalisation of performance is evidence that the productions (If—Then) at this stage of skill acquisition are resident in procedural memory as singular entities, unable to be separated consciously by the expert practitioner. However, the lack of verbalisation is also consistent with Dreyfus and Dreyfus' (1986) proposal in which expert performance is primarily dependent upon the activation of one or more of thousands of individual experiences which remain resident within long-term memory. These experiences are presumed to be classified in memory, such that the expert develops a relatively deep understanding of the inter-relationship between the features that comprise the problem environment. This, in turn, allows the expert the capability to perform successfully during relatively novel tasks, and in the absence of productions: a process that is consistent with case-based reasoning (Leake, 1996).

In contrast to Dreyfus and Dreyfus (1986), the cognitive mechanisms described by Anderson (1983, 1987) suggest that experts' performance during novel tasks is based upon the assumption that productions become more or less applicable within a variety of related domains. For example, the mechanism of generalisation involves the combination of related productions given evidence that one strategy is applicable during a variety of situations. Conversely, discrimination involves the decomposition of a production given evidence that the strategy associated with the production is not necessarily applicable within all environments.

While a number of differences exist between the various perspectives associated with the origin of expertise, it might be argued that these are more a reflection of differences in methodological approaches, rather than a substantive difference in terms of theoretical perspectives and constructs. Despite the methodological differences that exist between different analyses of skill acquisition leading to expertise, surprisingly consistent results have emerged in terms of the cognitive processes involved in the transition towards expertise. In particular, there appears to be consistent support for the notion that expertise comprises a vast series of individual experiences which are applied spontaneously to direct performance during task-related situations (Palmeri, 1997). This might explain the observation that expertise is highly task-specific (Glaser & Chi, 1988), although the limits of expertise have yet to be identified precisely (Clark, 1992). More importantly, the outcomes of the theoretical approaches outlined above appear to suggest that the origin of expertise lies in the prolonged exposure to an extensive series of task-specific situations (i.e., practice and/or experience) through which associations are developed between the various cues and particular outcomes.

Practical Considerations to Skill Acquisition in Aviation

As mentioned in the introduction, workplaces such as the aviation context rely on the use of highly skilled behaviour and on the need to facilitate the transition towards expert performance. Therefore, training in this area must focus on teaching strategies that will produce trainees able to reproduce skilled performance not only in the training environment, but also under the varying conditions of the workplace. Some of the important features of skilled performance of which aviation trainers must be aware are that skills in the aviation context are:

- often attempted under time constraint and, therefore, need to be practised for speed as well as accuracy;
- concerned with sequential behavioural responses and, therefore, the links in the sequence must be clear to learners;
- replicable under a range of conditions and requirements and, therefore, must be practised in a range of situations and under different environmental conditions;
- reliant upon sound judgement and effective decision-making skills;
- performed with the need for constant vigilance because of safety considerations;
- specific to the trainee's work environment; and
- assessed according to industry pre-established standards (adapted in part from Pithers, 1998, p. 170).

According to Cornford (1999), there are also nine defining characteristics of skilled performance that are typical of all skills:

- *Skill is learned* – skill is not an innate ability or a reflex;
- *Skill involves motivation, purpose and goals* – motivation, purpose and goals are important both for initial skill acquisition and for subsequent practice to attain expert performance;
- *Schemata are required* – skilled performance is based upon a series of related steps. Therefore, a schema (a mental model or representation) which includes the components, procedures and processes; the correct sequencing of these components, procedures and processes; and the time relations between the components, procedures and processes needs to exist in memory before the skill can be performed;
- *Skills are context-specific* – skilled performance requires specific cues that are linked to a specific context. This explains why trainees sometimes fail to apply their skills in appropriate ways in new settings (lack of generalisation);
- *Skills involve problem solving relevant to the context* – most skills involve particular principles that subsequently need to be applied to

settings or contexts that are often different in various ways from those in the initial training environment;

- *Skills involve relative judgements with individual differences in skilled performance evident* – different levels and standards of performance exist among practitioners in any specialist skill, that is, some individuals are more skilled than others;
- *Standards of excellence* – standards of excellence are an integral aspect of individuals making judgements about their own performance which may vary from time to time or from one situation to another and about the performance of others;
- *Skill involves comparable replication* – skilled performance can be replicated to similar standards on different occasions; and
- *Considerable periods of time are required to achieve high levels of skill* – skilled performance requires time and repeated practice (adapted from pp. 266-268).

The training implications of these features and characteristics of skilled performance will be discussed below.

Application of the Stages of Skill Acquisition to the Training Context

According to Ackerman and Kyllonen (1991), most current theorising into the process of skill acquisition is generally in agreement about the nature of the learning process involved. As was seen in the first part of this chapter, skill acquisition theorists usually describe this type of learning process as being broken into at least three phases: the cognitive (novice) stage, the practice-fixation (associative) stage, and the autonomous (proficient or expert) stage. However, it should be noted that these stages are not discrete or self-contained, rather they flow into each other and can overlap (Fitts, 1964).

Cognitive (Novice) Stage

The cognitive stage, as was explained in the first part of this chapter, begins with the learner's initial confrontation with the task or the manoeuvre to be performed. "During this phase, the learner begins to understand the basic task requirements.... In other words, the learner formulates a general idea of what is required of him or her" (Ackerman & Kyllonen, 1991, p. 203). This stage involves the building of a mental model or schema (an overall picture) of the skill, which will later serve as a standard of performance to guide behaviour and assist in detecting errors (Schmidt, 1975). In gaining an overall picture of what the skill involves, the student will also need to identify the separate steps or subskills, which constitute the skill, as well as the appropriate sequence in which they occur and the timing involved (Cornford, 1999).

The role of the teacher or trainer during this stage is to concentrate on building trainee understanding of the importance of the skill to job performance, its links with other skills, and how it is comparable to, or distinguished from, the other skills the student already has acquired. This is achieved through the use of clear verbal explanations and the provision of specific cues, and through a step-by-step physical demonstration of the skill by the teacher. This demonstration needs to emphasise the appropriate sequence of subskills as well as the elements of timing. In many situations, the explanation accompanies the demonstration. However, when the skills to be acquired are complex and/or need to be demonstrated and practised in a dynamic and often threatening environment, such as the cockpit of a small aircraft, most of the explanation should be presented in a classroom before the practical demonstration is performed. This will allow students to devote all their attention to the explanation being provided. The skills one is required to teach and the environment where the training takes place will determine the best approach to be used. The basic nature of the equipment that might be used and how the skilled performance is to be attempted are also important aspects of this stage of the training.

Practice-Fixation (Associative) Stage

The second phase of skill learning is the practice-fixation phase, also referred to as the associative phase by Fitts and Posner (1967). Cornford (1999) provides a clear description of this next important stage:

> What happens at this phase is that the steps or subskills become closely associated.... Each previous step or subskill serves as the stimulus for the next step and thus associative links are formed. Remembering the next steps in a sequence, and the requirements of successful performance of this, are often quite difficult and require considerable concentration and help from a teacher or trainer in the early stages of learning. It is through repetition that these associations and links become fixed in memory. We generally call this repetition practice, and there is little doubt that practice and feedback are the key elements at this stage of skill learning. (p. 270)

Clearly, this second stage of skill acquisition is typified by a laborious accumulation of knowledge, procedures and processes (Gordon, 1991). As a result, decision making and problem solving are slow and tedious and performance is often awkward and faltering. It is through repetition and practice that the associations and links between the steps and subskills needed to perform the skill successfully will be formed and become fixed in the student's memory. The more complex the skill, the longer it will take to acquire the accurate sequence of steps and the appropriate timing needed for successful performance. Therefore, the role of the trainer during this stage is to provide ample opportunity for the type of practice described above by Cornford (1999). This involves, on the part of the trainer, sound planning skills in terms of not only adequate resources, equipment and facilities, but also in time management. A demonstration segment of a lesson that is too long, due to a lack of preparation or recent rehearsal, will cut into valuable practice and feedback time. Given that many trainees may have their only exposure to specific equipment and materials during training sessions, this would certainly have a negative impact on effective skill acquisition.

Observation of student performance and coaching are other important skills that trainers in most aviation education settings need to develop, as will be explained later. Due to their importance during the practice-fixation phase of skill acquisition, the provision of correctives and feedback techniques are also discussed in some detail following this section. Finally, the importance of the debriefing period after the practical training session is underlined and some useful guidelines are suggested.

Autonomous (Proficient or Expert) Stage

The chief signal of achievement of this stage is that skilled performance requires minimal attentional effort. The skill can be performed at speed while remaining accurate (Ackerman & Kyllonen, 1991, p. 204). It is almost as if the trainee does not even have to think about the performance and the steps it involves. However, Fitts (1962, 1968) seems to have deliberately chosen the word *autonomous* rather than automatic because autonomous suggests that the skilled performer is very much in control of the performance of the skill (Cornford, 1999). In reality, during this phase, even though procedures have become second nature, the expert or proficient performer constantly monitors his or her performance and compares it to the schema or mental model that exists in memory. "Although there may not appear to be obvious concentration on performance, the skilled performers know when they have made an error" (Cornford, 1999, p. 271). However, because the trainee who has reached the autonomous stage of skill acquisition now has a certain amount of residual attention, he or she can even perform the task while experiencing distraction (Pithers, 1998, p. 173) or while concentrating on other aspects of the task.

Once again, Cornford (1999) stresses the importance of the gaining of feedback through practice and self-monitoring to reach this level of performance (p. 278). This is the stage where trainees can also apply self-evaluation and critical reflection skills through appropriate facilitation by the instructor. Therefore, students should be afforded practice in a variety of circumstances and under varying conditions and they should be encouraged to reflect on their performance and their practice (see Chapter 8 for suggestions on how to help students become reflective practitioners).

The Demonstration Approach

For a skill to be mastered, students need methods and strategies that are appropriate for the tasks and the setting, as well as ample opportunity to practise the new skills. Appropriate strategies will enhance skill acquisition by facilitating cognition, controlling affect, and directing psychomotoric execution (Pressley & Woloshyn, 1995). For example, one can improve retention and recall by using acronyms and integrative imagery or visualisation. As students become more proficient, their strategies must be adjusted in order for them to continue to improve and to remain motivated and challenged.

As stated previously, the demonstration approach is considered the most appropriate teaching strategy for psychomotor skill acquisition (Cross, 1982; Pithers, 1998). Briefly, in the demonstration approach, the teacher presents a skill, a strategy or a process to students, shows them how to do it, provides assistance as students initiate attempts to perform the task, and ensures that they practise so that they can be successful in performing the task on their own in the future.

Brewer (1997) suggests that there are four basic steps that should be followed when using the demonstration approach. These are preparation, presentation, application, and testing and follow-up. Similarly, Kroehnert (2000) indicates that a complete demonstration involves four phases: preparation, demonstration, student practice, and assessment. Within the aviation industry, and more specifically within the flight training context, similar steps are usually suggested for the instructor who is planning to use the demonstration teaching strategy. These usually include preparation, ground briefing, demonstration, practical application and feedback, and debriefing.

The following discussion pertaining to each step of the demonstration approach draws on the guidelines presented by Brewer (1997), Pithers (1998), Daines, Daines, and Graham (1995), and on practical experience gained in the field of flight training. However, even though the examples presented here are mainly derived from the flight training context, the guidelines should prove useful to any instructor using the demonstration teaching strategy within any of the aviation training settings.

Preparation

In planning a demonstration, Daines et al. (1995) emphasise the importance of breaking down the activity or skill into its basic steps or movements. They remind trainers to consider the position of the learners in that what is easy and quite comprehensible to an experienced practitioner will probably be completely new and barely understandable to novices. This may, at times, present problems to new trainers. Having reached the expert or at least the proficient stage of skill acquisition, they perform the skills with minimal attentional effort or concentration. Therefore, it may prove difficult initially to bring back all the steps and subskills into consciousness. Indeed, Bargh and Chartrand (1999) indicate that experts or experienced practitioners are not conscious of the cues that control a great deal for their behaviour. However, as instructors, we need to bring those cues back into consciousness and verbalise or identify the specific steps involved in performing complex psychomotor skills in order to help students understand how to perform the skills. Simplifying a complex task to the students' level of comprehension will, therefore, require considerable planning and some reflective practice on the part of new trainers.

Brewer (1997) also suggests that the "demonstration should be physically or mentally rehearsed to check for mistakes and create continuity of ideas" (p. 51). Indeed, the first demonstration of a new skill should be performed as near as possible to the ideal performance, since the things students learn first are normally learned best. This is called the "primacy effect" in the training literature (Kroehnert, 2000; Pithers, 1998; Transport Canada, 1998; Walker, 1995). Therefore, the critical importance of the initial practical demonstration students get from their instructor cannot be overstated.

Part of the planning phase may also involve the preparation of teaching aids and support material, such as diagrams and models, that will help students understand what is involved in performing the skill (Daines et al., 1995, p. 43). Furthermore, drafting a plan to be followed during the demonstration is also important so that all the relevant steps are covered in a logical sequence (Kroehnert, 2000). A written plan with clear and specific instructional objectives can also be a useful guide for trainees to follow during the explanation and the demonstration. It is also important to determine which material (i.e., concepts, procedures, etc.) should be explained *before* the practical lesson so that during the practical demonstration, explanations can be limited to essential verbal cues and to pointing out specific visual cues.

Time management is another essential factor to be considered. As mentioned previously, a demonstration that is too long, due to inadequate preparation, will deny the student valuable practice time and reduce learning effectiveness, and could also lead to student frustration. Similarly, during the ground briefing, if too much time is spent on providing "nice to know" details, students will be overwhelmed, possibly confused, and many will get bored. In contrast, if too little time is afforded to explaining the connections between the theory and the skill, learning gaps could be created and understanding will diminish.

Finally, a vivid and interesting introduction and an effective closure to the lesson are also important elements to bear in mind during the preparation phase. A good introduction will captivate students' attention and arouse interest in the explanation and the demonstration, while an effective closure should link the lesson to future lessons and motivate students by showing them how the lesson will contribute to the acquisition of skills needed in the workplace.

In summary, one cannot overstate the importance of the preparation phase of the demonstration approach. The analogy used by Kroehnert (2000) in discussing this major phase of skill training is very pertinent. He notes that "practical instruction is like an iceberg, in that the part that is visible (conducting a demonstration and supervising student practice) makes up only a small part of the effort. The bulk of the work goes on beforehand" (p. 56).

Ground Briefing or Explanation Phase (Cognitive Stage)

During this phase of skill acquisition, an explanation of the background theory and the concepts essential to understand the new skill is provided, and the procedures and steps required to perform the skill are clearly outlined. Indeed, an important part of the learning that takes place during skill acquisition has its beginning in teachers talking, that is, in teachers providing clear and accurate explanations and presenting relevant information needed to understand how the skill will be performed and how it fits in the big picture. Therefore, effective presentation skills, including the skill of explaining, are essential skills that teachers need to develop.

Effective presentation If teaching strategies provide the framework for learning, then effective presentation skills are the means by which the framework is held together — the "nuts and bolts" of good teaching and training, if you will. Without effective presentation skills, even the most sound teaching strategies will be compromised to some extent. In addition, good presentation skills allow the trainer to maximise the return on the careful planning that has taken place before the presentation. Presentation skills achieve this outcome by helping the trainer gain and maintain the attention and interest of participants while optimising learner motivation and involvement. According to Sir John Lubbock, English astronomer and mathematician, "if we succeed in giving the love of learning, the learning itself is sure to follow".

Approaching a classroom presentation or even a one-on-one briefing for the first time can be a daunting experience for beginning trainers. Popular wisdom tells us to be mindful of first and last impressions. Pithers (1998) supports this view stating that "it is important for the presenter to think about the clarity and impact of the introduction and the conclusion of any presentation and deliver these parts of the presentation with force" (p. 131). As mentioned previously, research into learning has shown that learners tend to remember the first (primacy effect) and last (recency effect) part of a lesson with greater frequency than content within the body of the presentation (Kroehnert, 2000; Walker, 1995). Due to the importance of these two phases of a lesson, the following discussion focuses briefly on the techniques of introduction and closure, and suggest some guidelines for trainers to follow. In addition, the skill of explaining will be examined because of the critical role that it plays as a complement to an effective demonstration.

Skills for beginning a lesson (Introduction) Daines et al. (1995) warn that the actions of a trainer in the first few minutes of a session "will critically affect what happens thereafter" (p. 106). Therefore, according to Pithers (1998), "the presenter should really rehearse what is to be said so that it can be delivered in a strong, energetic way with the learners, maximising eye contact" (p. 116). Thorough preparation is again the key to getting through this early part of the session. Confidence in your planning and organisation will mean that you can concentrate on delivery skills and on providing an effective demonstration. Several useful approaches have been proposed in the training literature. The following suggestions provide a summary that could be used as a checklist the day before training is to take place:

- Is the training room ready and reasonably comfortable?
- Is the equipment set up and are all the necessary tools, materials and supplies available and functioning?
- Is the seating arranged so that all the students can easily see you and your equipment?
- If a step involves small parts or intricate processes, do you have visual aids or models to clarify each step?
- Have you ensured that safety practices specific to the operation will be adequately covered?
- Are you still familiar with the material and the procedures to explain and perform the demonstration of the skill correctly?

Furthermore, instructors should consider the following suggestions before commencing a training session:

- get there early;
- recheck the equipment. Beware of Murphy's Law;
- set up your training aids so that they are at hand when needed;
- organise and place your notes so that headings can be seen at a glance and so that the notes can be readily consulted if needed; and
- mentally recall or visualise the sequence of events.

Perhaps one of the key functions of an effective introduction is to increase learner motivation by appealing to individual interest. One of the advantages of being a trainer in most aviation training contexts is that the majority of students are keenly motivated to learn the skills being presented. However, Pithers (1998) notes that motivation "is useful in learning only because it leads to practice.... Motivation is a necessary but not a sufficient condition for learning" (p. 33). Clearly then, motivation is desirable in learners, however, the skills of the presenter still have an important role to play in maximising this potential and providing the all important opportunities for practice.

Skills for concluding a lesson (Closure) As stated above, an effective ending to a training session can have positive outcomes for learning. Once again, there are no formulas for this aspect of the presentation. However, there are guidelines that can be useful to trainers. Pithers (1998) suggests the following approaches:

- a brief review or summary of the major teaching points, steps and procedures, preferably obtained by questioning the students, may prove a useful conclusion for at least two reasons. One is that this technique involves repetition or further practice of key points, processes or issues and so promotes learning. Secondly, it helps the instructor to evaluate whether the students have gained the knowledge, skills or attitudes of importance;
- the trainer should link the session to the "big picture" and conclude by connecting what has been learned during the session to the preceding session's learning and to learning to be experienced in future sessions;
- the instructor should also link what has been learned in the training room to the skills needed in the practical lesson and in the workplace;
- possibly an exercise or topic could be set to promote learner reflection, which involves further practice outside of session time and before the next training session (adapted from p. 131).

Turney, Eltis, Hatton, Owens, Towler, Wright, Cairns, and Williams (1983) suggest that the key components of effective session closure are reviewing and evaluating (p. 89). The reviewing aspect of the closure is based on the principle that material covered in a teaching session requires consolidation. To achieve this, trainers can, as mentioned above, have students reiterate key points, so that the main facts, skills, or ideas covered are recalled, highlighted again, and re-emphasised.

Evaluation, the second skills component of closure, forms part of a larger cyclical process where information about the outcomes of teaching and learning are fed back into the planning process for the purposes of fine-tuning training. At this point, evaluation basically requires trainers to check whether understanding or learning has taken place. If learning has occurred, then participants will be able to apply their new knowledge and demonstrate newly acquired skills. The previous step of "reviewing" is certainly a key approach in collecting information for this final step. The information gained will then "serve as a basis for future courses of action to consolidate ideas, remedy diagnosed weaknesses, supply omissions or reteach unassimilated material" (Turney et al., 1983, p. 91. Also see Chapter 15 for a more detailed examination of assessment and evaluation).

Explaining According to Brown (1978), "explaining is at the heart of teaching ... just as its obverse, understanding, is at the heart of learning" (p. 2). Pithers (1998) notes that "explanation has been defined in various ways but usually, it refers to a means of initiating and extending relationships, connections or links between concepts or things" (p. 150). Indeed, explanations can aid student understanding of important concepts, essential procedures and processes, clarify cause–effect relationships, purposes and desired outcomes. The following is a set of guidelines that could be useful to ensure that effective explanations precede a demonstration:

- clearly state the activity involved and indicate its purpose and outcome and explain why the new skill is important for aviation professionals;
- arouse and maintain student interest by relating the skill to previous relevant skills and by showing how it fits into the big picture — links and connections between a student's current knowledge and skills can be an effective tool in maintaining student attention and in encouraging student participation;
- clearly outline the main steps of the activity and identify the likely problem areas or the common errors, and suggest possible solutions to overcome these problems or common errors;
- provide standards of performance so that students can aim for specific goals. However, it is important to make sure that the standards are realistic for the student's level of learning; and
- aim to inspire confidence in students so that they will be willing and ready to try the skill themselves (adapted in part from Daines et al., 1995, p. 42, and Turney et al., 1983, p. 21).

Since explanations are intended to give students crucial information needed to understand and perform the skill that will be demonstrated, the explanations need to be clear and presented in a logical sequence. New terms need to be clearly defined and the language used during the explanation should be precise, explicit and devoid of unnecessary jargon. Using comparisons (similarities and differences), examples taken from real life experience, and providing reasons, testimonies (quotes from experts), and statistics can also help clarify explanations and make them more vivid and interesting to students.

The Demonstration

The demonstration is used as a means of illustrating the correct procedure to be followed in performing a task so that modelling and subsequent meaningful practice can take place. It is usually accompanied by a clear and concise oral explanation of what is going on. Demonstrations help establish the correct process and procedures in the student's mind. Therefore, it is crucial that the demonstration be crystal clear and correct in its approach and content. Once engraved in a student's mind, a poor or inaccurate demonstration is difficult to erase. In the case of complex skills, students will need to be exposed to the skill performance in an articulated or "broken down" version so that the simpler skill subsets can be learned effectively prior to being integrated into the complete activity.

The demonstration should include specific verbal and visual cues that focus the student's attention on the critical elements of the skill being performed and also could include helpful hints or "tricks of the trade". Such cues give learners a picture to recall later on during mental rehearsal and/or during self-talk in the practical application phase. When teachers provide explanations and demonstrations they develop an image of correctness so that while students observe the demonstration, they can develop an accurate mental model of the skill to be performed.

The demonstration–explanation phase is then followed by supervised imitation or modelling by the student. During practice, students compare the image of correctness presented by the instructor to their actual performance. However, initially students will need teachers to give them information about the correctness of their performance in the form of feedback since they are still unable to accurately self-assess. By closely watching the student's performance, the instructor can determine how well the student has understood and how accurately he or she can perform the skills involved. This observation of the student's practice provides immediate feedback and enables the instructor to come to a close understanding of the student's particular difficulties and thus be in a position to provide helpful correctives and valuable feedback.

Principles of demonstration Pithers (1998) states that a demonstration has the following advantages over verbal instruction alone as a teaching strategy for skill acquisition because it:

- involves more than one sense modality;
- relates verbal explanation to a workplace object, process or equipment;
- provides for active hands-on practice which can be followed by reinforcement and feedback;
- motivates or energises, that is if the steps are developmentally sequenced, appropriate to the learners and are within the learner's capability (p. 177).

Daines et al. (1995) concur with these statements and reiterate the importance of "a combination of seeing, hearing and doing" (p. 42). They suggest the following steps for carrying out the demonstration:

> Begin the demonstration, accompanying each step of your activity with a verbal description. Your words should marry your actions though you do not have to talk all the time or give a continuous running commentary. The more involved manoeuvres might be best completed in silence anyway. Language is important: use words that people understand (p. 43).

The suggestion by Daines et al. (1995) that complex skills might best be conducted without commentary needs emphasis because during the initial stage of skill acquisition students can easily become overloaded. They need to concentrate on each step and may not have any residual attention left to listen to explanations concurrently. That is also one of the reasons why, when teaching complex psychomotor skills, the steps need to be described and the procedures need to be explained before the practical demonstration actually takes place.

Brewer (1997), like Daines et al. (1995), stresses the importance of simple language that utilises basic vocabulary and short sentences to maximise understanding (p. 51), while Pithers (1998) notes the importance of presenting the procedure slowly, providing opportunities for trainees to comment or ask questions (p. 178). It is also important that the demonstration be appropriately timed so that students can undertake individual practice under close supervision immediately afterwards and receive feedback on their performance and appropriate correctives to help them improve later performance (Daines et al., 1995).

Checking for understanding is the final and most crucial stage of a demonstration. Daines et al. (1995) suggest one way that this can be facilitated is by asking "students to check with one another that they do know what to do by showing each other or talking it through" (p. 44). In this way, trainers can overcome the reticence of some trainees to ask questions in front of the whole class. If points have been missed, it may be necessary to go over aspects of the demonstration or even to carry out the demonstration again if this is possible.

However, it is important to stress that if students explain or talk through the skill with one another, the instructor still needs to provide close supervision, otherwise students could be developing an incorrect mental model of the skill to be performed. On the other hand, when training is conducted on a one-on-one basis, as in flight training, the student can be asked to talk through the manoeuvre while the instructor performs the demonstration again. Finally, asking participants to outline the main steps of the demonstration will allow the trainer to check understanding and the process of verbalising will help reinforce participant learning (Daines et al., 1995, p. 44).

Practical Application or the Practice-Fixation Stage

The practical application phase focuses on student practice under the close supervision of the instructor, who provides timely feedback and clear correctives to guide student performance toward mastery. For learners, this is the most important part of any training session since practice is at the heart of all skills learning. At this stage, the role of the trainer changes somewhat, from that of presenter to that of observer and guide or coach. This is where the skills of feedback and reinforcement will be most crucial. As students attempt to perform the task for the first time, the instructor will need to help and guide them both to perform the skill and to avoid making errors. However, Brewer (1997) underlines the importance of knowing when and how to act. He states that "it is important not to take over through suggestion or criticism and remove the participant's sense of importance or responsibility.... The presenter should point out errors and help the participant think of ways to correct the mistake, then let the participant fix it" (p. 52).

Research conducted by Henley (1987), who examined the association between student pilot stress and flight instructor training, confirms that instructors who appear nervous and take over too early during the approach to landing increase the student's stress level and fail to instil confidence. For example, one student reported: "*My instructor for my private [pilot licence] was fairly new at it [instructing] and on the landing he wouldn't let me do it — as soon as I'd do something wrong, he'd take over. It got very frustrating. I'd get a bit angry and thinking: How am I ever gonna learn to land if he always takes over?*" (cited in Henley, 1987, p. 80). Similarly, another student noted:

> *I think that [landing] was the most frustrating part of my training. For some reason, as soon as we'd start setting up for the approach, I would tense up and it just felt as if things were happening too fast. I was miles behind the aircraft.... My instructor would raise his voice saying: "Watch your airspeed! Keep it straight. Throttle back". I'd be ready to pull my hair! In fact, I think he was nervous himself. The closer we got to the ground, the more fidgety he'd get; he'd talk more and his comments would be snappy. I could see him move his hands ready to grab the controls and I found it very distracting. The fact that he seemed nervous or ready to jump on the controls to take over made me nervous.* (cited in Henley, 1987, p. 79)

Learning all the steps and subskills involved in a skilled performance requires considerable effort on the part of the student and patience on the part of the instructor. Practice takes time. Furthermore, slow practice is essential in the early stages of skill acquisition to ensure that all the relevant steps are accurate and that the sequence and timing are correct. According to Cornford (1999), accuracy is more important than speed at this stage of skill acquisition.

Many students talk to themselves (think aloud) as they recall each step in the correct sequence. Prompts and hints derived from the instructor are also often built into this self-talk. Such verbalisation can be an important practice in the process of integrating and consolidating the subskills needed to perform the skill accurately. Student self-talk can also be extremely helpful to the instructor because it lets the instructor know what the student is thinking and what mental model of the skill the student has developed.

The danger at this stage of skill acquisition is that students could incorporate incorrect procedures into their performance or omit important steps. Therefore, it is important to correct errors as soon as possible after they happen because if students practice incorrect steps and procedures, they will tend to remember them and repeat them (James, 1995). This is why student performance and instructor supervision are carried out concurrently during the initial stage of skill acquisition. Accordingly, the key tasks for the trainer during the practical application stage is to closely observe student performance and provide positive reinforcement and timely feedback as well as clear indications as to how to improve performance. These important teaching skills are discussed in some detail below.

Observation Discerning observation is a very important skill that teachers develop over time. With practice, the process of observing and making decisions on those observations in order to provide appropriate feedback and correctives should become easier and easier. It is important to remain objective during observation, to put aside biases and frame feedback from the view of a professional practitioner. Accurate observation is essential to provide feedback that is based on solid evidence, not on feelings and vague impressions. Objective observation requires that the instructor look at student performance with a lack of preconceived notions about what will be seen and why the student performed the way he or she did, and is able to record specific examples of behaviours.

Developing useful and meaningful observational competence is also part of the process of building reflective skills, which is addressed in detail in Chapter 8. Good observers think about and try to be aware of their own feelings and beliefs, and how these impact on what is seen during student performance. They know what stimuli lead to personal reactions. Once this awareness has been developed, instructors are able to use this knowledge to help them build and maintain objectivity when providing feedback on performance and correctives to improve future practice. This guidance is akin to that provided by coaches.

Guidance and coaching As mentioned in Chapter 2, coaching is described as "observing students while they carry out a task, providing a 'guide on the side' who intervenes and provides a scaffolding for learning to progress when necessary, but otherwise fades into the background, providing learning with opportunities for initiative and self-directed problem solving" (McLellan, 1996, p. 11). The coach is usually an expert practitioner who assists or guides the student in developing a skill and provides a safe learning environment that is supportive yet challenging. In most aviation learning contexts, the instructor is in a one-to-one relationship with the student. This allows the teacher to provide personalised and individualised training.

Once students can perform the skill to an acceptable standard, the coach–instructor will then move them to independent practice where students will be able to perform with less direct guidance. However, it should be stressed that students practicing alone may be less engaged than when they practice under close supervision. Therefore, independent practice needs to be well managed and closely monitored by the trainer.

Feedback and correctives Providing appropriate and timely feedback and correctives to learners has come to be seen as crucial to effective learning. As was seen in Chapter 1, this training skill has its theoretical basis in the behaviourist orientation to teaching and learning. Knox (1986) defines feedback as

> all ways by which learners obtain information about how their learning activities are progressing. It includes self-assessments as well as reactions by the instructor, peers, and other people. The feedback can be oral, written, or directed observation of an example. Feedback is important to learners because knowledge about the progress they are achieving and what affects it aids their decisions about persisting and adjusting their efforts. (p. 159)

According to Brookfield (1990),

> Giving evaluations is one of the most difficult, demanding, and complex tasks teachers have to face; yet, done well, it is also one of the most significant spurs to learning. Because teachers' evaluations are invested with enormous significance by students, many of whom have built up confident social faces and strong protective walls around their fragile egos, a critical comment from a teacher can be psychologically devastating. (p. 132)

As mentioned previously, it is essential that feedback be objective and constructive. Feedback must focus on the performance of the task and not on the person performing the task. The terms feedback and reinforcement are often found together in discussions of this teaching skill. However, Pithers

(1998) provides a useful guideline in differentiating between these two terms but also demonstrates how they work together to create effective learning. He states that feedback is an "inclusive term which refers to the provision of information or knowledge of results [following practice]" (p. 38). This information, put simply, lets learners know about what they may be doing correctly and what aspects of their performance need improvement. Reinforcement is a term drawn more from the behaviourist tradition and refers to positive or negative stimuli that occur after a student trials a new behaviour or skill (Pithers, 1998, p. 37). For example, positive reinforcement usually comes in the form of praise ("well done", "great work"). Research shows that positive reinforcement strengthens learning, although theorists still debate the reason for this (Pithers, 1998).

Knox (1986) suggests that there are a number of ways in which trainers can create conditions that contribute to useful feedback. These include:

1. *Receptivity* — if participants are involved in the evaluation process and understand the rationale for the use of feedback, they are likely to be receptive to the feedback, which should minimise self-protective mechanisms, such as defensiveness, distortion, and denial.
2. *Organisation* — as mentioned above, it takes planning and experience to become a discerning observer who can make judgements based on the evidence and report conclusions in useful ways.
3. *Judgements* — evaluative feedback about student performance entails explicit or implicit judgements. Therefore, it is important that students understand the basis on which such judgements are made. The most useful judgements interpret student progress in relation to clearly stated objectives, expectations, standards, or desired competencies and outcomes.
4. *Quality* — the amount and quality of feedback should be sufficient and timely so that learners have something with which to work. A list of strengths and weaknesses of a trial performance is far more useful than a summary critique such as "satisfactory" or "unsatisfactory". Receiving a critique right after some of the trials so the conclusions can be applied to the next trial is also far more useful than receiving a critique a week later when no more trials are scheduled.
5. *Use* — it is helpful to suggest to learners specific uses they can make of evaluative feedback and to provide opportunities for them to do so (adapted from Knox, 1986, pp. 159-160).

To these helpful guidelines, Turney et al. (1983) and Turney, Eltis, Hatton, Owens, Towler, Wright, and Williams (1993) would add the importance of peer support in the feedback process, advising that teachers adopt patterns of reinforcement which encourage participants to offer praise and support of each other. Of equal importance is the skill of self-assessment since the aim is to help students become independent learners. Therefore, instructors also need to encourage students to evaluate their own performance and assist them in developing that skill.

Testing and follow-up The testing or assessment of skill acquisition is the final step in this process toward mastery. This may not take place on the day of initial demonstration, especially if a complex skill is being taught. At this time, the trainer measures the participant performance against the assessment criteria, which will have been clearly communicated to the trainee previously. Competent performance under assessment formalises the acquisition of the skill and this is recorded according to the requirements of the training. If competent performance is not demonstrated, then diagnostic feedback should be provided to the trainee so that corrections can be made to performance before undergoing subsequent assessment (see Chapter 15 for more information on assessment and evaluation).

The Debriefing

During the debriefing, the instructor and the student critically reflect on the lesson in order to improve future performance. Although instructor analysis of student performance is initially helpful to the trainee, the goal is to develop in the student the habit of self-analysis. Amundson (1995), in his discussion regarding Line-Orientated Flight Training (LOFT) facilitation, states that "the most important aspect of LOFT is the facilitator's ability to get [flight] crew members to assess and discuss their performance" (p. 84). Self-assessment strategies are acquired through the instructor's guidance of student reflection on performance.

It should be stressed that a debriefing or critique is not a step in the grading process. Rather, it is a step in the learning process. Furthermore, a critique is not necessarily negative in content. It considers the good points as well as the weak ones. A critique or debriefing should help students improve their performance and provide them with something constructive with which to work and on which they can build. It should provide direction and guidance to help raise their performance level.

One practical method of developing essential self-assessment skills required by professionals within the aviation industry is for the instructor to let the student identify correct or incorrect actions carried out during the practical application of the new skill. Students should be encouraged to reflect on what happened and why it happened since self-discovery is an important component of self-analysis. According to Rogers (1969), independence and self-reliance are both facilitated when self-assessment is primary and evaluation by others is secondary. The reflective process needed for students to critically analyse and assess their performance can best be developed through student-centred debriefings.

Student-centred debriefing means that the instructor focuses the debriefing and the discussion on student self-learning rather than on "telling" the student what was performed incorrectly. Students learn more and remember better when they are active participants in the learning process (Rogers, 1969). When students make their own analysis of a situation rather than listen passively to the instructor, they become better at self-analysis and should function more effectively in new situations (Henley, Turney, & Schick, 2001).

The role of the instructor in student-centred debriefings is akin to that of a facilitator. Generally, skilful facilitators make sure that students do most of the talking. They enable students to figure out things for themselves. The more expertise that the facilitator develops, the more students can be assisted to understand the points or the issues that they have missed.

Instructors need to set expectations with students at the beginning of training. The expectation should be stated that students are encouraged to participate in evaluating their performance. Initially, students will need guidance in doing an analysis of what has taken place, but with encouragement from the instructor students will develop their self-assessment skills. The quiet student should be encouraged to voice at least some thoughts. It is the instructor's responsibility to be sure that all areas of training are covered in the conversations and that the positive aspects of the lesson are reinforced. This is usually necessary since most students tend to be self-critical and focus on errors, rather than on what has been done well.

Although it may sometimes be difficult, the instructor should avoid lecturing, or letting the discussion become instructor-centred. The instructor does not want to give an analysis of the student performance BEFORE the student does a self-analysis. This will stifle the student's thinking processes. It is important for the instructor to resist interrupting and "correcting" the student while the student is describing what he or she observed because this could give the student the feeling that only the instructor's observations count. Finally, the instructor should avoid a rigid agenda of what must be discussed (Henley, Turney, & Schick, 2001).

High level facilitation takes place over a period of many hours of instruction and practice. It is reached when the student discovers and talks about important issues and aspects of the lesson with minimal guidance from the facilitator. According to McDonnell, Jobe, and Dismukes (2000), "debriefing at this level provides the deepest and most enduring learning because crewmembers guide their own self-discovery" (p. 30). Medium- or intermediate-level facilitation can be achieved fairly quickly if the instructor subtly leads the student to practice analysis of what has just occurred during the practical lesson. Low-level facilitation occurs when students show little initiative and respond superficially. The instructor should continue to guide these students step-by-step to analyse critical issues and to consider appropriate solutions to problems. Although it will be necessary to provide some assistance, the instructor needs to resist the temptation to "tell" the student what is important.

The best way to assist students to self-assess is through the use of questioning. Questions that promote thinking and analysis, such as: What went well during your performance? or How do you think you handled the pressure? should be selected. For deeper analysis, it is best to ask questions that require the student to consider "what if" situations. Another way to promote deeper analysis could be to ask questions such as: What would you do differently next time?

The instructor can also use silence very effectively to encourage students to self-analyse. Most students are uncomfortable with silence and will make some response if the instructor waits quietly. Another benefit of silence is that students ask more questions and quiet students respond more often. Silence also allows students to formulate thoughts and to be proactive rather than reactive.

Finally, good listening skills are as important to effective instruction as good speaking skills. Active listening means paying attention to students, nodding, smiling, and making eye contact. The instructor may want to echo what has been said by repeating part of it and adding a comment or a question. This indicates serious attention to the student's thoughts. Restating the student's ideas in one's own words is another way to demonstrate active listening (Henley, Turney, & Schick, 2001).

Conclusion

Choosing teaching strategies should be an informed process. Teaching vocational skills requires a set of specialised teaching strategies. The process of demonstration, practice and feedback form the bases of these strategies and are generally agreed to be the most effective teaching approach for skill acquisition.

All trainers should aspire to develop effective presentation skills. However, the range of presentation skills can seem overwhelming to new trainers and indeed, the complexity of effective training cannot be overstated. In this chapter, key teaching skills were examined in terms of their main components, and hopefully, some of the mystery was stripped away. The importance of effective introductions and closures to training sessions was outlined with reference to research which suggests that these are optimum times to stimulate and reinforce learning. The issue of trainer anxiety in the beginning of a session was addressed in terms of the importance of planning and preparation.

Other important teaching skills, such as preparation, explanation, demonstration, practical application, the use of effective and constructive feedback were also highlighted. Each of these skills was discussed in relation to how they are thought to create conditions for effective skill acquisition and how they can be utilised in any training context.

References

Ackerman, P.L., & Kyllonen, P.C. (1991). Trainee characteristics. In J.E. Morrison (Ed.), *Training for performance: Principles of applied human learning.* (pp. 193-230). Fort Knox, KY: John Wiley and Sons.

Adelson, B. (1984). Problem-solving and the development of abstract categories in programming languages. *Memory and Cognition, 9,* 422-433.

Amundson, J.M. (1995). Line oriented flight training (LOFT) to improve cockpit-cabin communications. In N. Johnston, R. Fuller, & N. McDonald (Eds.), *Applications of psychology to the aviation system*, (pp. 62-66). Aldershot, UK: Avebury.

Anderson, J.R. (1983). *The architecture of cognition.* Cambridge, MA: Harvard University Press.

Anderson, J.R. (1987). Skill acquisition: Compilation of weak-method problem-solutions. *Psychological Review, 94,* 19-21.

Anderson, J.R. (1990*). The adaptive character of thought.* Hillsdale, NJ: Lawrence Erlbaum.

Bargh, J.A., & Chartrand, T.L. (1999). The unbearable automaticity of being. *American Psychologist, 54,* 462-479.

Brandt, R. (1992, April). On research on teaching. *Educational Leadership, 49*(7), 14-19.

Brewer, E.W. (1997). *13 proven ways to get your message across.* Thousand Oaks, CA: Corwin Press.

Brookfield, S. (1990). *The skilful teacher.* San Francisco: Jossey-Bass.

Brown, G. (1978). *Lecturing and explaining.* London: Methuen.

Charness, N. (1979). Components of skill in bridge. *Canadian Journal of Psychology, 33,* 1-6.

Chase, W.G., & Ericsson, K.A. (1982). Skill and working memory. In G.H. Bower (Ed.), *The psychology of learning and motivation* (Vol. 16, pp. 1-58). New York, NY: Academic Press.

Chi, M.T.H., Glaser, R., & Rees, E. (1982). Expertise in problem-solving. In R.J. Sternberg (Ed.), *Advances in the psychology of human intelligence* (Vol. 1, pp. 7-76). Hillsdale, NJ: Lawrence Erlbaum.

Clark, R.E. (1992). Facilitating domain-general problem-solving: Computers, cognitive processes and instruction. In E. DeCorte, M.C. Linn, H. Mandl, & L. Verschaffel (Eds.), *Computer-based learning environments and problem-solving* (pp. 265-285). Berlin: Springer.

Cooke, N.J. (1990). Modeling human expertise in expert systems. In R.R. Hoffman (Ed.), *The psychology of expertise* (pp. 29-60). New York: Springer-Verlag.

Cornford, I.R. (1999). Skill learning and the development of expertise. In J.A. Athanasou (Ed.), *Adult educational psychology* (pp. 263-290). Katoomba, Australia: Social Science Press.

Cratty, B.J. (1964). *Movement behavior and motor learning.* Philadelphia: Lea and Febiger.

Cronbach, L.J. (1954). *Educational psychology.* Chicago: Harcourt, Brace.

Cross, P. (1982). *Adults as learners.* Washington, DC: Jossey-Bass.

Daines, J., Daines, C., & Graham, B. (1995). *Adult learning, adult teaching.* Chippenham, UK: Antony Rowe.

Dreyfus, H.L., & Dreyfus, S.E. (1986). Why skills cannot be represented by rules. In E. Sharley (Ed.), *Advances in cognitive science 1* (pp. 315-335). Chichester, UK: Ellis Harwood.

Dunkin, M.J. (1987). Teaching: Art or science? In M.J. Dunkin (Ed.), *The international encyclopedia of teaching and teacher education* (p. 19). New York: Pergamon Press.

Eraut, M. (1987). Inservice teacher education. In M.J. Dunkin (Ed.), *The international encyclopedia of teaching and teacher education* (pp. 730-743). New York: Pergamon Press.

Ericsson, K.A., & Polsen, P.G. (1988). An experimental analysis of a memory skill for dinner-orders. *Journal of Experimental Psychology: Learning, Memory and Cognition, 14,* 305-316.

Fitts, P.M. (1962). Factors in complex skill training. In R. Glaser (Ed.), *Training, research and education*. Pittsburgh: University of Pittsburgh Press.

Fitts, P.M. (1964). Perceptual skill learning. In A.W. Melton (Ed.), *Categories of skill learning* (pp. 243-285). New York: Academic Press.

Fitts, P.M. (1968). Factors in complex training. In G.. Kuhlen (Ed.), *Studies in educational psychology*. Waltham, MA: Blaisdell Publishing.

Fitts, P.M., & Posner, M.I. (1967). *Human performance.* Belmont, CA: Wadsworth.

Foley, G. (1995). *Understanding adult education and training.* Sydney, Australia: Allen & Unwin.

Ghilsen, B. (1952). *The creative process.* New York: Mentor.

Glaser, R., & Chi, M.T.H. (1988). Overview. In R. Glaser, M.T.H. Chi, & M.J. Farr (Eds.), *The Nature of expertise* (pp. xv-xxvii). Hillsdale, NJ: Lawrence Erlbaum.

Gordon, D.R. (1986). Models of clinical expertise in American Nursing Practice. *Social Science and Medicine, 22,* 953-961.

Gordon, S.E. (1991). Implication of cognitive theory for knowledge acquisition. In R.R. Hoffman (Ed.), *The psychology of expertise* (pp. 99-120). New York: Springer-Verlag.

Gott, S.P., Lajoie, S.P., & Lesgold, A. (1991). Problem solving in technical domains: How mental models affect performance. In R.F. Dillon & J.W. Pellegrino (Eds.), *Instruction: Theoretical and applied perspectives* (pp. 107-117). New York: Praeger.

Henley, I. (1987). *The association between student pilot stress and flight instructor training: A study of perceptions*. Unpublished Master's thesis, University of Manitoba, Winnipeg, Manitoba.

Henley, I., Turney, M., & Schick, D. (2001). Practical strategies for aviation educators: Reflective practice and student-centred debriefing. In R.S. Jensen (Ed.), *Proceedings of the 11th International Symposium on Aviation Psychology* (CD-ROM). Columbus, OH: Ohio State University.

Hoffman, R.R. (1987). The problem of extracting the knowledge of experts from the perspective of experimental psychology. *AI Magazine, 8,* 53-67.

Hoffman, R.R., Shadbolt, N.R., Burton, A.M., & Klein, G. (1995). Eliciting knowledge from experts: A methodological analysis. *Organisational Behaviour and Human Decision Processes, 62,* 129-158.

James, R. (1995). *The techniques of instruction.* Hampshire, UK: Gower Publishing.

Klein, G.A. (1989). Recognition-primed decisions (RPD). *Advances in Man-Machine Systems, 5,* 47-92.

Knox, A.B. (1986). *Helping adults learn.* San Francisco: Jossey Bass.

Kroehnert, G. (2000). *Basic training for trainers: A handbook for new trainers* (3rd ed.). Sydney, Australia: McGraw-Hill.

Leake, D.B. (1996). CBR in context: the present and the future. In D.B. Leake (Ed.), *Case-based reasoning: Experiences, lessons, and future directions* (pp. 3-30). Menlo Park, CA: AAAI Press/ MIT Press.

Leinhardt, G. (1990, March). Capturing the craft knowledge in teaching. *Educational Researcher, 19*(2), 18-25.

Lesgold, A. (1988). Problem solving. In R.S. Sternberg & E.E. Smith (Eds.), *The psychology of human thought* (pp. 188-213). New York: Cambridge University Press.

McDonald, B. (1991). Motivation and learning: Prescriptions for change. In R.F. Dillon & J.W. Pellegrino (Eds.), *Instruction: Theoretical and applied perspectives* (pp. 118-139). New York: Praeger.

McDonnell, L.K., Jobe, K.K., & Dismukes, R.K. (2000). Facilitating LOS debriefings: Synopsis of a training manual. In R.K. Dismukes & G.M. Smith (Eds.), *Facilitation and debriefing in aviation training and operations* (pp. 26-49). Aldershot, UK: Ashgate.

McLellan, H. (1996). *Situated learning perspectives.* Englewood Cliffs, NJ: Educational Technology Publications.

Palmeri, T.J. (1997). Exemplar Similarity and the Development of Automaticity. *Journal of Experimental Psychology: Learning, Memory and Cognition, 23,* 324-354.

Patel, V.L., & Groen, G.J. (1991). The general and specific nature of medical expertise: A critical look. In K.A. Ericsson & J. Smith (Eds.), *Toward a general theory of expertise* (pp. 93-125). New York: Cambridge University Press.

Pithers, R.T. (1998). *Improving learning through effective training.* Katoomba, Australia: Social Science Press.

Pressley, M., & Woloshyn, V. (1995). *Cognitive strategy instruction that really improves children's academic performance.* Cambridge, MA: Bookline Books.

Rasmussen, J. (1983). Skills, rules, and knowledge: signals, signs, and symbols, and other distinctions in human performance models. *IEEE Transactions on Systems, Man, and Cybernetics, SMC-13,* 257-266.

Rogers, C.R. (1969). *Freedom to learn.* Columbus, OH: Charles E. Merrill.

Rouse, W.B., & Morris, N.M. (1986). On looking into the black box: Prospects and limits in the search for mental models. *Psychological Bulletin, 100,* 349-363.

Schmidt, R.A. (1975). A schema theory of discrete motor skill learning. *Psychological Review, 82,* 225-260.

Telfer, R.A. (1993). What makes aviation instruction different? In R.A. Telfer (Ed.), *Aviation instruction and training* (pp. 209-218). Aldershot, UK: Ashgate.

Terranova, M., Snyder, C.E., Seamster, T.L., & Treitler, I.E. (1989). Cognitive task analysis: Techniques applied to airborne weapons training. *Proceedings of the Human Factors Society 33rd Annual Meeting* (pp. 1309-1313). San Diego, CA: Human Factors and Ergonomics Society.

Transport Canada (1998). *Flight instructor guide.* Hull, Canada: Author.

Trappl, R. (1985). Artificial intelligence: A one hour course. In R. Trappl (Ed.), *Impacts of artificial intelligence* (pp. 5-30). North Holland, NL: Elsevier Science Publishers.

Turney, C., Eltis, K.J., Hatton, N., Owens, L.C., Towler, J., Wright, R., Cairns, L.G., & Williams, G. (1983). *Sydney micro skills: Series 1 handbook.* Sydney, Australia: Sydney University Press.

Turney, C., Eltis, K.J., Hatton, N., Owens, L.C., Towler, J., Wright, R., & Williams, G. (1993). *Sydney micro skills: Series 2 handbook.* Sydney, Australia: Sydney University Press.

Voss, J.F., Vesonder, G.T., & Spilich, G.J. (1980). Text generation and recall by high-knowledge and low-knowledge individuals. *Journal of Verbal Learning and Verbal Behaviour, 19,* 651-667.

Walker, J.T. (1995). *The psychology of learning: Principles and process.* Saddle River, NJ: Prentice Hall.

10 Implementing CRM in Initial Pilot Training

Mary Ann Turney

Air carriers initiated and implemented Crew Resource Management (CRM) more than 25 years ago. In spite of the importance placed on CRM training and human factors' skills by the aviation industry, the integration of CRM in initial pilot training is still nearly nonexistent. Most initial pilot training remains focused on the individual pilot, rather than on the pilot as a member of a team and of an aviation community of expertise. Indeed, the individualistic approach remains common in initial flight training programs. This chapter will address several important questions:

- what CRM skills are required of professional pilots?
- how does a curriculum incorporating CRM differ from the current initial training curriculum?
- what subject matter and teaching strategies can be implemented in initial flight training to include CRM skills? and
- what are the essential elements of a plan that will integrate CRM into initial flight training?

CRM Skills Requirements

As CRM programs developed, a variety of non-technical skills were identified as necessary skills for professional pilots. The number of CRM skills identified in the human factors literature ranged from as few as four to as many as 24 defined skills (Antersijn & Verhoef, 1995; Faulkner, 1996; Houle, 1995; Orasanu, 1994; Smith & Hanebuth, 1996; Young, 1995). However, the following skills were always included as essential CRM skills:

- crew coordination or teambuilding,
- communication,

- problem solving and decision making, and
- leadership and followership.

The United States Navy and the International Civil Aviation Organization (ICAO) identified these skills in the curricula of major air carriers, including United Airlines, American Airlines, Southwest, KLM, Northwest, and TWA. Two additional skills appeared in all but two of the above named air carrier curricula. These were situation awareness and workload management. The labels used to identify CRM skills sometimes varied. For example, Northwest dubbed the term “workload management” (Amundson, 1995, p. 83) while KLM used the term “stress management” (Antersijn & Verhoef, 1995, p. 243) to describe the need to avoid overloading a pilot crew member. Four other CRM skills named were planning, briefings, crew inquiry/advocacy, and conflict resolution (Antersijn & Verhoef, 1995; Houle, 1995; Maschke, Goeters, Hörmann, & Schiewe, 1995). The need for the professional pilot to integrate these CRM pilot skills with technical skills is an important goal of future pilot training (Johnston, 1993; Maurino, 1996).

Curricula Requirements for CRM

Current Curriculum

The traditional curriculum used for training flight students has been linear, sequential, and oriented toward technical proficiency alone. In the United States, for example, flight training is based on the Federal Aviation Administration’s (FAA) *Practical Test Standards* (PTS) manuals which focus on technical performance parameters, such as, “maintains heading plus or minus 10 degrees” (FAA, 1997). These manuals, which serve as the norm for flight tests and the granting of FAA certificates and ratings, have become the bases of training course designs.

Commercially produced syllabi based on the PTS manuals have become the norm for initial pilot training. Generally, the syllabi are divided into “stages” of training for each pilot certificate or rating (Jeppesen, 1996). Each stage of training provides a “stage objective”, as a guide for the student and the flight instructor, and a “stage completion standard” which will be used by the FAA Designated Examiner to assess the student throughout the training process. Although the FAA (1997) *Practical Test Standards* mentions crew resource management in its introduction to the flight examiner, there is little or no delineation of CRM skills in the examiner's checklist for the flight test. Similarly, the standard flight curricula mention skills such as communication, problem-solving, and situation awareness, but there are no specific objectives related to these skills, and no outcome measures related to CRM skills. Instead, the outcomes listed in the syllabi are strictly based on technical performance, suggesting to the instructor and student alike that the student need only acquire technical proficiency.

The situation in general aviation training contrasts dramatically with the norm for the professional pilot operating commercial aircraft. Thus, primary learning in general aviaiton does not lay a foundation for the professional pilot role.

CRM-Based Pilot Training Curriculum

After the analysis of a number of serious aviation accidents in which it was clear that human factors were a significant element, the aviation industry began to consider important training revisions. In 1989, the Flight Safety and Human Factors Study Group (a division of ICAO) published a digest entitled *Flight Crew Training: Cockpit Resource Management and Line Oriented Flight Training*. The publication was a guide for the introduction of CRM into flight training and was applicable for "all forms of flight crew training" (ICAO, 1989, p. 5).

Since the early ICAO publications, industry training initiated a number of modifications to traditional flight training. These modifications have gradually introduced non-technical skills into the flight training process. Line-Oriented Flight Training (LOFT) offered a significant opportunity to analyse technical proficiency in a real-time simulation that includes communication, decision making, leadership, and other important professional skills that are commonly referred to as CRM skills. However, as Maurino (1996) suggested, "technical and CRM skills may perhaps be separated for research purposes, but in the real world they never live an independent existence" (p. 101).

An example of a curriculum strategy that incorporates CRM is the LOFT scenario. A team of three trainers from United Airlines and the FAA, Hamman, Seamster, and Edens (1995), developed a framework for the development of LOFT scenarios. The framework included a group of related events inserted into a training session for specific CRM objectives, and a scenario that produces an operationally realistic environment that gives the crew the opportunity to combine CRM and technical skills. The event sets are complex enough to require coordinated action of all crew members for successful completion, but not complex enough to induce failure. LOFT methodology is non-linear and non-sequential. Rather, it blends non-technical and technical skills in an ongoing process.

Training the trainer is an important aspect of the LOFT methodology (Maschke, Goeters, Hörmann, & Schiewe,1995). Amundson stated: "The most important aspect of LOFT is the facilitator's ability to get the crew members to assess and discuss their performance as a crew in the LOFT and transfer that learning back to line operations. The debrief is where the real learning takes place" (p. 84).

Indeed, it is in the debriefing that pilots have an opportunity to reflect on what has taken place and to consider decisions, communications and actions in the light of alternatives and options. Hackman (1993) agreed, starting that "together, a well-designed scenario, video feedback, and an expert facilitator can provide a pilot with precisely the kind of experience that is needed to hone a new [CRM] skill and to become comfortable using it" (p. 7).

Another example of incorporating CRM in pilot training is a strategy recommended by Biegalski (1995) who developed training for the United States Air Force. The strategy involves the use of videotaped re-creations of illustrative accidents (p. 9). These videos are meant to present pilots with a "real-time" opportunity to assess their methods of enhancing situation awareness.

Videos should include humour and ought to be short. It is also more important to use examples of good performance rather than bad ones. Finally, for facilitators and evaluators, it is essential to have learners analyse negative situations — especially those which are subtle — and then follow up with a segment that shows appropriate behaviour.

Subject Matter and Training Strategies for CRM

A number of CRM strategies can be effectively implemented at the initial training levels. First, student pilots can become aware of the necessity of CRM skills through the distribution of informational articles (Cruse, 1995). Second, experiential exercises can be developed to underscore the synergy which develops from group problem solving (Young, 1995). Incorporating airline models, Young developed CRM training at Purdue University. He recommended that the curriculum include a focus on such skills as interpersonal communication, situation awareness, problem solving/decision making/judgement, leadership/followership, stress management, and critique. The four major course design areas which he developed were:

- teambuilding,
- psycho- and aero-medical factors affecting pilots,
- crew coordination and standardisation (use of checklists), and
- conflict resolution.

According to Young (1995), lectures are useful to introduce CRM concepts. However, experiential exercises are more meaningful. Synergy exercises and role-playing demonstrate the benefits of group versus individual problem solving.

Anderson and Henley (1995) advocated a curriculum design which involved the development of team skills through a problem-based learning (PBL) approach. The results of their research indicated that students felt that they had developed skills in excess of the technical knowledge and had also developed skills incorporated within the course. All the students indicated that they felt it was important to learn how to work as a team member and the majority understood the significance of having been exposed to small group interaction in order to adapt more readily to multi-crew operations in the aviation industry. The observations of the groups involved in the research project revealed clear differences between the levels of teamwork displayed by each group at the introduction of PBL compared to team performance at the conclusion of PBL. The results also showed the developmental process of some groups into effective teams (see Chapter 11 for a further discussion on the implementation of PBL to develop essential skills in aviation).

Role-playing and low-cost simulations are other effective teaching strategies for student pilots to develop CRM skills (Biegalski, 1995; Petrin, 1995). The use of LOFT in which the "student" pilot crew flies a trip in real time, then critiques its performance during a debriefing session, is an effective tool at any stage of flight training. The use of videotaping in conjunction with a LOFT scenario is a particularly effective teaching strategy, allowing the pilot crew to view their own performance and engage in self-assessment (Amundson, 1995; Biegalski, 1995; Cruse, 1995; Hackman, 1993). Today's technology provides the trainer with inexpensive videotaping equipment that can easily be combined with desktop simulation.

Essential CRM Elements in Initial Training

A Team-Based Program Design

The work of Mohrman, Cohen, and Mohrman (1995) reveals how to reconfigure an organisation "to perform work in teams when their strategy and the nature of their work call for such a design" (p. 1). In *Designing Team-Based Organizations*, the authors describe a team-based design that would accommodate important CRM program elements advocated by ICAO (1989) in *Flight Crew Training: Cockpit Resource Management and Line-Oriented Flight Training.* The team-based program design includes laying a foundation of acquired knowledge, diagnosing the current situation, agreeing on values, establishing criteria, and then creating a program design, implementation, and assessment components. A team-based design accommodates important CRM program elements, such as staff involvement in the development of training.

Changes take place in three stages according to Mohrman et al. (1995). Figure 10.1 illustrates these stages. The initial activity involves the "laying of a foundation for change through activities that get people involved in developing a shared vision of what the organization is trying to accomplish" (p. 28). Laying the foundation means identifying the expected outcomes and values which will drive the design, learning what design strategies will promote these outcomes, and diagnosing the current situation to determine the extent to which these strategies currently are in place.

The next step in the development of a team-based program design involves generating broad outlines for new design features. These features include structures, processes, systems, people practices, and roles that differ from those currently in place. For example, when teams initiate tasks, the teams should include employee input, so that team output is more likely to be shared by its members (Mohrman et al., 1995, p. 335). Highly specialised individuals can find it difficult to work in teams because they are steeped in their own knowledge and may lack some basic collaborative skills. They often apply their technical knowledge implicitly and find it hard to exchange views. The team approach supports these individuals in developing better communication skills.

Mankin, Cohen, and Bikson (1996) suggest the need to test change. They state that "the real test of the system comes within the context of its intended use — the pilot test" (p. 165). It is through the pilot test that change should initially be implemented, and then carefully evaluated with the support and involvement of management until the complete design has been refined and is ready for implementation.

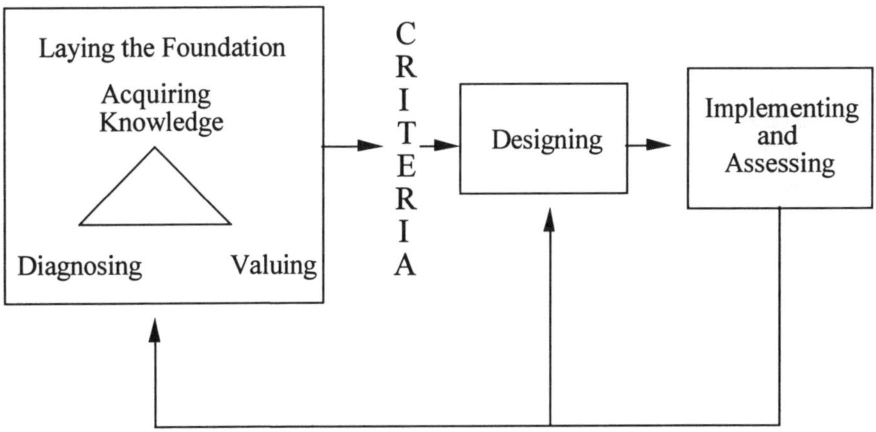

Figure 10.1 Self-design strategy for team-based organisations (Mohrman et al., 1995, p. 28)

Implementation and Assessment

The final step is the implementation and assessment process. This is a distinct activity from the design stage, yet can be done simultaneously. Important issues include deciding what sequence is best and what activities are required to start the process. "The culture of the organization may limit the speed of the transition" (Mohrman et al., 1995, p. 341). Ongoing assessment is a key learning activity. It should take place throughout the implementation process, and it should question whether or not the outcomes are occurring. Another important part of the Mohrman et al. (1995) model is the concept that implementation of new programs is a learning process. When design decisions are made by teams, the teams have an important stake in working through obstacles that may arise throughout implementation.

Critical Factors

Hayward (1995), who developed CRM training for Qantas Airlines, outlined some critical factors affecting the success of any CRM program design. These factors included "motivation for the program introduction, the quality of training design, training objectives, the budget for development of materials, training methods, training facilities, principles, ongoing reinforcement of those principles via recurrent training and the amount of real support the program receives from company management" (p 239). Biegalski (1995) advocated "a systems approach to CRM course design" and suggested that the designers "take care to design into the program only what can be received and internalized within the allotted time" (p. 6). It is obvious that these critical factors require a program design that involves considerable organisation and advanced planning.

In terms of sequence, Taggart (1997) stated that CRM program development should begin with assessment of the existing CRM culture, followed by setting objectives, and assessing limitations. Finally, research indicated that training of the trainers is critical to CRM program design (Jensen, Chubb, Adrion-Kochan, Kiekbride, & Fisher, 1995; Maschke, Goeters, Hörmann, & Schiewe, 1995). Indeed, an initial investment in training of instructional personnel is one of the most essential elements of the program design.

Guidelines for Implementing CRM in Initial Pilot Training

The following guidelines have been developed, based on the organisational design model presented by Mohrman et al. (1995), to assist course and curriculum developers in creating program designs that implement CRM in initial pilot training. The guidelines aim to provide a checklist and examples to assist in ensuring that programs are inclusive of all critical factors.

Lay a foundation for transition —

- motivate and involve all essential personnel, including top management, instructional staff, dispatchers and support staff;
- assess the existing culture and identify limitations;
- agree on important CRM skills; and
- identify expected outcomes and agree on values.

Design strategies for implementing CRM —

- identify any CRM strategies currently in place;
- list specific CRM training objectives;

- generate outlines for new training strategies and for training the instructional staff;
- consider costs and budgetary factors;
- define how instructional roles may differ;
- establish a time frame for transition; and
- consider recurrent training needs.

Assess the program design and implementation process —

- identify whether desired outcomes are occurring early in the implementation phase;
- assess obstacles, materials, equipment, and facilities (an ongoing process);
- adjust strategies, materials, equipment, sequence of activities to promote expected outcomes; and
- be sure all stakeholders are involved in the assessment process.

References

Amundson, J.M. (1995). Line Oriented Flight Training (LOFT) to improve cockpit-cabin communications. In N. Johnston, R. Fuller, & N. McDonald (Eds.), *Aviation psychology: Training and selection* (pp. 81-86). Aldershot, UK: Avebury.

Anderson, P., & Henley, I. (1995). The development of team skills using a problem-based learning approach in aviation studies. In N. Johnston, R. Fuller, & N. McDonald (Eds.), *Aviation psychology: Training and selection* (pp. 308-315). Aldershot, UK: Avebury.

Antersijn, P.A.M., & Verhoef, M.C. (1995). Assessment of non-technical skills: is it possible? In N. McDonald, N. Johnston, & R. Fuller (Eds.), *Applications of psychology to the aviation system* (pp. 243-250). Aldershot, UK: Avebury.

Biegalski, C.S. (1995). Techniques and media for CRM training: WEAAP '94 workshop report. In N. Johnston, R. Fuller, & N. McDonald (Eds.), *Aviation psychology: Training and selection* (pp. 5-10). Aldershot, UK: Avebury.

Cruse, G. (1995). CRM for the smaller operator. In N. Johnston, R. Fuller, & N. McDonald (Eds.), *Aviation psychology: Training and selection* (pp. 57-61). Aldershot, UK: Avebury.

Faulkner, J.P.E. (1996). Evaluating crew resource management in an ab-initio pilot training: The first seventy hours. In B.J. Hayward & A.R. Lowe (Eds.), *Applied aviation psychology* (pp. 143-147). Aldershot, UK: Avebury.

Federal Aviation Administration (1997). *Practical test standards for the commercial pilot.* Washington, DC: US Government Printing Office.

Hackman, J. (1993). Communication and CRM. In E. Wiener, B. Kanki, & R. Helmreich (Eds.), *Cockpit resource management* (pp. 51-55). San Diego, CA: Academic Press.

Hamman, W.R., Seamster, T.L., & Edens, E.S. (1995). LOFT/LOE in air carrier training. In N. Johnston, R. Fuller, & N. McDonald (Eds.). *Aviation psychology: Training and selection* (pp. 87-92). Aldershot, UK: Avebury.

Hayward, B. (1995). Extending crew resource management: An overview. In N. McDonald, N. Johnston, & R. Fuller (Eds.), *Applications of psychology to the aviation system* (pp. 237-242). Aldershot, UK: Avebury.

Houle, T.W. (1995). An operational model for the evaluation of Crew Resource Management (CRM) skills in Line Operations Simulations (LOS). In N. McDonald, N. Johnston, & R. Fuller (Eds.), *Applications of psychology to the aviation system* (pp. 251-255). Aldershot, UK: Avebury.

International Civil Aviation Organization. (1989). *Flight crew training: Cockpit Resource Management (CRM) and Line-Oriented Flight Training (LOFT).* (ICAO Circular 217-AN/132). Montreal, Canada: Author.

Jensen, R.S., Chubb, G.P., Adrion-Kochan, J., Kiekbride, L.A., & Fisher, J. (1995). Aeronautical decision making in general aviation: New intervention strategies. In R. Fuller, N. Johnston, & N. McDonald (Eds.), *Human factors in aviation operations* (pp. 5-10). Aldershot, UK: Avebury.

Jeppesen-Sanderson. (1996). *Private pilot syllabus*. Englewood, CO: Author.

Johnston, N. (1993). *LOS/LOFT debriefing skills: An introduction.* Aer Lingus, Flight Operations Training & Standards Division, Dublin, Ireland.

Mankin, D., Cohen, S.G., & Bikson, T.K. (1996). *Teams and technology: Fulfilling the promise of the new organization.* Boston, MA: Harvard Business School Press.

Maschke, P., Goeters, K.M., Hörmann, H.-J., & Schiewe, A. (1995). The development of the DLR/Lufthansa Crew Resource Management Training. In N. Johnston, R. Fuller, & N. McDonald (Eds.), *Aviation psychology: Training and selection* (pp. 23-31). Aldershot, UK: Avebury.

Maurino, D.E. (1995). The future of human factors and psychology in aviation from ICAO's perspective. In N. McDonald, N. Johnston, & R. Fuller (Eds.), *Applications of psychology to the aviation system* (pp. 9-15). Aldershot, UK: Avebury.

Maurino, D.E. (1996). Eighteen years of the CRM wars: A report from headquarters. In B.J. Hayward & A.R. Lowe (Eds.), *Applied aviation psychology* (pp. 99-109). Aldershot, UK: Avebury.

Mohrman, S.A., Cohen, S.G., & Mohrman, A.M., Jr. (1995). *Designing team-based organizations: New forms for knowledge work.* San Francisco, CA: Jossey-Bass.

Orasanu, J.M. (1994). Shared problem models and flight crew performance. In N. Johnston, N. McDonald, & R. Fuller (Eds.), *Aviation psychology in practice* (pp. 255-281). Aldershot, UK: Avebury.

Petrin, D.A. (1995). Human factors in aviation: A course description. In N. Johnston, R. Fuller, & N. McDonald (Eds.), *Aviation psychology: Training and selection* (pp. 258-263). Aldershot, UK: Avebury.

Smith, G.M., & Hanebuth, C.E. (1996). Differences in CRM evaluation between LOFT facilitators and line check captains. In B.J. Hayward & A.R. Lowe (Eds.), *Applied aviation psychology* (pp. 148-155). Aldershot, UK: Avebury.

Taggart, W.R. (1997). *The NASA/UT/FAA Line/LOS checklist: Assessing system safety and crew performance.* Retrieved May 24, 2001 from University of Texas, NASA/University of Texas/FAA Aerospace Crew Research Project Web site http://www.psy.utexas.edu/psy/helmreich/nasaut.htm.

Young, J.P. (1995). Cockpit (Crew) resource management: Development of a collegiate course. In N. Johnston, R. Fuller, & N. McDonald (Eds.), *Aviation psychology: Training and selection* (pp. 11-16). Aldershot, UK: Avebury.

11 Using Problem-Based Learning to Develop Essential Skills in Aviation

Irene Henley

Introduction

The importance of critical thinking, problem solving, decision making, and effective teamwork in aviation cannot be over emphasised. While the failure of teams in many fields may simply result in reduced effectiveness and cause frustrations and/or delays, the failure of a flight crew or an air traffic controller team may lead to a loss of life. Hackman (1993) notes that "it is the team, not the aircraft or the individual pilot, that is at the root of most accidents and incidents" (p. 49). Indeed, over two-thirds of all accidents in commercial aviation have been attributed to human factors (Hunter, 1995; O'Hare, 1999). Some of the problems identified in accident investigations include inadequate leadership, inappropriate distribution of tasks, poor workload management, ineffective communication between crew members, limited use of available resources, and the failure to create an environment that would promote the development and maintenance of effective teamwork (Helmreich & Foushee, 1993; Prince, Chidester, Bowers, & Cannon-Bowers, 1992). In spite of these findings, little attention has been given to the development of crew resource management and team skills in *ab initio* (primary) pilot training.

Until recently, pilot training, especially at the *ab initio* stage, has concentrated mainly on the acquisition of factual knowledge and the development of the piloting skills of the individual pilot, that is, on individual performance. However, with the reassessment of Crew Resource Management (CRM) in the airline industry, both researchers and practitioners are arguing that this "single-pilot tradition" (Helmreich & Foushee, 1993), or "individualistic ethos" (Johnston, 1993) in flight training needs to be reconsidered. Further, the development of CRM and team skills should not be limited to multi-crew operations because even in single-pilot operations pilots do not work in isolation; they need to interact effectively and efficiently with a host of people, such as other crew members, managers, maintenance personnel, passengers, air traffic controllers, etc. Clearly, more emphasis needs to be placed on the performance of pilots as team members and on factors that affect crew coordination and teamwork (Hackman, 1993). Johnston (1993) stresses that, "if we want pilots to perform as a crew – as team members – we should train them as a crew throughout" (p. 16).

Furthermore, it has increasingly been implied that upon entering the workforce, graduates for professions are inadequately prepared when it comes to the practical application of the knowledge gained for their degree or in vocational courses (Barrows & Abrahamson, 1964; Barrows & Mitchell, 1975; Barrows & Tamblyn, 1976; Boud & Feletti, 1998; Diehl, 1991; Henley, 1995; Ross, 1992). This could suggest that the teaching approaches adopted may not adequately prepare students for all the demands of professional life (Gibbs, 1992; Harden, Sowden, & Dunn, 1984).

The aim of traditional training programs has generally been to impart large bodies of "need-to-know" information at the expense of higher-level competencies (i.e., inquiry skills, critical thinking, decision making, self-assessment, and team skills), with more emphasis on memorisation and the regurgitation of material presented in class than on analytical reasoning and professional judgement. This factual information may prove helpful to pass examinations but it may not necessarily be useful in practice because it may have a limited "shelf life" (Adriaanse, Broekhuizen, Geerligs, van Oorde, & Zaremba, 1998; Anderson, 1993; Anderson & Henley, 1994). Once qualified, however, students are generally expected to be able to intuitively apply this knowledge in solving workplace problems. However, the expanding knowledge base and the rate of change and technological advancement experienced in professions such as aviation make self-directed learning skills a more desirable qualification than the acquisition of large amounts of factual knowledge which quickly become redundant or outdated. According to Boud and Feletti (1998), "it is more important for students to be able to learn quickly, effectively and independently when they need it, than it is for them to have assimilated (at graduation) all the information which their teachers believe is desirable" (p. 4).

Problem-Based Learning (PBL) is an alternative teaching strategy designed to facilitate learning and skill development using problematic scenarios based on realistic workplace situations. The problem or key issue of practice that serves as a trigger for research and learning is intended to actively engage students in the learning process by allowing them to identify the limits of their current skills and knowledge and then devise strategies whereby they gain the knowledge necessary to solve the problem or satisfactorily address the key issue. This chapter defines problem-based learning, examines some of the educational objectives of a PBL approach, and discusses their application to the aviation training context. The PBL process is also described and steps for designing a PBL course are suggested.

Defining the Terms

There is often some confusion regarding the difference between problem-solving learning and problem-based learning (Boud & Feletti, 1998). Therefore, a brief definition of these terms seems in order.

Problem-Solving Learning

Problem solving is a technique where a problem is presented and a plan of attack or a strategy to solve the problem is given to the students by the teacher. The teacher guides the students through the problem to arrive at a set solution. Exercises are then provided for the students to apply the plan of attack or strategy in solving similar problems.

The problem scenarios in problem-solving learning normally are set within and bounded by a specific subject or discipline area. The focus is largely on acquiring the answers expected by the teacher, and these answers are generally rooted in the information provided to the students by the teacher in lectures and/or reading material (Barrows & Tamblyn, 1980; Savin-Baden, 2000; Woods, 1985, 1994). In order to discover the solutions to the problems being presented, the students are not generally expected to explore or research much further than the material or texts provided for the subject area. As a result, the solutions are usually "linked to a specific curricula content, which is seen as vital for students to cover in order for them to be competent and effective practitioners" (Savin-Baden, 2000, p. 2).

Problem-Based Learning (PBL)

Problem-based learning also predominantly uses problems, however, it is not simply adding problem-solving activities to an otherwise discipline-centred curriculum. PBL is a way of integrating and coordinating subject disciplines, where the curriculum is centred on key issues or problems of professional life (Boud & Feletti, 1998). PBL is defined as an activity in which the trigger for learning is a key issue, a query or a problem the learner wishes to solve (Barrows, 1985; Barrows & Tamblyn, 1980; Boud, 1985; Morgan, 1983). Further, in PBL, a "real-life" situation or problem is presented to students *before* any factual information is given (Woods, 1985, 1994). The learners must conduct the research necessary to obtain all the pertinent information needed to resolve the problem. Once the relevant information is acquired, it is applied back into the problem to arrive at a reasonable and justifiable solution.

During the research and problem-resolution process, learners broaden their understanding of content (expand their knowledge base) through direct involvement with problems of professional realism and of relevant context, and at the same time they develop higher-level competencies, such as problem definition, inquiry skills, analytical reasoning, critical judgement, self-assessment, and self-directed learning. Furthermore, "students are in control because they must select the knowledge needed to solve the problem, learn that knowledge and relate it to the problem" (Woods, 1985, p. 19). In short, PBL encourages learners to be actively involved in learning and to be responsible for solving problems relevant to professional practice (Anderson & Henley, 1994; Henley, Anderson, & Wiggins, 1999).

Finally, PBL also provides the opportunity to integrate knowledge and skills from various related disciplines because scenarios based on problems that arise in professional practice are not compartmentalised into distinct subject areas. For example, the resolution of a problem encountered during a night charter flight would require the integration of knowledge and skills gained in flight training, flight rules and procedures, principles of flight, engines, systems and instrumentation, aircraft performance and operations, navigation, meteorology, and human factors.

Problem-Based Learning in Context

Although PBL is not new, it resurfaced close to four decades ago because of the realisation that traditional teaching often falls short of the fulfilment of many educational objectives, such as the development of analytical reasoning, critical thinking, decision making, and team skills (i.e., higher-level competencies). In 1965, in an attempt to overcome discrepancies between tertiary education and the requirement of the professions, the Faculty of Health Sciences at McMaster University in Canada implemented what became known as a problem-based approach to learning (Barrows & Tamblyn, 1980). Since then, problem-based learning has been adopted by numerous educators in teaching the professions (e.g., medicine, nursing, architecture, engineering, etc.) in many countries around the world (Boud & Feletti, 1998).

Traditional courses in aviation have also been criticised for their lack of attention to human factors, for placing little emphasis on teamwork, and for poor attempts at developing critical thinking, decision making, research skills and self-assessment in students (Diehl, 1991; Henley, 1995; Johnston, 1993; Ross, 1992). Based on research evidence, curriculum changes have occurred at a few colleges and universities worldwide and PBL has been implemented as an alternative teaching approach because of the outcomes suggested by PBL research and the identification of the needs of adult learners and of graduates in the workplace (Adriaanse et al., 1998; Anderson, 1993; Anderson & Henley, 1994, 1995).

Problem-Based Learning versus Traditional Teaching

The respective advantages and disadvantages of PBL and traditional teaching have frequently been compared in the literature dealing with PBL (Anderson, 1993; Boud & Feletti, 1998; Harden et al., 1984; Schwartz, 1998; Todd, 1991). Proponents of PBL argue that a PBL approach allows for a broader, more relevant and integrated knowledge base than the traditional, lecture-based approach (Boud, 1988; Coles, 1985; Engel, 1998; Margetson, 1998). While traditional teaching tends to burden students with large amounts of facts to be assimilated (memorised) and subsequently regurgitated for the purpose of passing examinations, PBL emphasises knowledge which is pertinent to the profession by placing it in a practical context (Birch, 1986; Boud & Feletti, 1998).

Traditional teaching approaches also tend to focus more on the teachers and what they teach, whereas in a facilitative, student-centred approach such as PBL students have more responsibility for their learning (Harden et al., 1984). Furthermore, PBL provides an excellent medium for developing team skills because it is based on simulated, but true-to-life problems that are usually best solved through the interactions and efforts of a small group. As mentioned previously, aviation is an industry where effective, coordinated teamwork is essential. Accordingly, both researchers and practitioners are suggesting that training in team skills should be commenced during the *ab initio* stage of flight training, since team skills, like any other skill, will only fully develop with continued practice (Anderson, 1993; Diehl, 1991; Johnston, 1993; Sams, 1989). In this respect, PBL can be seen as a precursor to CRM and Line-Oriented Flight Training (LOFT).

PBL, according to its advocates, also takes account of how students learn and encourages a deep approach to learning (Boud & Feletti, 1998; Coles, 1998). A study by Coles (1985) compared the learning styles of PBL medical students with those of traditionally taught medical students both prior to and following their first year of study. Coles found that, although both groups of students showed similar approaches to studying prior to entering medicine (low reproduction, high meaning and high versatility), by the end of their first year the traditional students appeared to be detrimentally affected. Their studying approaches declined towards greater reproduction, lower meaning and lower versatility. However, PBL students maintained their desirable approaches to learning, even leaning towards less reproduction by the conclusion of their first year. Coles concluded that traditional teaching appeared to have detrimental effects on learning approaches, while PBL seemed to be "creating an educational climate [enabling] ... students to learn ... in a desirable manner" (p. 309). Similarly, Albanese and Mitchell (1993) found that PBL students were "more likely to study for understanding or to analyse what they need to know and study accordingly" (p. 52) which implies a deep rather than a surface approach to learning and the use of more effective learning strategies (see Chapters 4 & 5 for more information on approaches to learning and learning strategies).

Despite the claimed advantages of PBL, not all teachers and students readily or easily embrace PBL. There are many challenges to PBL (Margetson, 1998). Most teachers or trainers in the aviation training context, for example, lack expertise in the development and planning of a PBL curriculum and many may be reluctant to change their teaching approach since the lecture is more comfortable for them. It is the method with which they are most familiar, therefore they may not adapt easily to other teaching strategies. People generally resist new approaches because they fear change.

Like teachers, most students are used to subject-based learning and didactic approaches, such as the lecture. Some students may feel threatened by a PBL approach and it may conflict with their preferred learning style. For example, some students find PBL more demanding and challenging than a lecture because they have to determine the information needed and conduct the research to resolve problems; some learners would prefer to be more passive (Anderson & Henley, 1994). Furthermore, some students feel uncomfortable because there are no straightforward answers to most problems or they may feel insecure not knowing exactly what the teacher expects as a correct answer (Anderson, 1993; Schwartz, 1998).

Another challenge stems from those who "conceive education as a largely one-way process of direct information transmission from the knowledgeable teacher to the ignorant student" (Margetson, 1998, p. 39). In most traditional teaching contexts, the teacher is seen as a content expert and teaching is viewed mainly as a process of information transmission with "learning merely an information-absorption process" (p. 39). As a result, undue emphasis is placed on didactic forms of teaching. The ill-founded assumption in this case is that unless the teacher covers "need-to-know" material in class, students will not acquire what is considered essential knowledge. Unfortunately, this narrow association of teaching with learning has been adopted unreflectively and uncritically by many educators and practitioners alike. Margetson (1998) argues that PBL, unlike the conventional approach, "pays due respect to both student and teacher as persons with knowledge, understanding, feelings and interests who come together in a shared educational process" (p. 39). He also stresses that PBL practitioners do not deny the importance of content, quite the contrary. However, "problem-based learning requires a much greater integration of knowing *that* with knowing *how*" (p. 38).

Objectives of Problem-Based Learning

According to Engel (1998), some of the objectives of a PBL course are to develop high professional competency, foster reasoned decisions in unfamiliar situations, deal with ill-defined problems, adapt to and participate in change, develop self-evaluation skills, appreciate another person's point of view, and work productively as a team member — all of which are of paramount importance in the education and development of aviation professionals. Barrows (1985, 1986) also identified a number of key educational objectives considered relevant to medical education which hold similar value for other professions, including aviation. They are practical application, analytical reasoning, team skills, motivation for learning, and self-directed learning and self-assessment skills. These objectives, Barrows (1986) suggests, could best be achieved using a PBL approach.

Practical Application

An essential objective of education for the professions, according to Barrows (1986), involves the structuring of knowledge for use in practical applications. Tulving and Thomson (1973) found that the more realistic the learning situation is to its practical application, the better the performance of students. Supporting this perspective, research by Barrows and Tamblyn (1976) and Willems (1981) revealed that many students familiar with content knowledge were incapable of applying their knowledge given a practical situation. These students wrote highly knowledgeable reports and solved complex problems but appeared unable to take the next step and use their new knowledge in a practical context. Based on the findings of these studies, Schmidt (1983) concluded that "people can possess knowledge which they seem unable to apply. They know information but can not use it" (p.11). In fact, the way knowledge is stored and retrieved depends on how it is acquired (Barrows, 1986). In a PBL approach, the authenticity of the learning scenarios ensures that the knowledge and skills gained will be readily available in the kinds of situations that graduates will face once in the workplace.

Furthermore, adult educators have repeatedly stressed the importance of both practical application and relevancy for adult learners (Brookfield, 1990, 1995; Cross, 1982; Houle, 1980; Kidd, 1973; Knowles, 1980). Hence, there needs to be an integration of practical skills, higher-level competencies, and theoretical knowledge in teaching the professions. A PBL approach would seem more apt to facilitate such an integration of skills and knowledge than would a traditional lecture-based approach because PBL is triggered by realistic situations where practical skills and theoretical knowledge are closely intertwined (Anderson, 1993; Henley, Anderson, & Wiggins, 1999). According to Barrows (1986), learning that is driven by the challenges of practice and integrated into the reasoning required to evaluate and resolve problems experienced in the workplace promotes structuring of knowledge to support future practical application and ensures that graduates are better prepared to enter practice with strong attributes of self-directed learning and well-grounded knowledge, attitudes and procedural skills.

Analytical Reasoning Process

Proponents of PBL argue that as students acquire knowledge, they should also be empowered to think and reason (Barrows, 1986; Boud, 1988; Engel, 1998; Savin-Baden, 2000; Woods, 1994). Barrows (1986) agrees, stressing that factual knowledge is not the most valuable contribution to be made to a doctor's education. He suggests instead that problem-solving skills required in analytical reasoning processes are of paramount importance to the student doctor. According to Barrows, the problem-solving skills involved in clinical reasoning include hypothesis generation, inquiry, data analysis, problem synthesis, and decision making. These skills must be developed in association with the acquisition of theoretical knowledge and the practical skills inherent to the profession to ensure that problem solving and knowledge will function together in the workplace. Research has shown that PBL is an instructional method that can provide students with knowledge and skills suitable for problem solving or analytical reasoning (Barrows & Tamblyn, 1980; Engel, 1998; Schmidt, 1983; Woods, 1994).

Aviation, like medicine, presents situations that consistently require the use of effective problem-solving skills. Every contingency cannot be planned for or foreseen in aviation's complex and ever-changing environment (Foley, Anderson, & Henley, 1995). For example, there was no specific training for the crew of United Airlines Flight 232 that crashed at Sioux City on July 19, 1989, where the pilots managed to crash land the jet airliner using differential power when all the hydraulic primary control systems had failed (Job, 1996; National Transportation Safety Board, 1990). Nor was there specific prior training for the two pilots of an Aloha Airlines B737 on April 28, 1988, who found themselves with a unique situation when the forward upper section of the fuselage tore off in-flight causing multiple systems to fail (Job, 1996). These predicaments were unprecedented and, as such, the pilots had no formal training in how to deal with systems failures of such magnitude. However, the pilots used refined problem-solving skills and effective team skills to improvise and adapt their collective knowledge and experience to a unique and unpredictable situation. These skills enabled the crew in both cases to handle multiple systems failures and unusual aerodynamic characteristics of a crippled aircraft so that they could land the aircraft with a minimum number of casualties (Henley, Anderson, & Wiggins, 1999). As in clinical reasoning, the skills used by the pilots included identifying the problem, diagnosing its likely cause(s), generating hypotheses, data analysis, problem synthesis, decision making, managing the condition as effectively as they could, and continuously monitoring and assessing their actions.

Through the presentation of problem simulations and real-life scenarios, PBL has the potential to refine the critical reasoning processes and the decision-making skills of students so that the knowledge and skills gained can be applied successfully in a practical context. According to Savin-Baden (2000),

> problem-based learning can help students to *learn with complexity*, to see that there are no straightforward answers to problems scenarios, but that learning and life take place in contexts, contexts which affect the kinds of solutions that are available and possible. Learning such as this is not just a straightforward method of problem-solving, but it helps people to learn how to learn and to link learning with their own interests and motivations (p. 5).

Team Skills

Since the flightdeck is a time-critical environment where complex aircraft systems have to be continuously monitored and interpreted and masses of information need to be analysed and processed in-flight, safe operations require highly coordinated team interactions and performance (Foushee, 1984; Diehl, 1991; Hackman, 1993; Johnston, 1993). According to Boud and Feletti (1998), having students work cooperatively as a group, exploring information in and out of class, can facilitate the group's learning process. Within a small group, the exploration of the problem from various perspectives helps stimulate the curiosity of students to find an explanation or to fill in knowledge gaps. Furthermore, because the hypotheses and solutions proposed have to be explained and defended, critical thinking, analytical reasoning, communication skills, and conflict resolution skills are also developed. Teamwork divides the tasks among group members but it often doubles the success due to the synergy within the group.

Research by Anderson (1993) has shown that a PBL approach using small tutorial groups could enhance the development of effective team skills.

A problem-based learning approach was implemented in a multidisciplinary subject in aviation studies (*Light Twins and Specialised General Aviation Operations*) at an Australian university to investigate whether a PBL approach would facilitate the development of team skills in *ab initio* students enrolled in an undergraduate aviation studies program. Since the development of a team is a dynamic process, team effectiveness or specific team skills can be difficult to quantify (Baker & Salas, 1992). Therefore, to obtain a holistic insight into the developmental process and the performance of an effective team, Anderson (1993) adopted a mixed-method approach that allowed for the incorporation of both quantitative and qualitative sources of data (structured observations, video recordings, and participants' perceptions through the use of interviews and a questionnaire). The analysis of each block of data sought to enhance, elaborate and illustrate the others, so that overlapping yet different facets of the problem could be addressed. The multi-method approach enabled the measurement of team skills, in addition to providing an insight into factors relating to the successful development of a collection of individuals into either a group or a team (Buchholz & Roth, 1987; Woodcock, 1989).

The structured observations and video recordings revealed clear differences between the levels of teamwork displayed by each group at the introduction of PBL compared to team performance at the conclusion of PBL. The observations also showed the developmental process of some groups into effective teams. The data revealed that all the groups displayed some behaviours critical for effective teams, however, not all groups developed to the same extent (Anderson, 1993; Anderson & Henley, 1994, 1995). These findings were supported by the interviews with students.

Generally, students indicated that as a result of the PBL course they had a better understanding of how to work more effectively with the members of their groups, they had gained "group skills" which participants described as tact, effective listening, better communication skills, and cooperation (Anderson, 1993). All the students also noted that since they planned to work in the aviation industry they recognised the importance of learning how to work as a team member. The majority indicated that they valued the opportunity of having been exposed to small-group interactions because it would help them adapt more readily to multi-crew operations once they started working in the aviation industry (Anderson & Henley, 1994).

Many participants also indicated that by working with problems in small groups they were applying their inquiry skills, problem solving skills and critical thinking skills more explicitly. One student echoed the responses of many, saying that he had learned "*team skills, research skills, working to a deadline, submitting group work and being able to look at a scenario and make decisions*" (cited in Anderson & Henley, 1994, p. 333). Other students reported that the course had made them more sensitive of their behaviour within a group and reinforced their awareness of how they reacted and dealt with people. Participants also indicated that through exposure to group work and by having to reflect on the group's progress on a regular basis, they learned "first hand" about the process and development of their group and about themselves.

According to most students, an added benefit of the PBL approach using tutorial groups was that they felt more confident about their understanding of the content for the course. Many said that because they had to research the material themselves, discuss their findings with other group members, and even defend the result of their research by providing convincing evidence to support their views, they did not have to study for this subject before the final assessment because they were confident that they understood the material well (Anderson, 1993). The final results on the essay-type examination at the end of the semester supported their self-assurance. Overall the results of the PBL students were better than the marks obtained by groups taught using a traditional approach and the essays of the PBL group showed a more in-depth understanding of critical concepts regarding light twin-engine aircraft performance and a better grasp of significant human factors related to multi-engine operations.

Motivation for Learning

Because students are actively involved in the learning process and the problems to be solved are realistic, the PBL approach usually generates more interest than the traditional lecture-based approach which, according to research on learning styles, appeals to only 22 percent of learners. Group work and problem solving, on the other hand, are preferred by well over 70 percent of learners (Burns, 1989; Kolb, 1984, 1999. Also see Chapter 4). The perceived relevance of working with true-to-life problems and the challenge of solving these problems provide strong motivation for learning (Anderson, 1993; Barrows, 1986). Students are generally stimulated to extend their knowledge in the context of meaningful problems (Boud & Feletti, 1998; Margetson, 1998). According to Barrows (1986), the increased motivation to learn about the subject drives students and facilitates the extraction and more in-depth understanding of information from learning resources; as a result, students improve both their inquiry skills and their knowledge base.

Self-Directed Learning and Self-Assessment Skills

Technological advances in aviation are forcing increasingly rapid changes with more far-reaching implications than ever before. Information is quickly rendered redundant; as a result, professionals must continually update their knowledge and their skills. Thus, the aviation professional must be a perpetual student. Accordingly, some educators have recognised that their primary goal is not to teach specific amounts of knowledge, but rather to foster the development of self-reliance, self-assessment skills, self-directed learning, and professional competence (Barrows, 1985; Foley, Anderson, & Henley, 1995; Henley, Anderson, & Wiggins, 1999; Woods, 1994). As Engel (1998) notes, accelerating and massive changes within the professions "make self-directed learning throughout [a professional's career] a *sine qua non*" (p. 17).

PBL students are required to independently locate information, critically assess material, and use a broad range of resources in order to understand new concepts and solve problems, just as professionals are required to do when working in the industry. Furthermore, skills of reflection and self-assessment allow students to become more critical and sensitive to their personal learning needs, and motivate them to locate appropriate resources to broaden their knowledge base. In short, research has shown that PBL is a teaching approach that can foster the development of open-mindedness, inquiry skills, self-reliance, self-assessment, and self-directed learning by making students responsible for their own learning in preparation for that time when they join the workforce and become truly independent learners (Anderson, 1993; Barrows, 1986; Barrows & Abrahamson, 1964; Barrows & Tamblyn, 1976, 1980; Barry, 1996; Engel, 1998; Woods, 1994).

The Problem-Based Learning Process

Several authors and researchers have discussed the use of PBL in an industry training context and have described key steps involved in the PBL process (Barrows, 1985; Barrows & Tamblyn, 1980; Barry, 1996; Boud, 1985, 1988; Boud & Feletti, 1998; Chappell, Gonczi, & Hager, 1995; Engel, 1982; Woods, 1994). The following steps are common to most PBL approaches:

- *The learning process begins with participants encountering a problem or a key issue for practice.* According to Chappell et al. (1995), "the problems posed in PBL courses are central to practice and, therefore are said to avoid many of the difficulties which are found in more traditional vocational education programs. These difficulties include issues of relevance, learner experience, and the difficulties associated with the 'front-end' approach to many courses, whereby all the knowledge is provided first" (p. 184).
- *The participants encounter the problem or key issue in the same way they would in the workplace.* In this way, the learning is contextualised and seen by the participants as directly relevant to their working lives. This clearly enhances motivation on the part of participants, especially adult learners, because the problem forms the stimulus and reason for deciding what should be learned.
- *The participants work with the problem or the key issue and apply critical reasoning and existing knowledge. The material is sufficiently challenging to cause the participants to evaluate personal levels of knowledge.* First, students analyse the problem and begin to generate broad hypotheses in relation to the underlying cause(s) for the problem. Important issues are isolated. Second, participants identify what further information is required in order to challenge and evaluate the hypotheses and ultimately solve the problem. Limitations or gaps in current knowledge and understanding set the starting point for the acquisition of new knowledge and skills, and are used as a guide for individualised study and subsequent group discussions.
- *The participants identify what they don't know and develop strategies to acquire the knowledge they will need to solve the problem or address the key issue that was raised.* Information is gathered and organised in some way that guarantees that important issues are not overlooked. Students apply an inquiry strategy to conduct research and individualised study in order to formulate logical and rational answers to the questions raised by the problem that was presented. Effective decision making is the art of asking the right questions. According to proponents of PBL, this is an important workplace skill since "knowledge is never static and can never be totally accommodated in pre-employment programs. [They] therefore view the purpose of pre-employment courses as developing the skills to learn new knowledge quickly, effectively and independently" (Chappell et al., 1995, p. 184). The benefits of these skills can also be seen in relation to continual professional development long after initial training

content has become outmoded or superseded.

- *The skills and knowledge acquired are then related back to the problem which has formed the basis of the learning.* The students apply the newly acquired knowledge back to the initial problem to evaluate the effectiveness of learning and to reinforce learning. During this phase learners need to articulate their thinking and clarify what they have learned. They discuss how they approached the problem and attempt to resolve any remaining difficulties or misconceptions. Articulation and clarification are the keys to the transfer of knowledge to new situations. As students re-examine the problem and evaluate the hypotheses that were formulated, they can also monitor their own progress and, as a result, they also improve their critical reflection skills.
- *The learning that has occurred in working with the problem or the key issue is summarised and integrated into the participants' existing knowledge.* This integration process can be facilitated through student-centred debriefing where students are encouraged to critically reflect on what they have learned, on group dynamics, on their progress as a team, and on their personal contribution to the group. By reflecting periodically on their learning processes, learners can consider how their approaches are effective or ineffective (see Chapter 8 for more information on critical reflection). Video recordings of the group tutorials can become powerful tools in assisting students to self-assess during this reflective phase. Seeing their behaviour as a group member can be quite revealing for many participants and assist in the development of unbiased self-assessment. Entries in reflective journals can also be used as a source for discussion and further reflection. Critical reflection is the key to continuous improvement and to constructing a more holistic, in-depth understanding of one's learning approach and of the requirements of professional practice (adapted from Barry, 1996, pp. 15-16; Barrows & Tamblyn, 1980, pp. 38-56, 71, 191-192; Chappell et al., 1995, p. 184).

Designing a PBL Course

It is important at the outset to understand that PBL is not limited to one specific learning or teaching approach. PBL is an umbrella term widely used to describe an array of problem-centred programs. Barrows (1986) developed a taxonomy of PBL methods to identity and create an awareness of the spectrum of PBL approaches. PBL methods, for example, vary greatly in their degree of teacher or student direction. PBL can be situated anywhere from a fairly high teacher-directed approach using lecture-based cases to *closed loop* PBL, where students learn completely independently, as aviation professionals would do once in the industry (Henley, Anderson, & Wiggins, 1999). Accordingly, the role of the facilitator progresses on the continuum from modelling, to coaching, and finally to fading (Barrows, 1985).

In designing a PBL course, the first step is to develop clear educational objectives to ensure that all the course content is covered and that the desired skills are developed through the use of appropriate triggers (see Figure 11.1). Having established educational objectives, the degree of teacher direction should be considered. For example, implementing a closed loop PBL approach with first year students might be unreasonable and ineffective, since these students may not have the necessary skills, experience, or resources to be fully independent learners. However, designing PBL so that it becomes increasingly student-directed as students near graduation is one way in which the knowledge and skills gained in the educational environment and those needed in the workplace can be effectively integrated (Henley, Anderson, & Wiggins, 1999).

The degree of integration of disciplines also will need to be considered. For example, will the course be interdisciplinary or intradisciplinary? In most aviation programs, subjects such as flight rules and procedures, meteorology, navigation, aircraft performance and operations, engine systems and instrumentation, and human factors are segregated from one another as individual subjects during the learning process. In practice, however, the professional pilot does not work in a compartmentalised environment, but integrates knowledge acquired from all the aviation disciplines (Anderson & Henley, 1994, 1995). Pilots conducting night freight operations, for example, are a long way from segmented courses in which they learned meteorology, human factors, aircraft systems and performance as separate units (Foley, Anderson, & Henley, 1995). Therefore, something must be done to link what is learned in the classroom to what is needed in the working environment.

PBL is rooted in problems taken from the working context in an attempt to bridge that gap. A PBL approach can facilitate the integration of knowledge and practical skills by eliminating course segmentation to create a more practical and integrated approach to the study of aviation. Knowledge, skills, and attitudes to be learned are embedded in tasks and settings that reflect the uses of these competencies in the workplace.

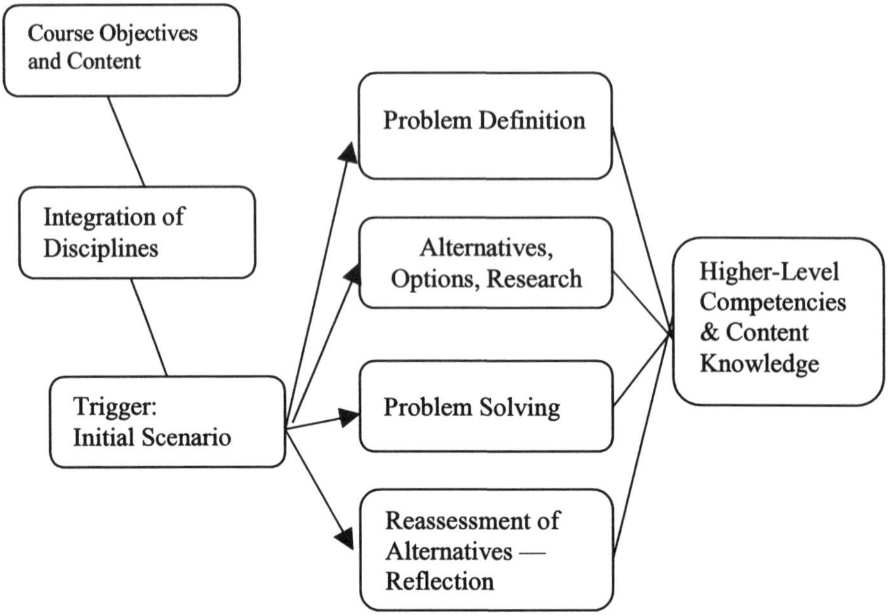

Figure 11.1 PBL scenario development

In PBL, scenarios act as the catalyst for free inquiry and learning in any relevant subject area and, therefore, act as a stimulus for learning the basics of any discipline. In aviation studies, for example, the appropriate choice of simulated situations could encourage students to learn all the relevant issues pertaining to meteorology, navigation, aerodynamic, aircraft performance and operations raised in the scenario. Thus, in the process of solving a problem, students would review prior knowledge and learn a range of skills, in addition to gaining new technical knowledge. In this approach, the emphasis is on the students' gathering and sharing task-relevant information and making logical and rational decisions to address a realistic situation or solve a problem that could be encountered in the work environment (Henley, Anderson, & Wiggins, 1999). According to Margetson (1998), PBL reflects the true nature of knowledge, that is, knowledge is complex and changes as a result of people's responses to problems. Therefore, a multidisciplinary approach would provide more realism than a subject-based curriculum.

The next step in designing a PBL course is to develop scenarios, situations or problems that will serve as catalysts for learning. The problem or situation used as a trigger for learning can take any form, being limited only by the creator's imagination. For example, it might be an audio-visual presentation, such as the "Time for Decision" (Lynch, 1988) exercise where a tape recording of a cockpit situation presents the problem to be solved. Similarly, a video covering any major accident or incident can be the medium used to develop student inquiry, knowledge and problem-solving skills. Other triggers include card games, accident investigations, unexplained situations or phenomena, or tasks such as preparing a charter flight, or an article to be written, a job interview or a debate.

The Night Visual Flight Rules (NVFR) syllabus, for example, could be suitably covered by using an accident investigation or a night charter operation which would prompt students to research the legal requirements, the rules and procedures, and the visual and sensory illusions associated with NVFR operations. A planned flight scenario could encourage the study of navigation, meteorology, flight rules and procedures, aviation medicine, aerodynamics, and aircraft systems and performance. An accident report could serve as the catalyst for an investigation into aviation safety, human factors, and forced landing procedures. Similarly, a video-taped presentation or role-playing scenario could be used as the basis for a study of multi-crew operations, in particular with respect to personality traits, diplomacy, conflict resolution, and crew coordination. However, it should be stressed that, despite its name, PBL should not be limited to problems, but should also address key issues relevant to professional practice.

Well-designed PBL scenarios or problems are essential to ensure the successful achievement of educational objectives. As with any skill, writing scenarios takes practice and considerable effort. To infuse as much realism as possible, a recommended procedure is to use the present tense when writing the scenario and provide specific context, such as time, place, and the specific role of the professional involved. It is also important to include objective rather than interpretive data and ensure that the problem or key issue to be addressed will lead students to conduct research and then make a response (i.e., make a decision or take an action).

A common problem experienced by neophyte scenario writers is to include too much information. Only essential information should be provided. The problem or key issue needs to be presented to students in the same way it would present itself in reality. True-to-life problems or issues are generally confusing and ill-defined when first encountered. It will be the task of the group to explore the problem and define it further. Above all, the aim of a problem or scenario is to arouse specific questions in the minds of the students, the answering of which will cause learning to take place and new skills to be developed. Barrows and Tamblyn (1980) have compiled a number of items to consider when designing problems or scenarios. They stress the need to:

- consider the graduate, not the student. Think of the "end product" who will graduate from the course, not what you want the student to know now;
- ensure the problem is one frequently experienced or poorly handled by pilots, air traffic controllers, aircraft maintenance engineers, flight attendants, dispatchers, and/or managers, or the issue is essential to professional practice;
- ensure the problem or the issue is of vital importance to aviation;
- ensure the problem is one in which intervention (preventative or corrective) can significantly alter the outcome;
- decide on achievable specific learning objectives for each session;
- decide how much information to give in the initial trigger (scenario) and ensure it will naturally elicit the learning issues or questions planned;
- ensure the trigger is on-going and can be continued in another session; and
- arrange appropriate resource material (maps, videos, articles and manuals) to be made available at the correct time (adapted from p. 159).

Another important step to consider while designing a PBL curriculum is the implementation phase. It is crucial to carefully plan the introduction of PBL. A convincing explanation of the relevance and benefits of a PBL approach and a clear outline of the PBL process should be provided to students. It is also unrealistic to expect groups initially to either generate an effective problem-solving strategy themselves or work systematically and constructively through problem-solving processes without guidance. The exploration and definition of the first problem will, therefore, require careful facilitation to allow students to experience the problem-solving process and develop the necessary skills to successful cope with future scenarios or problems on their own. If there is more than one facilitator, it is also advisable to write a facilitator's guide and train the facilitators.

Further, if small groups are used, it is also critical to train students to work in groups. Groups provide a rich variety of people with diverse talents, learning styles, and points of view. As a result, learning can greatly be enhanced through group interactions; however, that variety can also breed conflicts. Team skills, like any other skills, need to be developed; they do not magically happen as soon as a group comes together. Therefore, it is essential that students learn how to handle group dynamics effectively (Woods, 1994). According to Woods, there are seven fundamental personal rights that are the foundation of effective group interactions: "We all have the right to choose, to have opinions, to be respected, to have needs, to have and express feelings, to make mistakes (and be forgiven) and to accept these rights in others" (p. 5-3). These fundamental underpinnings of interpersonal relationship will help build trust and encourage group members to communicate empathetically with each other.

Although problem solving and working in groups may not be the preferred learning style of some students, if they understand that analytical reasoning, decision making, and team skills are essential competencies for professionals working in the aviation industry, they will appreciate the need to develop these skills (Anderson, 1993). An effective way of addressing this issue is to have students themselves list the essential traits of aviation professionals. Problem solving and team skills, among other competencies, will inevitably surface as vital characteristics of aviation professionals.

The final step to be considered in designing a PBL course is the assessment of students and the evaluation of PBL units. Albanese and Mitchell (1993) identified one of the major difficulties in evaluating the success of PBL as being the unit assessment. To be effective, the assessment of PBL, as with any other course, must assess the specific objectives of the unit. Since PBL objectives emphasise a deep rather than a surface approach to learning, this should be reflected in the assessment process. It is argued that multiple-choice examinations promote a surface rather than a deep approach to learning (Albanese & Mitchell, 1993; Moore & Telfer, 1993) since students are only required to recognise or recall the material which is one of the lowest levels of cognition (Bloom, 1956). Therefore, the assessment of PBL should encompass more holistic forms of assessments, such as a personal interview, a group assignment, an essay or an examination which measures the student's critical analysis, synthesis, evaluation and application of PBL.

As mentioned earlier, PBL is more than the teaching of specific content. It aims to develop higher-level competencies, such as analytical reasoning, critical thinking, self-assessment, effective communication, and team skills.

The development of these skills should also be measured for the assessment to truly reflect the objectives of the course. This aspect of the assessment process could include a group project or a more formative assessment such as the ongoing review of team work and communication skills as witnessed by the teacher throughout the tutorial sessions. Another form of assessment could be the group performance in a time-limited activity where team skills and communication styles could be assessed. In short, the assessment should mirror the workplace context and engage students in demonstrating the complex capabilities and competencies of the profession (Anderson, 1993; Anderson & Henley, 1995; Foley, Anderson, & Henley, 1995).

Finally, it is essential that any problem-based learning course or unit be evaluated at many points during its development and implementation. When first designed, the evaluation of the effectiveness of the PBL course can take several forms including reflection-on-action by the facilitator(s) involved, student reviews, and comments of colleagues, all of whom can point out problems and inconsistencies that may have escaped the designer's notice. Throughout the course there should also be a mechanism whereby students and staff can provide feedback on their use of the units and material, what works, problems encountered, and suggestions for improvement (see Chapter 15 for a more detailed discussion of the assessment and evaluation processes).

Conclusion

This chapter has outlined how PBL has the potential of developing self-directed learning skills and contextual knowledge which combine to develop a graduate who is an independent learner, and a problem-solver who is better prepared for the industry. Furthermore, it was argued that PBL using small tutorial groups can enhance team skills and increase intrinsic motivation. For these reasons, PBL has the potential, if effectively implemented in aviation education, to yield definite benefits for future graduates and for the industry.

However, the potential uses of a PBL strategy in the aviation context will clearly vary depending on the immediate setting of the training and on the learning outcomes to be achieved. In deciding on a teaching strategy, instructors need to consider whether problem-based learning is the best way to arrange a curriculum to provide the kind of learning required for professional practice in aviation. As mentioned in Chapter 7, teachers should first decide on the desired educational objectives and then select the teaching method that will be best suited to attain the desired results. Learners with some knowledge of the work context will perhaps have more to gain from a PBL strategy given that it is very useful in extending knowledge and skills in a known environment.

Another consideration may be the amount of time allotted to achieving certain educational objectives. The development of PBL skills may require more time than resources permit. Conventional courses still have their place, but in some areas they could be redesigned in line with the principles of problem-based learning and experiential learning. Research has shown that PBL can help students develop higher-level competencies that are considered to be extremely beneficial to professionals working within the aviation industry (Anderson, 1993; Anderson & Henley, 1994, 1995; Henley, Anderson, & Wiggins, 1999). As Kolb (1984, 1999) suggests, the ideal learning cycle should proceed from concrete experience (contextual learning) to observations and reflections, then through the formulation of abstract concepts and generalisations. This is followed by testing implications of these concepts and generalisations in new situations, and as a result learners gain additional skills and experience that will be of benefit to them in the workplace.

References

Adriaanse, H., Broekhuizen, P., Geerligs, T., van Oorde, K., & Zaremba, T. (1998, June). Problem-based learning. *Civil Aviation Training, 9*(3), 40-43.

Albanese, M.A. & Mitchell, S. (1993). Problem-based learning: A review of literature on its outcomes and implementation issues. *Academic Medicine, 68*(1), 52-81.

Anderson, P.M. (1993). *Problem-based learning and the development of team skills in Aviation Studies.* Unpublished Honours Thesis, University of Newcastle, Newcastle, Australia.

Anderson, P.M., & Henley, I. (1994). Problem-based learning and the development of team skills in Aviation Studies. In S.E. Chen, R. Cowdroy, A. Kingsland, & M. Ostwald (Eds.), *Reflections on problem based learning* (pp. 319-345). Sydney, Australia: Australian Problem Based Learning Network.

Anderson, P.M., & Henley, I. (1995). *Assessing the development of team skills using a time-limited exercise.* In R.S. Jensen & L. Rakovan (Eds.), *Proceedings of the*

Eighth International Symposium on Aviation Psychology (pp. 625-630). Columbus, OH: Ohio State University.

Baker, D.P., & Salas, E. (1992). Principles for measuring teamwork skills. *Human Factors, 34*(4), 469-475.

Barrows, H.S. (1985). *How to design a problem-based curriculum for the preclinical years.* New York: Springer Publishing.

Barrows, H.S. (1986). A taxonomy of problem-based learning. *Medical Education, 20*, 481-486.

Barrows, H.S., & Abrahamson, S. (1964, August). The programmed patient: A technique for appraising student performance in clinical neurology. *Journal of Medical Education, 39*(8), 802-805.

Barrows, H.S., & Mitchell, D.L.M. (1975, December). An innovative course in undergraduate neuroscience experiment in problem based learning with "problem boxes". *British Journal of Medical Education, 9*(4), 223-230.

Barrows, H.S., & Tamblyn, R.M. (1976, January). An evaluation of problem based learning in small groups utilizing a simulated patient. *Journal of Medical Education, 51*(1), 52-54.

Barrows, H.S. & Tamblyn, R.M. (1980). *Problem-based learning: An approach to medical education.* New York: Springer Publishing.

Barry, M. (1996, May). Evaluation of problem-based learning strategies applied to short courses with Pacific Power. *Probe, 15*, 13-16.

Birch, W. (1986). Towards a model of problem-based learning. *Studies in Higher Education, 11*(1), 73-82.

Bloom, B.S. (Ed.). (1956). *Taxonomy of educational objectives. Handbook 1: Cognitive domain.* New York: David McKay.

Boud, D.J. (Ed.). (1985). *Problem-based learning in education for the professions.* Sydney: Higher Education Research and Development Society of Australasia.

Boud, D.J. (Ed.). (1988). *Developing student autonomy in learning,* (2nd ed.). London: Kogan Page.

Boud, D.J., & Feletti, G. (1998). Changing problem-based learning. In D.J. Boud & G. Feletti (Eds.), *The challenge of problem-based learning* (2nd ed., pp. 1-14). London: Kogan Page.

Brookfield, S. (1990). *The skilful teacher.* San Francisco: Jossey-Bass.

Brookfield, S. (1995). *Becoming a critically reflective teacher.* San Fransisco: Jossey-Bass.

Buchholz, S., & Roth, T. (1987). *Creating the high performance team.* New York: John Wiley & Sons.

Burns, S. (1989). There's more than one way to learn. *Australian Wellbeing, 33*, 42-44.

Chappell, C., Gonczi, A., & Hager, P. (1995). Competency-based education. In G. Foley (Ed.), *Understanding adult education and training* (pp. 175-187). Sydney: Allen and Unwin.

Coles, C. (1985). Differences between conventional and problem-based curricula in their students' approaches to studying. *Medical Education, 19*, 308-309.

Coles, C. (1998). Is problem-based learning the only way? In D.J. Boud & G. Feletti (Eds.), *The challenge of problem-based learning* (2nd ed., pp. 313-325). London: Kogan Page.

Cross, P. (1982). *Adult learners*. San Francisco: Jossey-Bass.

Diehl, A.E. (1991, Fall). Cockpit decision making. *FAA Aviation Safety Journal, 1*(4), 14-16. Washington, DC: US Department of Transport.

Engel, C.E. (1982). Problem-based learning. In K.R. Cox & C.E. Ewan (Eds.), *The medical teacher* (pp. 94-101). Edinburgh, Scotland: Churchill-Livingstone.

Engel, C.E. (1998). Not just a method but a way of learning. In D.J. Boud & G. Feletti (Eds.), *The challenge of problem-based learning* (2nd ed., pp. 17-27). London: Kogan Page.

Foley, S., Anderson, P., & Henley, I. (1995). Preparing student pilots for their profession by implementing problem-based learning in Aviation Studies. In R.S. Jensen & L.A. Rakovan (Eds.), *Proceedings of the Eighth International Symposium on Aviation Psychology* (pp. 1177-1182). Columbus, OH: Ohio State University.

Foushee, H.C. (1984). Dyads and triads at 35,000 feet: Factors affecting group process and aircrew performance. *American Psychologist, 39,* 886-893.

Gibbs, G. (1992). *Improving the quality of student learning*. London: Technical and Education Services.

Hackman, R.J. (1993). Teams, leaders, and organizations: New directions for crew-oriented flight training. In E.L. Wiener, B.G. Kanki, & R.L. Helmreich (Eds.), *Cockpit resource management* (pp. 47-70). San Diego, CA: Academic Press.

Harden, R.M., Sowden, S., & Dunn, W.R. (1984). Educational strategies in curriculum development: The SPICES model, *Medical Education, 18*, 284-297.

Helmreich, R.L., & Foushee, H.C. (1993). Why crew resource management. In E.L. Wiener, B.G. Kanki, & R.L. Helmreich (Eds.), *Cockpit resource management* (pp. 1-45). San Diego, CA: Academic Press.

Henley, I. (1995). *The quality of the development and evaluation of flight instructors in Canada and Australia*. Unpublished doctoral dissertation, University of Newcastle, Newcastle, Australia.

Henley, I., Anderson, P., & Wiggins, M. (1999). Integrating human factors into general aviation: Issues and teaching strategies. In D. O'Hare (Ed.), *Human factors in general aviation* (pp. 89-117). Aldershot, UK: Ashgate.

Houle, C.O. (1980). *Continuing learning in the professions*. San Francisco: Jossey-Bass.

Hunter, D.R. (1995). Aeronautical decision making: Historical results and a new paradigm. In R. Fuller, N. Johnston, & N. McDonald (Eds.), *Human factors in aviation operations* (pp. 11-16). Aldershot, UK: Avebury Aviation.

Job, M. (1996). *Air disaster* (Vol. 2). Weston Creek, Australia: Aerospace Publications.

Johnston, N. (1993). Integrating human factors training into ab initio airline pilot curricula. *ICAO Journal, 48*(7), 14-17.

Kidd, J.R. (1973). *How adults learn*. New York: Association Press.

Knowles, M. (1980). *The modern practice of adult education: From pedagogy to andragogy*. Chicago: Association Press/Follett Publishing.

Kolb, D.A. (1984). *Experiential Learning*. Englewood Cliffs, NJ: Prentice Hall.

Kolb, D.A. (1999). *Learning style inventory*. Boston, MA: Hay/McBer Training Resource Group.

Lynch, M. (1988). *Time for decision*. Sydney: Gower Northgate Training.

Margetson, D. (1998). Why is problem-based learning a challenge? In D.J. Boud & G. Feletti (Eds.), *The challenge of problem-based learning* (2nd ed., pp. 36-44). London: Kogan Page.

Moore, P., & Telfer, R.A. (1993). A comparative analysis of airline pilots' approaches to learning. *Journal of Aviation/Aerospace Education and Research, 4*(3), 17-23.

Morgan, A. (1983). Theoretical aspects of project based learning in higher education. *British Journal of Educational Technology, 14*(1), 68-78.

National Transportation Safety Board. (1990). *Aircraft accident report: United Airlines Flight 232, McDonnell Douglas DC-10-10, Sioux Gateway Airport, Sioux City, Iowa, July 19, 1989*. Springfield, VA: National Technical Information Service.

O'Hare, D. (1999). Introduction to human factors in general aviation. In D. O'Hare (Ed.), *Human factors in general aviation* (pp. 3-10). Aldershot, UK: Ashgate.

Prince, C., Chidester, T.R., Bowers, C., & Cannon-Bowers, C. (1992). Aircrew coordination: Achieving teamwork in the cockpit. In R.W. Swezey & E. Salas (Eds.), *Teams: Their training and performance* (pp. 329-353). Norwood, NJ: Ablex.

Ross, M.J. (1992). *The integrated commercial flying school: An evaluation of commercial flight training and flight simulator effectiveness*. Unpublished doctoral dissertation, University of Newcastle, Newcastle, Australia.

Sams, T. (1989, April). *Developing cockpit resource management training curricula for ab initio airline pilot training*. Paper presented at the 5th International Symposium on Aviation Psychology, Columbus, Ohio.

Savin-Baden, M. (2000). *Problem-based learning in higher education: Untold stories*. Buckingham, UK: Society for Research into Higher Education and Open University Press.

Schmidt, H.G. (1983). Problem-based learning: Rationale and description. *Medical Education, 17*, 11-16.

Schwartz, P. (1998). Persevering with problem-based learning. In D.J. Boud & G. Feletti, (Eds.), *The challenge of problem-based learning* (2nd ed., pp. 58-63). London: Kogan Page.

Todd, S. (1991). Preparing tertiary teacher for problem-based learning. In D.J. Boud & G. Feletti, (Eds.), *The challenge of problem-based learning* (pp. 130-136). London: Kogan Page.

Tulving, E., & Thomson, D.M. (1973). Encoding specificity and retrieval processes in episodic memory. *Psychological Review, 80*, 352-373.

Willems, J. (1981). Problem-based (group) teaching: a cognitive science approach to using available knowledge. *Instructional Science, 10*, 5-21.

Woodcock, M. (1989). *Team development manual* (2nd ed.). Worcester, UK: Gower.

Woods, D.R. (1985). Problem-based learning and problem solving. In D. Boud (Ed.), *Problem-based learning in education for the professions* (pp. 19-42). Sydney, Australia: Higher Education Research and Development Society of Australasia.

Woods, D.R. (1994). *Problem-based learning: How to gain the most from PBL.* Waterdown, Canada: Author.

12 Computer-Based Approaches in Aviation Education

Mark W. Wiggins

Introduction

The introduction of computer-based training systems within advanced technology environments is becoming increasingly prevalent as the costs associated with training and education increase. Computer-based systems are often perceived as a cost-effective means of providing information, since information acquisition is not necessarily limited to a particular time frame or venue. Furthermore, there are mechanisms that allow the evaluation of learning outcomes and the provision of feedback in the absence of human intervention.

The aim of this chapter is to consider some of the advantages and disadvantages associated with the development and introduction of computer-based systems for aviation training and education. In particular, the vast resources invested in advanced training technology, often for very little return, have implications for high reliability environments such as aviation where appropriate knowledge and skills are widely regarded as part of the process of productivity improvements and error management. As a result, there are likely to be significant costs associated with the development of systems that may fail to achieve their intended goals.

Throughout this chapter there is an emphasis on effective and appropriate design as an integral part of the process of developing successful computer-based systems. The notion of system design incorporates both the physical features of the system, such as usability, and the structure of the system in terms of the extent to which it enables users to acquire information in a form that is consistent with their preferred learning approach or learning style (see Chapter 4 for a description of learning styles). This discussion provides the basis for a consideration of flexible delivery approaches whereby learners have the capacity to acquire information at a time and at a place that suits their own particular needs. The aim of this strategy is to ensure that the opportunities for learning and the process of skill acquisition are maximised.

Computer-Based Instruction

Computer-based instructional systems are becoming increasingly prevalent within industrial environments as a means of augmenting and, in some cases, replacing more traditional teaching strategies. They offer a number of potential benefits, most significant of which is the opportunity for the development of individualised learning programs that meet the particular needs of employees (Jonassen, 1996). However, despite the potential afforded by computer-based instruction, the benefits have yet to be fully realised in terms of facilitating improvements in human performance (Light & Littleton, 1999).

Computer-based instruction is one of a number of terms that is used to encapsulate the broad aims and objectives of computer-based andragogical initiatives. Other terms include computer-based tutoring systems, Computer-Assisted Learning (CAL), and Computer-Based Training (CBT), and they each reflect slightly different roles that the computer undertakes as part of the learning process (Alessi & Trollip, 1991). For example, in the case of CAL, the computer adopts the role of a tutor that guides, rather than directs, the process of information acquisition (Yazdani, 1987). In contrast, CBT involves an approach whereby the computer directs both the sequence in which information is acquired, and the type of information acquired (Krendl & Lieberman, 1988).

Computer-based instructional initiatives range from relatively simplistic approaches, through to highly complex computer programs that incorporate "intelligent" feedback (Anderson, Boyle, & Reiser, 1985). The relative complexity of a computer-based system is likely to be dependent upon a number of factors including the costs involved, the time available for program development, the needs of the learners, and the type of material to be acquired. Each of these factors will impact upon the quality and the characteristics of the computer system that is ultimately developed and implemented.

The embrace of computer-based instruction within the industrial environment is testament to the perception that there are significant benefits associated with the communication of information using computers. However, it should be noted that the aim of computer-based instruction, like other andragogical initiatives, is to facilitate the efficient and effective acquisition of knowledge, attitudes, and/or skills amongst learners (Criswell, 1989; Roblyer, Edwards & Havriluk, 1997). Computers simply represent a mechanism whereby this process can be achieved.

The Advantages of Computer-Based Instruction

Clearly, one of the most significant advantages associated with computer-based instruction is the potential flexibility afforded by the acquisition of information at a time that is convenient for the learner (Newby, Stepich, Lehman, & Russell, 2000). This flexibility in delivering information has the concomitant effect of reducing the demand on an employer to ensure that employees are available simultaneously for purposes of education or training. Furthermore, there is also likely to be a reduction in the demand for training equipment and accommodation.

In addition to the flexibility associated with the delivery of information, computer-based instruction also enables the acquisition of information in a form that is consistent with the needs of learners (Goodyear, 1990; Sleeman, 1987). For example, learners who exhibit different learning styles can adopt different strategies for the acquisition of information, depending upon their particular needs (Jonassen, 1996). Similarly, individual learning approaches can also be accommodated by computer-based instruction systems.

An example of the accommodation of learning approaches and learning styles within a computer-based instructional system can be drawn from the development of a computer-based Transfusion Medical Tutor (TMT) that was designed to improve students' ability to identify antibodies in human blood (Obradovich, 1996). The TMT incorporated an expert system that was capable of detecting and responding to different types of errors that were made by students. These errors were categorised as either errors of commission, errors of omission, errors due to incomplete mental models, or errors due to a lack of consistency with the information available. The categories of error that occurred were used to provide feedback concerning the most appropriate remedial action. Evaluations of the performance of students prior to, and following exposure to, the TMT revealed a 64 percent reduction in the frequency with which antibodies were misidentified in human blood.

Within high reliability organisations such as aviation and nuclear energy production, there are considerable difficulties associated with the development of appropriate training scenarios for system failures. An alternative approach to training in the actual environment is to employ simulation as the basis for the development and application of knowledge, attitudes and skills (Thurman, 1994). Although there are a number of disadvantages associated with the use of simulation as the basis for skill acquisition, it provides a unique opportunity to test the outcomes of a range of strategies that may or may not be regarded as standard responses.

Where complex industries are involved, adaptation is often an important basis for the resolution of a system failure (Entin & Serfaty, 1999). The complexity of such failures, combined with the potential outcomes, is such that a range of responses need to be considered in addition to the responses to failures that might be regarded as standard. The relative appropriateness of the response to unusual system failures is likely to be dependent upon the knowledge of the operator concerning the relationship between system components (Takano, Sasou, & Yoshimura, 1997). One of the advantages associated with computer-based instruction is that it provides an opportunity to simulate and, thereby, examine the full ramifications of the inter-relationship between system components (Jones & Endsley, 2000).

The development of an understanding of the interrelationship between the components of a system is consistent with the development of a mental model of the system. An accurate and reliable mental model is the basis for effective and appropriate adaptation to the non-standard changes that might occur within a particular environment (Chi, Bassok, Lewis, Reimann, & Glaser, 1989). For example, Chandler (1996) describes a computer-based system for training aviation regulations, and notes that one of the requirements necessary to fully appreciate the implications of aviation regulations is to approach the same information from a number of different perspectives. This type of strategy enables the development of a broad understanding of the wide range of situations to which a particular regulation might be applied. In essence, the program appears to facilitate the development of a broad-based mental model of the regulatory environment.

From a management perspective, one of the main advantages associated with the development of computer-based instruction systems is the potential for a reduction in the long-term costs associated with training and education (Cannell, 1999). For example, Rockwell Collins anticipates savings of up to

US$14 million following the transition from a disparate classroom-oriented instructional strategy to a more centralised, technology-based approach to training its workforce (Fister, 2000). Similarly, organisations such as Apple Computer, Andersen Consulting, and Storage Technology reported savings of up to US$1.5 million following the introduction of computer-based training initiatives (Ellis, 1999). The reduction in costs was due to a number of factors including the reduction in face-to-face contact with an educator and the possibility that learning could occur outside the structured classroom environment (Wallace, 2000).

The Disadvantages of Computer-Based Instruction

Although there are a number of advantages associated with the development and implementation of computer-based instruction, there are also a number of disadvantages, the most significant of which is the potential lack of direct communication with an educator. The significance of direct communication with an educator is illustrated by the failure of a computer-based training system developed by Flightsafety for corporate pilot training. The system was originally developed as a "stand-alone" product whereby pilots could acquire information in the absence of interaction with an educator (Bovier, 1993). However, following discussions with pilots, it became clear that the training needs of pilots were not being met, and that this was due primarily to a perceived lack of direct interaction with an educator.

The failure of a particular type of instructional strategy to meet the learning needs of users is often dependent upon the previous experience of learners within the domain. For example, the introduction of a computer-based instructional strategy in the absence of any preparatory measures is likely to reduce the extent to which an instructional strategy meets the needs of users. This is especially the case if learners were previously exposed to a face-to-face teaching strategy where they had the opportunity to engage in a verbal dialogue with an educator. Evidence to support this perspective can be drawn from Watters (2000) who undertook a comparison between the perceptions of learners who were exposed to either computer-based instruction or a face-to-face technique that involved reciprocal questions. The results of the comparison indicated a relatively greater level of acceptance of computer-based instruction, and Watters (2000) interpreted this outcome as a product of the previous experience of the learners. In particular, he noted that the majority of the participants involved in the study had been taught using a non-interactive, lecture-based approach and they responded very positively to the statement "professors are supposed to lecture". This evidence, albeit limited in scope, suggests that the relative acceptance of a teaching strategy is dependent upon the previous experience and the resultant expectations of the learners.

Like any novel method of education, the level of acceptance of computer-based instruction amongst learners is likely to be dependent upon their expectations and previous experiences. Therefore, the successful introduction of computer-based instruction would normally be preceded by a strategy that prepares the learners for a change in the process of training and education. The need for the preparation of learners imposes an additional cost to the introduction of computer-based instruction, particularly during the initial stages of implementation. However, it will usually prove to be well-invested money in the long term.

The need to prepare learners for the introduction of computer-based instruction is normally established as part of the process of curriculum design. This process is consistent with the approach taken in the development of other training initiatives. However, the time required and the costs associated with the initial development of computer-based instructional initiatives can be relatively high in comparison to other teaching strategies (Criswell, 1989). This is especially the case where the system is developed for a unique environment and cannot be purchased off-the-shelf (Ellis, 1999).

Seidel and Wagner (1979) suggest that the costs associated with the development and introduction of computer-based instruction can be categorised according to:

- development,
- procurement and production,
- operations, and
- maintenance.

The costs associated with the development of the system include curriculum design, the development of a working model, programming, testing, and evaluation. Procurement and production costs are generally associated with the physical installation of the system within the operational environment and the purchasing of materials and hardware. The costs associated with the training of personnel and the preparation of learners are categorised as operations costs. Finally, the administration of the system and any costs associated with revisions to a system are classified as maintenance costs.

Criswell (1989) suggests that, as a general rule of thumb, roughly 200 hours are required for the development of approximately one hour of computer-based instruction. However, where a computer-based system requires a degree of "intelligence", such as the provision of feedback, the time involved in the development of the system can increase markedly beyond the standard rule of thumb. Nevertheless, it should be noted that there may be a number of andragogical benefits associated with the development of "intelligent" feedback, such that the benefits outweigh the costs involved.

Aspects of Computer-Based Instruction

The nature of computer-based instruction is such that systems can be classified along a number of continua. These continua include the level of interaction available, the degree of feedback provided, and the extent to which a program simulates the operational environment. Each of these features has the potential to impact upon the rate at which task-related information is acquired and the quality of the information acquired by learners.

Interaction and Computer-Based Instruction

The level of interaction associated with computer-based instructional initiatives can be classified according to two components: the complexity of the information processing required and the complexity of the response required. In the case of information processing, interaction can range from systems that involve a relatively linear progression of information acquisition, through to more complex systems in which a multitude of paths might be taken to reach the same destination (Chandler & Sweller, 1996). Increasing the numbers of paths available to reach a destination increases the cognitive load imposed upon the learner to acquire information and make selections (Martin-Michiellot & Mendelsohn, 2000).

The level of cognitive load imposed upon a learner is presumed to be related to the ease with which information is acquired (Sweller & Chandler, 1991). Intrinsic cognitive load relates to the nature of the information that is to be acquired. Information possesses an intrinsic level of difficulty, and the cognitive load imposed by the information is largely immutable. In contrast, extrinsic cognitive load is associated with the process through which information is presented to the learner. Where the presentation of information is complex, there is a concomitant increase in the cognitive load imposed upon the learner. Martin-Michiellot and Mendelsohn (2000) suggest that, in some situations, the combination of intrinsic and extrinsic cognitive load upon the learner reaches a threshold, beyond which there is an adverse impact upon the acquisition of information.

Despite the potential difficulties associated with increasing the extrinsic cognitive load imposed upon learners, it can have benefits in terms of the efficient and effective acquisition of information. For example, Rotheray and Sewell (1987) suggest that developers of computer-based instruction should avoid the perception of the learner as a passive recipient of information. Rather, the most appropriate strategy is one that is consistent with, and builds upon, the existing cognitive structures of the learner. This perspective is consistent with Norman's (1988) notion of transparent design, in which the optimal approach is one which is consistent with the user's mental model of the system.

Apart from information processing, the complexity of the response required also impacts significantly upon the efficacy of computer-based instruction. For example, where a response is consistent throughout a piece of software, there is the potential that information may not be processed to the extent that the learner develops an accurate and reliable model of the domain (Criswell, 1989; Jonassen, 1996). However, alternative strategies, such as answering questions or responding with detailed points of view, have the capacity to increase the level to which the information must be processed in an effort to formulate an appropriate response (Sibbald, 2000).

Feedback and Computer-Based Instruction

Consistent with other forms of education and training, the success of computer-based instruction is fundamentally dependent upon the appropriateness and the timeliness of the feedback provided. Appropriate and timely feedback provides the basis for reflection and the application of information (Van Hoven, 2000). Ideally, the feedback provided as part of computer-based instruction should be encouraging and, where necessary, should guide the learner towards an accurate response.

From a behavioural perspective, the modification of performance occurs through the provision of rewards for correct responses or punishers for incorrect responses (see Chapter 1 for an overview of the behaviourist theory of learning). The former are designed to increase the rate at which a behaviour occurs, while the latter are presumed to decrease the rate at which a particular response will occur (Foxall & Greenley, 1998). In the case of computer-based instruction, Criswell (1989) suggests that punishers for incorrect responses should be relatively mild and should avoid terms such as "error" or "wrong". Severe or repeated punishers can lead to the demoralisation of the user and may lead to a situation where the learners develop strategies to avoid using computer-based instruction.

Where possible, natural reinforcers and punishers should be employed as feedback, as a means of drawing the relationship between the computer-based environment and the operational setting (Criswell, 1989). Nevertheless, the feedback provided as part of computer-based instruction should always be oriented towards improving performance. This perspective is based upon the assumption that participation in computer-based instruction is likely to be voluntary, and learners need to be encouraged to acquire information, rather than be forced to do so.

Consistent with the need to encourage the acquisition of information, feedback concerning the accuracy of responses should incorporate suggestions for improvement, rather than simply respond on the basis of the relative accuracy of a response. "Intelligent" feedback also involves monitoring a series of responses, rather than assessing each response in isolation (Anderson et al., 1985). This information provides the basis for diagnosing cognitive errors, such as the failure to draw relationships between particular constructs (see Figure 12.1). The main advantage associated with the recognition and diagnosis of cognitive errors is that feedback can be directed towards identifying problems that might exist within the knowledge base. This type of feedback is presumed to provide the basis for a more efficient remedial strategy than simply responding on the basis of individual responses to questions.

The development and evaluation of an accurate and reliable knowledge base constitutes an important principle for the development of cognitive skills such as decision making and planning. For example, Busby and Payne (1999) used a computer-based instructional program to draw analogies between various constructs of planning as a means of increasing the accuracy of the knowledge base amongst engineers. Users were asked to compare pairs of behaviours that might be confronted within the operational environment and determine which is the more appropriate. Although the initial feedback provided simply referred to the relative accuracy of the response, there were two different strategies available to remedy the error.

The first of the remedies offered by Busby and Payne (1999) involved repeating the exercise in a simpler, analogous domain, and identifying the differences that may exist between the two behaviours proposed. It was assumed that the nature of the behaviour would be clearer in the analogous domain and, therefore, performance would be more accurate. The analogous domain employed by Busby and Payne (1999) involved journeys in motor vehicles, since this was a domain with which most users were familiar. Moreover, it was anticipated that more concrete examples would facilitate the retention of information within long-term memory. This assumption is consistent with the principles associated with case-based reasoning in which exposure to a series of carefully constructed cases has been demonstrated to facilitate the acquisition of knowledge, attitudes and skills amongst learners (Kolodner, 1996).

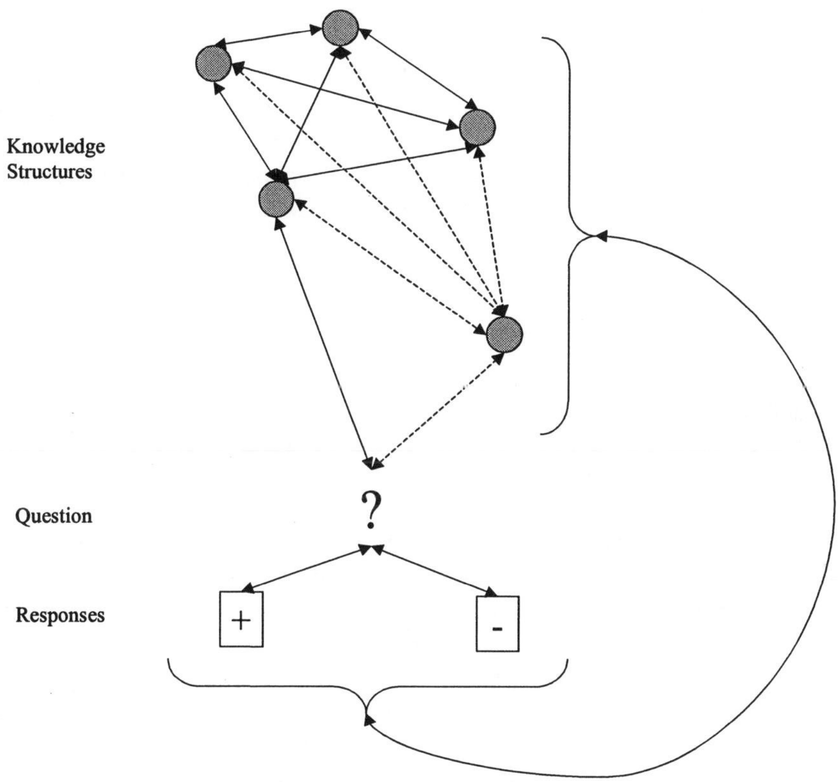

Figure 12.1 A conceptual illustration of the relationship between patterns of responses to questions and the extent to which they reflect deficiencies associated with the underlying knowledge base

The second remedial approach was more complex and drew upon additional information pertaining to the assumptions about the situation (Busby & Payne, 1999). In particular, each of the behaviours was associated with a number of assumptions pertaining to the situation, and the accuracy of these assumptions would determine the appropriateness of the behaviours. Therefore, by drawing the user's attention to the accuracy of the underlying assumptions, the relative appropriateness of the behaviour could be determined.

Although the evaluation of the tutoring system developed by Busby and Payne (1999) was limited to subjective responses, there was some indication that the provision of detailed feedback in the form of analogous examples and/or an evaluation of assumptions may facilitate the acquisition of complex cognitive skills amongst learners. Support for this conclusion can be drawn from Bassick (1990) and Novick (1988) who note the significance of analogy in the transition towards expertise. Therefore, it might be concluded that the development of cognitive skills, such as decision making and situational awareness, requires a detailed level of feedback that focuses upon the accuracy of the knowledge base, rather than simply indicating the relative accuracy of responses to a vast series of knowledge-based questions.

Virtual Reality and Computer-Based Instruction

The substantial increase in the capacity of contemporary computing systems to process information has resulted in the development of increasingly realistic computer-based systems that attempt to mimic the sensory responses that are triggered within the actual environment. Computer gaming organisations, in particular, have sought to develop programs that provided realistic feedback to users in domains such as motor car racing and flight simulation. The ultimate aim of these programs is to create an environment in which the user interacts with the features of the environment as if they actually existed (Thurman & Russo, 2000).

Virtual Reality (VR) is a term that is used to describe the process by which a user employs multi-sensory systems to engage with a simulated, computer-based environment and experience the responses from the environment as if they were real (Marran & Schor, 1997; Thurman & Russo, 2000). From the perspective of training and education, the main advantages associated with these systems include the degree of feedback afforded, and the level of situated cognition provided by the learning environment (Greeno, 1998; Wilson, 1999). In addition, VR provides an immersive environment and an opportunity for the learner to test the outcomes of various ideas and strategies that may or may not be applicable within a given situation (Salzman, Dede, & Loftin, 1995).

At present, there are a number of technical difficulties with VR that need to be overcome to ensure that it is a useful and reliable instructional tool (O'Neil, Mayer, Herl, Niemi, Olin, & Thurman, 2000). These difficulties include the potential for simulator sickness that may arise from the mismatch between the visual and the vestibular cues presented (Wann & Mon-Williams, 1996). Furthermore, it should be noted that the application of VR, in the absence of an appropriate instructional strategy, is unlikely to be particularly effective in terms of facilitating improvements in human performance (O'Neil et al., 2000; Wilson, Foreman, & Tlauka, 1997).

Computer-Based Intelligent Tutoring Systems

The efficiency with which aviation-related information is acquired by learners is largely dependent upon the extent to which a particular teaching strategy is consistent with the needs of the learner (Boles, Pillay, & Raj, 1999). Each individual perceives and processes information differently, to the extent that adopting a unitary approach to the acquisition of information can lead to a differential impact in terms of the rate and the quality of the information that is ultimately acquired by different learners (Pounds & Bailey, 1999). As a consequence, the goal of educators is to provide task-related information in a form that meets the needs of a majority of students within a domain.

In face-to-face teaching, the capacity to respond to the different needs of learners is due to the broad understanding of the domain amongst educators, and the capability to explain the same information in a number of different modes to facilitate understanding (Pressley & Woloshyn, 1995). For example, an educator may perceive that a particular learner tends to respond to concrete examples, rather than abstract definitions or concepts. In response, the educator may illustrate periodically the application of the concepts discussed as part of a lesson. The adaptation of the information represents the flexibility that is inherent in an effective educator (Barry, 1993).

In the absence of face-to-face communication between the educator and the student, the process of learning can occur through a number of strategies including reading books, watching video recordings, participating in simulations, and/or observing lectures (Ellis, 1999; Van Hoven, 2000). While there are a number of advantages associated with the application of these strategies, the main disadvantage is their inability to respond to differences in the learning needs of students (Anderson & Henley, 1994). In each case, the onus tends to rest with the learner to identify alternative approaches to facilitate understanding.

From the perspective of advanced technology, computer-based training systems offer some potential as a surrogate for face-to-face education. Referred to as Intelligent Tutoring Systems (ITS) or Intelligent Computer-Assisted Instruction (ICAI), they are distinguished from other computer-based training systems by their capacity to respond "intelligently" to the needs of the learner (Anderson et al., 1985). ICAI systems are characterised by a vast repertoire of knowledge concerning a particular domain, to the extent that they simulate the knowledge of the educator. It is this vast knowledge, in combination with the organisation of information, that enables ICAI to respond to the needs of a particular individual learner (Obradovich, 1996).

The development of ICAI has paralleled the construction of artificially intelligent systems that are capable of reasoning and developing novel solutions to problems in the absence of direct human intervention (Anderson, et al., 1985). This capacity to resolve novel problems is presumed to be due to the provision of a series of higher-order principles that specify the process through which different forms of information are related. According to production system theory, these higher-order principles are defined by rules that incorporate a condition statement (IF) and an action statement (THEN) (Anderson, 1993). By contrast, connectionist theory prescribes inherent relationships between the information located within long-term memory (Card & Newell, 1989). The pattern of activation that is triggered in response to a stimulus is presumed to direct the process of activation and subsequent responses.

ICAI responds to the needs of the learner by formulating inferences on the basis of previous responses to questions and problems (Howell & Cooke, 1989). A base level of knowledge can be determined at different levels of sophistication. For example, a learner may have difficulty in solving a problem due to a lack of background knowledge pertaining to a specific area. Answers to a series of questions can identify the absence of knowledge, and the learner can be directed to subsequently complete a program of study prior to continuing with the primary course of study.

At a higher level of cognition, difficulties can be identified at the perceptual stage, the data processing stage, and/or the response activation stages of information processing. In the case of perception, difficulties may be due to a lack of understanding concerning the significance of cues, and/or an inability to resolve the pattern of cues that is presented (Flach, 1995). During data processing, the difficulties that occur may be due to an inappropriate mental model, an incapacity to process information efficiently, and/or an inappropriate or inefficient organisation of information within long-term memory. Finally, problems associated with response activation may be due to a lack of understanding as to the point at which the behaviour is expected to occur.

The principle of responding to learners on the basis of the types of errors that they make is consistent with the approach taken amongst many face-to-face educators in attempting to facilitate an understanding of material (Gage & Berliner, 1998). However, in the case of ICAI, the material necessary to resolve the error is already encapsulated within the system, such that the information can be provided as part of a seamless approach to information acquisition (Glaser, Shauble, Raghavan, & Zeitz, 1992). The advantage for the learner is that the process of knowledge acquisition can be adapted to the individual needs of the student and provide the information necessary to resolve a problem. The outcome is likely to be an increase in motivation amongst learners, and a progressive improvement in their capacity to solve difficult problems.

Expert Systems and Aviation Instruction

Despite the promises afforded by ICAI, computer-based systems have yet to reach the stage where they are sufficiently sophisticated to respond spontaneously to the needs of learners. This is due primarily to the limitations associated with existing cognitive architectures, and the lack of understanding concerning the nature of human performance. As a consequence, the alternative approach to ICAI involves a less "intelligent" strategy whereby the final decision to undertake a particular strategy remains with the user. This type of strategy is typically embodied within an expert system.

An expert system is a computer-based program similar to decision aid, in which advice is provided to the user concerning the appropriateness of various options (Todd & Benbasat, 1994). From an andragogical perspective, the main aim associated with an expert system is similar to ICAI: to facilitate the acquisition of the knowledge, attitudes, and skill necessary to improve human performance. In an expert system, however, the selection of the process by which the information is acquired rests with the learner rather than with the system.

In comparison to ICAI, the main advantage associated with expert systems is that the responsibility for the learning process remains with the learner. The expert system is a tool, rather than a director of the learning process. Amongst adult learners, a degree of choice over the process of learning tends to be perceived positively, and appears to facilitate the development of intrinsic motivation towards the task (Knowles, Holton, & Swanson, 1998).

Where ICAI is likely to be effective amongst younger learners, expert systems are likely to be more effective amongst adult learners (Goodyear, 1990). Similarly, ICAI may be more appropriate for a subject-centred curriculum, where expert systems may be more effective as part of a student-centred curriculum. Nevertheless, the relative efficacy of the two approaches is likely to be dependent upon the extent to which the models are consistent with the expectations of learners.

From an aviation perspective, interaction with expert instructional systems will generally occur as a normal part of operational behaviour. For example, expert systems are currently being incorporated within aircraft, air traffic control systems, and nuclear energy processing to facilitate effective decision-making amongst operators (Endsley & Kris, 1994; Hopkin, 1995; Kang, Cheon, & Chang, 1992). An inevitable part of this interaction is the acquisition of information on the part of operators, and the development of strategies that will eventually render the expert systems obsolete.

Expert systems are designed to undertake analyses of information derived from the operational environment. In civil aviation, an expert system might be used to determine which of two alternatives is most appropriate for a passenger aircraft given a range of factors including distance, services, fuel remaining, maintenance facilities, and the needs of passengers. In the military environment, an expert system is already being tested in the form of the Pilot's Associate. This system is designed to perform a range of functions, including the management of warnings, and suggestions as to the most appropriate track to intercept an enemy aircraft (Banks & Lizza, 1991).

Consistent with other automated systems, the efficacy of an expert system is dependent upon the appropriateness of the design of the system. Errors that occur during programming development can result in errors that are incapable of being detected by the flight crew, especially within a time-constrained environment. In the case of the Pilot's Associate, the system can provide advice to the pilot concerning the most suitable manoeuvre to avoid an air-to-air or surface-to-air missile (Banks & Lizza, 1991). Any error that occurs within the system can delay the pilot's response, thereby increasing the probability of a missile strike.

In addition to errors associated with the process of system development, system failures can occur when operators develop a level of complacency concerning the accuracy of the system. Automation-induced complacency has been identified amongst flight crews and air traffic controllers who rely upon advanced technology to assist in the management of task-related information (Parasuraman & Riley, 1997; Singh, Molloy, & Parasuraman, 1993). Consequently, it might be expected that automation-induced complacency will also be evident amongst users of expert systems, particularly where the system incorporates a specific recommendation for action.

Although expert systems are likely to be employed within advanced socio-technological environments as a tool to assist operators, the designers of such systems need to be aware that they also offer a unique opportunity for the development of skills amongst operators. At the very least, there is a need to maintain the skills of operators in the case where an expert system fails or when there is a need to cross-check the outcomes that have been recommended. Ensuring that the algorithms employed in a particular analysis are available to operators for scrutiny will enable a level of redundancy and provide the basis for a useful learning experience.

The Design of Intelligent Tutoring Systems

One of the main issues that has directed the development of ICAI and expert systems has been the requirement to provide information that is applicable within an operational context (Herrington & Oliver, 2000). The notion of situated cognition is significant in this regard, since it has provided a theoretical framework that ensures that the basis of instructional systems is consistent with the requirements that exist within the operational environment (see Chapter 2 for a further discussion regarding situated learning). In particular, Herrington and Oliver refer to the principle of authentic activity as a strategy that provides the motivation for learners to engage the instructional process, develop their own strategies for knowledge acquisition, and maintain their attention to the task.

An authentic activity is somewhat consistent with a problem that might be employed as part of a problem-based learning initiative (Sibbald, 2000, also see Chapter 11 in this book). The activity is situated within the operational environment, and is sufficiently difficult and ill-defined so that the learner will be encouraged to access other resources (Herrington & Oliver, 2000). In the case of ICAI and expert systems, it is equally important to provide a context for the learning strategy and, thereby, encourage the use of the technology and the acquisition of information. Setting the context for learning represents the first stage in the design of an effective and efficient intelligent tutoring system.

In many cases, the design of an intelligent tutoring system will be based upon a detailed analysis of the information processing strategies employed by recognised experts in the field (Roth & Woods, 1989). A process consistent with cognitive task analysis is employed to 'capture' both the knowledge of experts, and the process by which this knowledge is integrated to formulate decisions (Prerau, Adler, & Gunderson, 1990). This information is subsequently embedded within a computer-based decision tree that facilitates the decision-making process. Essentially, the aim of this strategy is to emulate the accuracy of decision-making strategies employed by experts within the operational environment.

Having acquired the task-related information from experts, it is necessary to ensure that the program is constructed using an appropriate cognitive architecture that can facilitate the acquisition of knowledge, attitudes, and/or skills. Arruarte, Fernandez-Castro, and Greer (1996) posit a model of instruction that is specifically designed to assist the development of intelligent tutoring systems. The model is based upon the relationship between instructional actions and cognitive processes. In particular, it is assumed that appropriate cognitive processes can be initiated through exposure to one or more instructional actions. This process is designed to facilitate the acquisition of knowledge at an appropriate cognitive level and in the absence of face-to-face contact with an educator.

Arruarte et al. (1996) suggest that the relationship between instructional actions and cognitive processes can occur at a number of different levels. For example, in a one-to-one relationship, a single instructional action, such as a general explanation, may be associated with a cognitive process such as selective perception. More complex relationships can also occur in which multiple instructional actions influence a single cognitive process (many-to-one relationship), or a single cognitive process is triggered by a number of instructional actions (one-to-many relationship). The extent to which these effects occur will be dependent upon a number of factors, including the previous knowledge and experience of the learner.

The Cognitive Learning of Automated Instruction (CLAI) model of ITS instructional design is mediated through a central component that determines the instructional actions that ought to be triggered within a particular instance (Arruarte et al., 1996). This principle is consistent with the notion of an "executive function" that automatically assesses the characteristics associated with a specific situation and determines the most appropriate responses from the system (Perkins, 1990). Once the instructional actions have been determined, the related instructional events are initiated and may include tutorial programs, drills, and/or simulations (Arruarte et al., 1996).

Instructional events can each be associated with a specific teaching style, depending upon the aims and the objectives of the andragogical initiative (Srisethanil & Baker, 1995). For example, a tutorial program might involve a guided learning approach or a self-directed learning approach, depending upon whether the initiative is designed according to a subject-centred or student-centred curriculum. However, it should be noted that the main principle associated with CLAI model of instructional design is that instructional events, in combination with instructional actions, lead to specific cognitive processes. It is the nature of these cognitive processes that will determine the type of information acquired, and the rate at which this information is acquired. Specifying the nature of these relationships provides the basis for the development of an intelligent system that is capable of responding to the needs of individual learners.

Aviation and Intelligent Tutoring Systems

Once a learner has established the basic principles associated with performance within the operational environment, an ITS has the potential to facilitate the transition from the acquisition of knowledge to the development of skills. In combination with the principles of situated learning, an ITS also has the capacity to facilitate the transfer of training to the operational setting. Nevertheless, it should be noted that this type of education occurs in the absence of direct face-to-face contact with an educator.

The extent to which aviation-related education and training can occur in the absence of direct contact with an educator is unknown. Therefore, one of the issues that needs to be resolved during the progression towards the development of human factors-related ITS is whether or not learners require direct contact with educators for the development of skills. It might be argued that ITS offers greater advantages for skills, such as decision-making that do not necessarily require interaction with peers. However, advances in computer-based speech perception may eventually result in a situation in which an ITS can simulate contact with an educator or with other members of a group as a means of facilitating the development of skills such as communication.

Despite the potential difficulties associated with ITS, it offers the promise of a relatively cost-effective strategy that progresses beyond the development of knowledge and attitudes. The development of skilled performance is the ultimate goal of aviation instruction, and ITS may offer one of a number of strategies to meet this goal.

Human Factors and Computer-Based Instruction

From a human factors perspective, computer-based instruction has the capacity to facilitate the acquisition of human factors-related knowledge, skills and attitudes within a setting that is consistent with the demands imposed by the aviation environment. However, the success of this type of andragogical approach is fundamentally dependent upon the extent to which the training initiative meets the needs of the individuals for whom it was designed.

Consistent with other behaviours within the aviation environment, the application of human factors principles requires an appropriate combination of knowledge, attitudes and skills pertaining to the domain. These features of performance are necessary to recognise the requirement for an intervention, determine the most appropriate intervention under the circumstances, and apply the intervention as necessary to effect an appropriate outcome. The principles can be applied equally successfully at an individual or at an organisational level within advanced technology environments.

The Requirement for an Intervention

One of the primary difficulties associated with the appropriate application of human factors-related behaviours is the failure to recognise the requirement for an intervention. For example, in the aviation environment, there have been a number of aircraft accidents that can be attributed to a lack of awareness, on the part of the flight crew, concerning the state of the aircraft (Sarter & Woods, 1991). Similar difficulties have been noted by air traffic controllers, and amongst other operators who are required to interpret information on the basis of a complex set of stimuli (Durso, Truitt, Hackworth, Crutchfield, & Manning, 1998; Gaba, Howard, & Small, 1995).

Computer-based instruction offers a range of possibilities to increase both the accuracy and the timeliness in recognising the requirement for an intervention. Part of this process involves the development of an understanding of those patterns of cues that are indicative of a potential system failure. The recognition and response to cues is a feature of expertise, and has been demonstrated amongst expert fire fighters and engineers (Hammond, Hamm, Grassia, & Pearson, 1987; Klein, 1989)

Both Chappell (1991) and Mosier (1991) advocate repeated exposure to patterns of cues as the basis for timely responses from operators to changes that occur within a dynamic system. While evidence to support this perspective is limited, Wiggins (2001) has demonstrated the potential for this strategy amongst pilots using a computer-based instructional system for in-flight weather-related decision making (also see Wiggins, O'Hare & Hunter, 2000). The results of an evaluation of the strategy indicated that pilots who had been taught the cues necessary to recognise a deterioration in in-flight weather conditions were more likely to initiate a timely diversion than those pilots who had not been exposed to the cues.

In addition to highlighting the cues associated with changes in the system state, computer-based instruction also provided the basis for the development of a knowledge base of the relationships that exists between the various components of a system. This knowledge base constitutes a mental model and is likely to facilitate the generalisation of skills to the extent that novel events are recognised as efficiently and as effectively as those events that are more familiar (Resnick, 1987).

Determining an Appropriate Intervention

Having identified that a change in the system state has occurred, an operator is required to establish the most appropriate response from the repertoire of responses available. This process can involve a number of cognitive skills such as decision making, workload management, and/or stress management. Consequently, successful performance in this regard requires the development of a repertoire of strategies, the capacity and skills necessary to analyse each of the options available, and the ability to conduct this process while maintaining control over the system (Wickens, Gordon, & Liu, 1998).

One of the advantages associated with computer-based instruction is the capacity for the learner to use the system to test the outcomes of various initiatives that may or may not be appropriate. Active participation in the process of testing different strategies tends to reinforce the learning process, such that there is a greater level of discrimination between the relative efficacy of different strategies for different types of situations (Anderson, 1993). This capacity is especially significant in those situations where the time available for deliberation is severely restricted and the consequences of an error are significant.

From a human factors perspective, the selection of an appropriate strategy for the resolution of a problem depends upon both the acquisition of a suitable knowledge base, and the motivation to select the optimal strategy. In the aviation environment, inappropriate motivation has generally been referred to as a hazardous attitude towards the task, and is characterised by risk taking and a perception of invulnerability (Buch & Diehl, 1984). Hazardous attitudes have been implicated in poor decision making amongst pilots, and there has been a significant attempt to redress the problem through education and training (Harris, 1994; Murray, 1999). Although this approach has met with some success, Wiggins and O'Hare (1993) suggest that a lack of personal involvement in the consequences of decisions may have limited the overall efficacy of the approach.

Computer-based instruction offers an opportunity for learners to experience the consequences of hazardous attitudes within the confines of a simulated environment. Anecdotal evidence arising from crew resource management initiatives suggests that personal involvement in the outcomes of a task has the capacity to increase the motivation to apply appropriate strategies in the future (Mecham, 1994). Therefore, it might be argued that, in combination with an accurate and reliable knowledge base, personal experience arising from the application of hazardous attitudes can increase the probability of selecting an appropriate intervention in response to changes in the system state.

Since the process of selecting an appropriate intervention is reliant upon working memory, there are a number of limitations that are likely to impact upon the process of information retrieval and analysis. For example, working memory is both capacity-limited and is subject to transcience, such that only a limited amount of information can be processed at any one time (Shiffrin & Schneider, 1977). While the amount of information processed per unit time can be increased through strategies such as chunking, the limitation imposed upon the allocation of working memory resources may reduce the capacity for an operator to manage a number of tasks simultaneously (Duffy, 1993; Tsang & Voss, 1996).

The successful management of concurrent tasks while selecting an appropriate intervention can be difficult in the absence of extensive, task-related experience. One of the advantages associated with experience is the extent to which information can be processed more efficiently than might otherwise be the case. For example, experts are often characterised by their capacity to manage a number of tasks simultaneously in the absence of any deterioration in performance (Cooke, 1990; Ye & Salvendy, 1994). By providing novices with task-related experience, attentional capacity can be released for the management of other tasks such as selecting an intervention in the face of an unusual or ambiguous system failure. O'Hare and Batt (1999) refer to the process as artificial seasoning, whereby a novice is provided with a repertoire of experiences through a computer-based instructional environment.

Applying an Intervention

Once an appropriate human factors-related intervention has been selected, the process of applying the intervention occurs, wherein the operator physically responds using the appropriate controls. Where the intervention requires a complex series of responses, there is a potential for errors as controls are either overlooked (error of omission), or controls are initiated in an incorrect sequence (error of commission). Decreasing the potential for such errors requires practice and feedback concerning the accuracy of performance: a process that is afforded by computer-based instruction.

Where a response is required within a relatively limited period of time, an increase in the perceived level of stress is likely to occur to the extent that errors are more likely to occur (Kontogiannis, 1996). Part of the process of reducing the potential for errors involves managing the stress associated with the performance of the task. Computer-based instruction has the advantage of enabling learners to practice the implementation of an intervention such that the process becomes more familiar and, thereby, less stressful.

Usability and Computer-Based Instruction

Computer-based instruction occurs through a conduit (the computer) and thereby increases the complexity of the process of interaction between the educator and the learner. The efficiency with which computer-based instruction occurs is likely to be dependent upon a number of factors including the usability of the system, the teaching strategies incorporated within the program, and the accuracy of the conceptual map upon which the program was developed (see Figure 12.2). Where a system lacks sufficient usability either to facilitate the acquisition of information or to engage and motivate the user, the system is unlikely to yield successful outcomes.

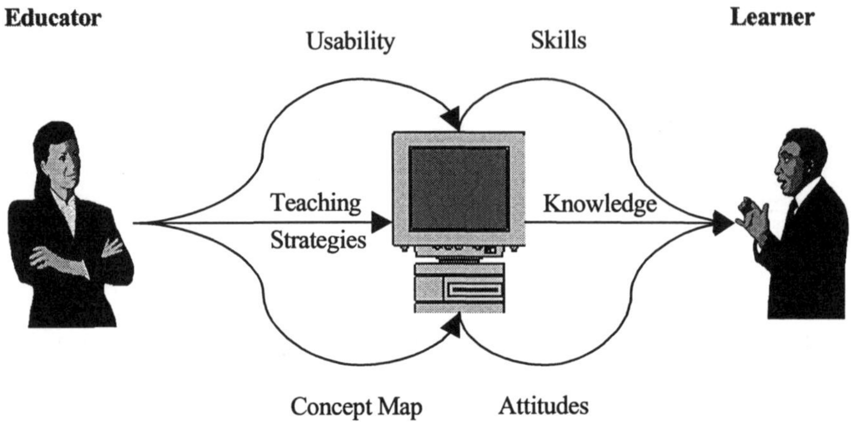

Figure 12.2 A conceptual diagram of the relationship between the educator and the learner in computer-based instruction

The evaluation of the usability of a system is normally conducted as part of the process of product development and implementation. Therefore, an important part of the process of usability assessment involves establishing the needs and the capabilities of potential users prior to the development of the system (Spyridakis, Miller, & Barfield, 1998). For example, the age of potential users or their previous experiences may limit the extent to which a particular type of software interface is likely to be appropriate.

Once the needs and abilities of the potential users have been established, the system can be constructed. However, Savage and Pearsall (1998) recommend the involvement of subject matter experts as part of this process to ensure that the system remains consistent with user needs. This type of iterative approach is an important part of the principles of human-centred approach to system design (Billings, 1997).

Consistent with the principles of human factors, a system is most effective when it is capable of being used optimally in the absence of specialist education and training (Norman, 1988). Computer-based instruction often suffers from the problem of a learner having to acquire the knowledge and skills necessary to use the system before there is an opportunity to acquire the knowledge and skills that is expected to be imparted by the system. The development of an overly complex interface or navigation process can reduce the learner's motivation to interact with the system and, thereby, decrease the probability that task-related information will be acquired.

Summary

The aim of this chapter was to explore some of the principles associated with computer-based instruction, particularly in terms of the extent to which they relate to the acquisition and retention of aviation-related information. The first part of the chapter explored the advantages and the disadvantages associated with computer-based instruction. These contrasting perspectives were designed to illustrate the potential for computer-based instruction, and the limitations of this type of andragogical strategy.

As part of this chapter, three main aspects of computer-based instruction were introduced: interaction, feedback, and virtual reality. Interaction is the primary means through which information is acquired by the learner, and the type of interaction employed within a training initiative has important implications for the development of aviation-related skills. Similarly, the level and quality of the feedback provided is likely to impact upon the quality of information acquisition. Finally, the future of computer-based interaction was examined in the form of virtual reality, and some of the principles were discussed in terms of their implications for aviation instruction.

The chapter also included a consideration of the development of computer-based instruction from an information-processing perspective. Specifically, the role of computer-based instruction was examined in terms of recognising the need for a human factors-related intervention, determining the type of intervention required, and the application of the intervention within the operational environment.

The potential application of intelligent tutoring systems was also considered at various stages throughout the chapter. In addition, the distinction was made between intelligent systems and expert systems. The design of these systems was explored using a model that purports to guide the process of cognitive learning within intelligent instructional environments. Finally, some of the difficulties associated with the application of intelligent systems were examined from a human factors perspective.

References

Alessi, S.M., & Trollip, S.R. (1991). *Computer-based instruction* (2nd ed.). Englewood Cliffs, NJ: Prentice-Hall.

Anderson, J.R. (1993). *Rules of the mind.* Hillsdale, NJ: Lawrence Erlbaum.

Anderson, J.R., Boyle, C.F., & Reiser, B.J. (1985). Intelligent tutoring systems. *Science, 228,* 456-462.

Anderson, P.M., & Henley, I. (1994). Problem-based learning and the development of team skills in aviation. In S.E. Chan, R.M. Cowdroy, A.J. Kingsland, & M.J. Oswald (Eds.), *Reflections on problem-based learning* (pp. 319-345). Sydney, Australia: Problem Based Learning Network.

Arruarte, A., Fernandez-Castro, I., & Greer, J. (1996). The CLAI model: A cognitive theory of instruction to guide ITS development. *Journal of Artificial Intelligence in Education, 7,* 277-313.

Banks, S.B., & Lizza, C.S. (1991). Pilot's associate: A cooperative, knowledge-based system application. *IEEE Expert, 6,* 18-29.

Barry, K. (1993). *Beginning teaching: A developmental text for effective teaching* (2nd ed.). Wentworth Falls, Australia: Social Science Press.

Bassick, M. (1990). Transfer of domain-specific problem-solving procedures. *Journal of Experimental Psychology: Learning, Memory, and Cognition, 16,* 522-533.

Billings, C.E. (1997). *Aviation automation: The search for a human-centred approach.* Mahwah, NJ: Lawrence Erlbaum.

Boles, W.W., Pillay, H., & Raj, L. (1999). Matching cognitive styles to computer-based instruction: An approach for enhanced learning in electrical engineering. *European Journal of Electrical Engineering, 24,* 371-383.

Bovier, C. (1993, August). How a high-tech training system crashed and burned. *Training,* 26-29.

Buch, G., & Diehl, A. (1984). An investigation of the effectiveness of pilot judgement training. *Human Factors, 26,* 557-564.

Busby, J.S., & Payne, K.H. (1999). A behavioural training system for planning judgement. *Journal of Computer Assisted Learning, 15,* 61-72.

Cannell, M. (1999, August). Trends in training. *The Safety and Health Practitioner,* 38-39.

Card, S.K., & Newell, A. (1989). Cognitive architectures. In J.I. Elkind, S.K. Card, J. Hochberg, & B.M. Huey (Eds.), *Human performance models for computer-aided engineering* (pp. 173-180). Washington, DC: National Academy Press.

Chandler, P., & Sweller, J. (1996). Cognitive load while learning to use a computer program. *Applied Cognitive Psychology, 10,* 151-170.

Chandler, T.N. (1996). System for Training Aviation Regulations (STAR): Using multiple vantage points to learn complex information through scenario-based instruction and multimedia techniques. *Journal of Educational Multimedia and Hypermedia, 5,* 349-378.

Chappell, S.L. (1991). Training and cockpit design to promote expert performance. In R.S. Jensen (Ed.), *Proceedings of the Sixth International Symposium on Aviation Psychology* (pp. 133-138). Columbus, OH: Ohio State University.

Chi, M.T.H., Bassok, M., Lewis, M.W., Reimann, P., & Glaser, R. (1989). Self explanations: How students study and use examples in learning to solve problems. *Cognitive Science, 11,* 145-182.

Cooke, N.J. (1990). Modeling human expertise in expert systems. In R.R. Hoffman (Ed.), *The psychology of expertise* (pp. 29-60). New York, NY: Springer-Verlag.

Criswell, E.L. (1989). *The design of computer-based instruction.* New York: Macmillan.

Duffy, L. (1993). Team decision-making biases: An information processing perspective. In G.A. Klein, J. Orasanu, R. Calderwood, & C.E. Zsambok (Eds.), *Decision making in action: Models and methods* (pp. 346-359). Norwood, NJ: Ablex.

Durso, F.T., Truitt, T.R., Hackworth, C.A., Crutchfield, J.M., & Manning, C.A. (1998). En route operational errors and situation awareness. *The International Journal of Aviation Psychology, 8,* 177-194.

Ellis, T. (1999, June). The future of training. *Occupational Health and Safety,* 34-43.

Endsley, M.R., & Kris, E.O. (1994). Information presentation for expert systems in future fighter aircraft. *The International Journal of Aviation Psychology, 4,* 333-348.

Entin, E.E., & Serfaty, D. (1999). Adaptive team coordination. *Human Factors, 41,* 312-325.

Fister, S. (2000, April). Reinventing training at Rockwell Collins. *Training,* 64-70.

Flach, J.M. (1995). Situation awareness: Proceed with caution. *Human Factors, 37,* 149-157.

Foxall, G., & Greenley, G. (1998). The affective structure of consumer situations. *Environment and Behaviour, 30,* 781-798.

Gaba, D.M., Howard, S.K., & Small, S.D. (1995). Situation awareness in anesthesiology. *Human Factors, 37,* 20-31.

Gage, N.L., & Berliner, D.C. (1998). *Educational psychology* (6th ed.). Boston, MA: Houghton Mifflin.

Glaser, R., Shauble, L., Raghavan, K., & Zeitz, C. (1992). Scientific reasoning across different domains. In E. DeCorte, M.C. Linn, H. Mandl, & L. Verschaffel (Eds.), *Computer-based learning environments and problem-solving* (pp. 345-371). Berlin, DE: Springer-Verlag.

Goodyear, P. (1990). The provision of tutorial support for learning with computer-based simulations. In E. DeCorte, M.C. Linn, H. Mandl, & L. Verschaffel. (Eds.), *Computer-based learning environments and problem-solving* (pp. 391-409). Berlin, DE: Springer-Verlag.

Greeno, J.G. (1998). The situativity of knowing, learning and research. *American Psychologist, 53,* 5-26.

Hammond, K.R., Hamm, R.M., Grassia, J., & Pearson, T. (1987). Direct comparison of the efficacy of intuitive and analytical cognition in expert judgement. *IEEE Transactions on Systems, Man, and Cybernetics, 17,* 753-770.

Harris, J.S. (1994). Improved aeronautical decision-making can reduce accidents. *Flight Safety Foundation: Helicopter Safety, 20*(2), 1-6.

Herrington, J., & Oliver, R. (2000). An instructional design framework for authentic learning environments. *Educational Technology, Research and Development, 48,* 23-49.

Hopkin, V.D. (1995). *Human factors in air traffic control.* London, UK: Taylor and Francis.

Howell, W.C., & Cooke, N.J. (1989). Training the human information processor: A review of cognitive models. In I.L. Goldstein (Ed.), *Training and development in organisations* (pp. 121-182). San Francisco, CA: Jossey-Bass.

Jonassen, D.H. (1996). *Computers in the classroom: Mindtools for critical thinking.* Englewood Cliffs, NJ: Prentice-Hall.

Jones, D.G., & Endsley, M.R. (2000). Overcoming representational errors in complex environments. *Human Factors, 42,* 367-378.

Kang, K-S., Cheon, S-W., & Chang, S-H. (1992). Development of an expert system for performance evaluation and diagnosis. In E.W. Hagen (Ed.), *Conference record for the fifth conference on human factors and power plants* (pp. 308-313). Monterey, CA: IEEE.

Klein, G.A. (1989). Recognition-primed decisions (RPD). *Advances in Man-Machine Systems, 5,* 47-92.

Knowles, M.S., Holton, E.F., & Swanson, R.A. (1998). *The adult learner.* Houston, TX: Gulf Publishing.

Kolodner, J.L. (1996). Making the implicit explicit: Clarifying the principles of case-based reasoning. In D.B. Leake (Ed.), *Case-based reasoning* (pp. 349-370). Menlo Park, CA: MIT Press.

Kontogiannis, T. (1996). Stress and operator decision-making in coping with emergencies. *International Journal of Human-Computer Studies, 45,* 75-104.

Krendl, K.A., & Lieberman, D.A. (1988). Computers and learning: A review of recent research. *Journal of Educational Computing Research, 4,* 367-389.

Light, P., & Littleton, K. (1999). Getting IT together. In K. Littleton & P. Light (Eds.), *Learning with computers* (pp. 1-9). London, UK: Routledge.

Marran, L., & Schor, C. (1997). Multiaccommodative stimuli in VR systems: Problems and solutions. *Human Factors, 39,* 382-388.

Martin-Michiellot, S., & Mendelsohn, P. (2000). Cognitive load while learning with a graphical computer interface. *Journal of Computer Assisted Learning, 16,* 284-293.

Mecham, M. (1994, January). Cathay refines approach to simulator training. *Aviation Week and Space Technology,* 35-36.

Mosier, K.L. (1991). Expert decision-making. In R.S. Jensen (Ed.), *Proceedings of the Sixth International Symposium on Aviation Psychology* (pp. 266-271). Columbus, OH: Ohio State University Press.

Murray, S.R. (1999). FACE: Fear of loss of face and the five hazardous attitudes concept. *The International Journal of Aviation Psychology, 9,* 403-411.

Newby, T.J., Stepich, D.A., Lehman, J.D., & Russell, J.D. (2000). *Instructional technology for teaching and learning* (2nd ed.). Upper Saddle River, NJ: Prentice-Hall.

Norman, D.A. (1988). *The design of everyday things.* New York: Doubleday.

Novick, L.R. (1988). Analogical transfer, problem similarity, and expertise. *Journal of Experimental Psychology: Learning, Memory and Cognition, 14,* 510-520.

Obradovich, J.H. (1996). The transfusion medicine tutor: A model for the design and use for expert systems for teaching. *Proceedings of the Human Factors and Ergonomics Society 40th Annual Meeting* (pp. 1055-1059). Santa Monica, CA: Human Factors and Ergonomics Society.

O'Hare, D., & Batt, R. (1999). A pilot for all seasons: Beyond simulation. In D. O'Hare (Ed.), *Human performance in general aviation* (pp. 173-192). Aldershot, UK: Ashgate.

O'Neil, H.F., Mayer, R.E., Herl, H.E., Niemi, C., Olin, K., & Thurman, R.A. (2000). Instructional strategies for virtual aviation training environments. In H.F. O'Neil & D.H. Andrews (Eds.), *Aircrew training and assessment* (pp. 105-130). Mahwah, NJ: Lawrence Erlbaum.

Parasuraman, R., & Riley, V. (1997). Humans and automation: Use, misuse, disuse, and abuse. *Human Factors, 39,* 230-253.

Perkins, D.N. (1990). *Person plus: A distributed view of thinking and learning.* Cambridge, MA: Harvard Graduate School of Education.

Pounds, J., & Bailey, L.L. (1999). *Cognitive style and learning: Performance of adaptors and innovators in a novel dynamic task.* Washington, DC: Federal Aviation Administration (NTIS DOT/FAA/AM-99/12).

Prerau, D.S., Adler, M.R., & Gunderson, A.S. (1990). Eliciting and using experiential knowledge and general expertise. In R.R. Hoffman (Ed.), *The psychology of expertise* (pp. 137-148). New York: Springer-Verlag.

Pressley, M., & Woloshyn, V. (1995). *Cognitive strategy instruction that really improves children's academic performance.* Cambridge, MA: Brookline.

Resnick, L.B. (1987, December). Learning in school and out. *Educational Researcher,* 13-20.

Roblyer, M.D., Edwards, J., & Havriluk, M.A. (1997). *Integrating educational technology into teaching.* Upper Saddle River, NJ: Prentice-Hall.

Roth, E.M., & Woods, D.D. (1989). Cognitive task analysis: An approach to knowledge acquisition for intelligent systems design. In G. Guida & C. Tasso (Eds.), *Topics in expert system design* (pp. 233-264). Amsterdam, Netherlands: Elsevier.

Rotheray, D., & Sewell, D. (1987). The release of cognitive resources: A unifying perspective on mainstream and special education. In J. Rutkowska & C. Crook (Eds.), *Computers, cognition and development* (pp. 69-83). Chichester, UK: John Wiley.

Salzman, M.C., Dede, C., & Loftin, R.B. (1995). Usability and learning in educational virtual realities. *Proceedings of the Human Factors and Ergonomics Society 39th Annual Meeting* (pp. 486-490). Santa Monica, CA: Human Factors and Ergonomics Society.

Sarter, N.B., & Woods, D.D. (1991). Situation awareness: A critical but ill-defined phenomenon. *The International Journal of Aviation Psychology, 1,* 45-57.

Savage, P., & Pearsall, S. (1998). A case study in iterative design. *Ergonomics in Design, 6,* 18-25.

Seidel, R.J., & Wagner, H. (1979). A cost-effectiveness specification. In H. O'Neill (Ed.), *Procedures for instructional system development* (pp. 233-251). New York: Academic Press.

Shiffrin, R.M., & Schneider, W. (1977). Controlled and automatic human information processing: II. Perceptual learning, automatic attending and general theory. *Psychological Review, 84,* 127-190.

Sibbald, D. (2000). Bridging the gap from classroom to practice: PBL students develop consumer website for nonprescription drugs. *American Journal of Pharmaceutical Education, 64,* 339-349.

Singh, I.L., Molloy, R., & Parasuraman, R. (1993). Automation-induced "complacency": Development of the complacency-potential rating scale. *The International Journal of Aviation Psychology, 3,* 111-122.

Sleeman, D. (1987). Cognitive science, AI and developmental psychology. Are there links? Could there be links? In J. Rutkowska & C. Crook (Eds.), *Computers, cognition and development* (pp. 235-254). Chichester, UK: John Wiley.

Spyridakis, J.H., Miller, A.E., & Barfield, W. (1998). Usability evaluation for intelligent transportation systems. In W. Barfield & T.A. Dingus (Eds.), *Human factors in intelligent transportation systems* (pp. 217-246). Mahwah, NJ: Lawrence Erlbaum.

Srisethanil, C., & Baker, N.C. (1995). Applications and development of multiple teaching styles to an engineering ITS. In J. Greer (Ed.), *Proceedings of AI-ED '95* (pp. 315-322). Charlottesville, VA: AACE.

Sweller, J., & Chandler, P. (1991). Evidence for cognitive load theory. *Cognition and Instruction, 8,* 351-362.

Takano, K., Sasou, K., & Yoshimura, S. (1997). Structure of operator's mental models in coping with anomalies occuring in nuclear power plants. *International Journal of Human-Computer Studies, 47,* 767-789.

Thurman, R.A. (1994). Instructional simulation from a cognitive psychology viewpoint. *Educational Technology Research and Development, 41,* 75-89.

Thurman, R.A., & Russo, T. (2000). Using virtual reality for training. In H.F. O'Neil & D.H. Andrews (Eds.), *Aircrew training and assessment* (pp. 85-104). Mahwah, NJ: Lawrence Erlbaum.

Todd, P.A., & Benbasat, I. (1994). The influence of decision aids on choice strategies under conditions of high cognitive load. *IEEE Transactions on Systems, Man and Cybernetics, 24,* 537-547.

Tsang, P.S., & Voss, D.T., (1996). Boundaries of cognitive performance as a function of age and flight experience. *The International Journal of Aviation Psychology, 6,* 359-377.

Van Hoven, C. (2000, August). Keys to effective training. *Occupational Health and Safety, 69,* 102-108.

Wallace, S. (2000, January). The benefits of interactive CBT. *Occupational Health and Safety,* 22-26.

Wann, J., & Mon-Williams, M. (1996). What does virtual reality need? Human factors issues in the design of three-dimensional computer environments. *International Journal of Human-Computer Studies, 44,* 829-847.

Watters, M. (2000). Reciprocal questions and computer-based instruction in introductory auditing: Student perceptions. *Journal of Education for Business, 75,* 263-266.

Wickens, C.D., Gordon, S.E., & Liu, Y. (1998). *An introduction to human factors engineering.* New York: Longman.

Wiggins, M.W. (2001). *Expertise and weather-related decision-making.* Unpublished doctoral dissertation, University of Otago, New Zealand.

Wiggins, M., & O'Hare, D. (1993). A skill-based approach to aeronautical decision-making. In R. Telfer (Ed.), *Aviation Instruction and Training* (pp. 430-475). Aldershot, UK: Gower.

Wiggins, M., O'Hare, D., & Hunter, D. (2000). *Weatherwise* [Computer-Based Training Program]. Washington, DC: Federal Aviation Administration.

Wilson, J.R. (1999). Virtual environments applications and applied ergonomics. *Applied Ergonomics, 30,* 3-9.

Wilson, P.N., Foreman, N., & Tlauka, M. (1997). Transfer of spatial information from a virtual to a real environment. *Human Factors, 39,* 526-531.

Yazdani, M. (1987). Artificial Intelligence, powerful ideas and children's learning. In J. Rutkowska & C. Crook (Eds.), *Computers, cognition and development* (pp. 99-114). Chichester, UK: John Wiley.

Ye, N., & Salvendy, G. (1994). Quantitative and qualitative differences between experts and novices in chunking computer software knowledge. *International Journal of Human-Computer Interaction, 6,* 105-118.

13 Internet-Based Education and Training: Using New Technologies to Facilitate Instruction and Enhance Learning

Matthew J.W. Thomas

Introduction: The Promise of Internet-Based Education and Training in Aviation

The recent rapid growth of the Internet as a global telecommunications medium has also led to significant innovations in relation to the delivery of education and training. Many aspects of the learning process, from the presentation of information and interpersonal communication to assessment of learning outcomes and performance, can now be supported using widely accessible Internet-based formats. Within the aviation industry, the Internet has recently been embraced as an innovative means of providing education and training. Stating a common perception of unlimited potential, Wilson (2000) suggests, that "the Internet is set to revolutionise certain portions of the training curriculum, as distance learning takes more and more pilots out of the classroom and on to the web" (p. 22). One typical vision of the future of aviation training involves flight-crew members on a long-haul stopover logging into the company's secure web site from a notebook computer in their hotel room and undertaking an Internet-based module on the A380 field-length requirements as part of conversion training. While this scenario might have been seen as an element of science fiction only a decade ago, today it is much more of a reality.

In an industry where staff who require regular training are frequently travelling across all corners of the globe, the benefits of an instructional delivery mode that transcends both space and time are obvious. Unlike classroom-based learning, or traditional Computer-Based Training (CBT), which both require groups of students attending a central training location, the Internet allows the delivery of instruction at any time and in any place (Walters, 1999). The benefits of Internet-based training are typically seen in relation to cost savings and improved training efficiency. As Earp (2000) suggests, the adoption of distance learning in the aviation industry offers savings in relation to travel costs and scheduling flexibility, and considerable improvement in operation performance by allowing people access to instruction on demand.

While prospective gains in efficiency, along with reductions in costs, are certainly attractive to any organisation, the shift to Internet-based education and training must also be driven by reference to the use of the Internet as an instructional medium. From the perspective of instruction, the Internet has a number of potential uses including the delivery of course materials, interpersonal interaction and dialogue between students and teaching staff, and the Internet can even be used as a medium for educational evaluation and assessment.

Perhaps the most common instructional use of the Internet currently involves the distributed provision of Internet-based hypermedia and multimedia courseware. Hypermedia and multimedia were developed as new computer-based forms of information representation and have been accepted as alternatives to the use of paper-based instructional materials. The term multimedia describes the storage and presentation of information in a combination of media forms: for instance audio, images, and text (Reushle, 1995). Hypermedia refers to the representation of multimedia in a hypertext system, where pieces of information are stored as nodes in non-linear networks, and connected through links. Together, these new forms of information representation can be used to build complex course materials. In order to understand better the potential of these new Internet-based forms of educational delivery, it is necessary to examine the use of these new information and communications technologies in relation to the principles of adult learning.

New Technologies and the Principles of Adult Learning

For decades computers have been used in instruction, and are now accepted as tools which can promote high-quality learning outcomes through the provision of integrated multimedia instructional programs. Historically, computer-based learning developed from within the highly didactic paradigm of behaviouristic models of learning. The early developments of teaching machines and programmed instruction embodied the very essence of the didactic transfer of knowledge and reinforced an overtly mechanistic and technical perspective of learning (Fry, 1961; Glaser, 1965; Skinner, 1968). In direct contrast to this legacy, Internet-based instruction offers means by which new modes of learning can be realised. In particular, a considerable amount of recent research has focused on how the principles of adult learning can be realised through the use of new information and communication technologies (Laurillard, 1993).

When thinking about the role of the Internet in relation to adult learning, it is useful to reflect on the principles of andragogy, or adult learning theory. As described by Knowles (1983), andragogy holds that adults learn in a manner quite different from children, and therefore require different modes of instruction. In contrast to the emphasis on the didactic transfer of knowledge which is central to traditional pedagogy (childhood learning), andragogy (adult learning) involves instructional modes and methods that emphasise aspects drawn from our understanding of the cognitive development of adults. Three basic tenets of andragogy are important to the increasing use of the Internet as an instructional medium. Each of these principles of adult learning provides a framework for appropriately embedding Internet-based instruction within a training curriculum and can offer considerable guidance for practice.

A primary principle of andragogy is that adults have considerable autonomy as learners. This is to say that adults in general have the ability to engage in self-directed learning, through evaluating their own knowledge deficiencies, identifying their own learning objectives and assuming certain responsibly for their own learning outcomes (Rowntree, 1992). To a large degree this means a shift away from overly prescriptive approaches to instruction typified by the traditional "chalk and talk" model. From the perspective of andragogy, the role of the instructor and thus by default, the instructional medium is to assist adults to learn in a way that enhances their capabilities to function as self-directed learners (Mezirow, 1981). The use of the Internet as an instructional medium has been shown to promote these important aspects of adult learning. Park (1991) discusses a number of research studies which have shown that hypermedia is ideal for adaptive student-centred instruction and increases learner control. Although using computer-based platforms that predate the Internet proper, these studies utilised a hypermedia format almost identical to that of the World Wide Web. In one such study it was argued that the potential of hypermedia systems lies in the provision of a non-constrictive information environment, where students choose the paths through information, and are engaged in a constant process of decision making and evaluation through navigation (Roselli, 1991). More recent research has reinforced these findings through the use of Internet-based instructional materials and suggests that the information structures in Internet-based materials can mirror the development of students' own knowledge structures (Barnard, 1997; Starr, 1997).

A second and allied principle of andragogy relates to the need for a practical and problem-based curriculum. Adult learners already have considerable experience and knowledge, and are most frequently engaged in learning tasks with practical implications for their everyday work tasks. Accordingly, the learning outcomes of adult learners can be greatly enhanced by promoting the integration of new knowledge within a framework of past experience and future practice. As Knowles (1983) suggests, andragogy involves a distinct shift away from techniques that involve the transmission of information, towards modes of instruction that are participatory and experiential. Moreover, it has been argued that in the aviation training environment a problem-based approach is essential to the development of more complex forms of knowledge and skills whereby a student is able to use acquired knowledge in a practical context (Henley, Anderson, & Wiggins, 1999). By constructing interactive learning environments in which students engage in the development of solutions to various problems, Internet-based learning can be enhanced. As recent research has shown, educational uses of hypermedia can support various forms of problem-based and exploratory learning (Jacobs, 1992). Furthermore, the development of Internet-based simulation and part-task training by major equipment manufacturers illustrates the potential for these types of Internet-based learning in the aviation context.

A third and final aspect of andragogy that relates particularly to the use of the Internet in instruction is the role of interaction and dialogue in the adult learning process. Laurillard (1988, 1999) has developed a "conversational" model of adult learning, in which the processes of internal and interactive dialogue are seen as essential to the formulation of new knowledge and abilities. In turn, this naturalistic view of the learning process highlights several other important aspects of adult learning, including critical reflection, learner engagement and the socialisation processes inherent in knowledge construction. As Laurillard (1999) states, "Essentially, a learning process complex enough to achieve the aims of academic learning must involve at least two participants, operating iteratively and interactively on two levels — practice and discussion — and connecting those two levels by the activities of adaptation and reflection" (p. 114).

According to this theory of adult learning, new knowledge is not merely acquired through passively receiving transmitted information. Rather, adults integrate new information into increasingly complex knowledge structures through a dialectic process in which dialogue is crucial. Through the provision of asynchronous, non-hierarchical and reciprocal communication environments, computer-supported collaborative work has been shown to increase participation, interaction and collaborative thinking (Boshier, 1990; Ruberg, Moore, & Taylor, 1996). Furthermore, Internet-based learning environments appear to provide a perfect forum for an academic discourse that promotes increased student engagement, critical analysis and reflection, and the social construction of knowledge. As Warschauer (1997) argues, computer-mediated communication forums allow for learning which integrates student interaction and critical reflection.

Each of these three aspects of andragogy: learner autonomy, practical and problem-based approaches, and the role of dialogue, are important to the development of Internet-based instruction. Moreover, the forms of non-linear information representation in hypermedia have been shown to create qualitative improvements in the construction of knowledge and understanding (Burbules & Callister, 1996). Recently, the *engagement theory* of information and communication technology use has been used to describe the potential benefits of Internet-based teaching and learning. By way of bringing together a number of different potential benefits, engagement theory encapsulates the means by which technology can provide student-centred learning. As Kearsley and Shneiderman (1998) describe, "the fundamental idea underlying engagement theory is that students must be meaningfully engaged in learning activities through interaction with others and worthwhile tasks.... We believe that technology can facilitate engagement in ways which are difficult to achieve otherwise" (p. 20).

Emerging from the theories of qualitatively improved learning, research studies have shown that the use of information and communication technologies can lead to increased levels of cognitive engagement with the learning task, increased motivation, deeper levels of processing, and the construction of personal meaning (Burbules & Callister, 1996; Land & Hannafin, 1996). However, while much of the recent research has focused on the role of new technologies in facilitating qualitatively enhanced forms of student learning, a considerable amount of practice in the area still fixates on the possible increases to productivity, efficiency and cost-effectiveness. This focus on the bottom line has the potential to detract from the production of effective instruction within the aviation industry.

Current Examples of Internet-Based Education and Training

Currently, there are considerable moves throughout the aviation industry towards the development of Internet-based training programs. In all areas of the industry, from university-based *ab initio* flight training, through to airline-based recurrency training, the Internet is being used to facilitate the delivery of distance education and training. To better understand both the diversity and potential of current practice, two innovative programs will be explored in detail. While these programs differ in educational objectives, curriculum development and instructional design, both use the Internet as an essential mode of delivery. Furthermore, through the provision of flexible learner-centred education, both programs embody the major principles of adult learning.

Case Study One: Air New Zealand Internet-based Airport Familiarisation Briefings

Project context: Training requirements and instructional objectives The Air New Zealand Internet-based Airport Familiarisation Briefings project was undertaken to facilitate the process of flight crew acquiring and maintaining route and aerodrome proficiency qualifications. As a regulatory requirement under international air transport regulations, all flight crew must undergo regular training and evaluation in relation to the routes and aerodromes they use. For instance, under JAR-OPS1.975 an operator must ensure that, prior to being assigned as commander or as pilot to whom the conduct of the flight may be delegated by the commander, the pilot has obtained adequate knowledge of the route to be flown and of the aerodromes (including alternates), facilities and procedures to be used.

Similar regulatory requirements exist worldwide and ensure that all flight crew are familiar with the particular characteristics and operational requirements of all airports serviced on specific routes, as well as the alternate airports with which crews might be required to be familiar given the need for diversion. Proficiency in relation to the specific requirements of a variety of unique airports requires flight crew to develop a significant body of knowledge relating to local conditions, airport layout, and airspace. Traditionally, this knowledge has been acquired by flight crew during ground and simulator training and from personal study of the company's operations manuals. Flight crew is then required to demonstrate its acquisition and effective use of this knowledge through adequate performance during recurrent route checks. While this process establishes a measure of operational proficiency, the overall training approach does little to assist flight crew in developing an integrated body of knowledge nor does it employ the principles of adult learning.

To create a more effective training system, Air New Zealand has recently undertaken a major project to develop Internet-based airport briefings for its international flight crew. This project involved the development of individual modules of study constituting a detailed briefing for each of the airports Air New Zealand services on its international routes. The overall objectives of this project were twofold. Firstly, the project sought to enhance the learning process in relation to flight-crew route and aerodrome proficiency by taking advantage of both the multimedia capabilities of CBT combined with the flexibility offered by Internet-based delivery. Secondly, the project sought to reduce training costs by removing the requirement to roster pilots to come to base specifically to complete the airport briefings.

Technical Details and Instructional Processes

The training program was constructed using Macromedia Authorware, a well-known industry standard software application used for the development of computer-based training packages. This software allows for the inclusion of multimedia content, and facilitates the development of an integrated learning package including elements such as the structured delivery of content and automated on-line assessment. Although used primarily for the development of stand-alone CBT packages, the software also allows for the development of Internet-based packages. Using a standard Web browser and a common software plug in, users are able to login to a secure web site from any computer with Internet access and work through the CBT package when required. Each airport briefing covers a variety of topics including general details concerning the geographic location of the airport and its surroundings as well as issues relating to terrain and local weather patterns (see Figure 13.1).

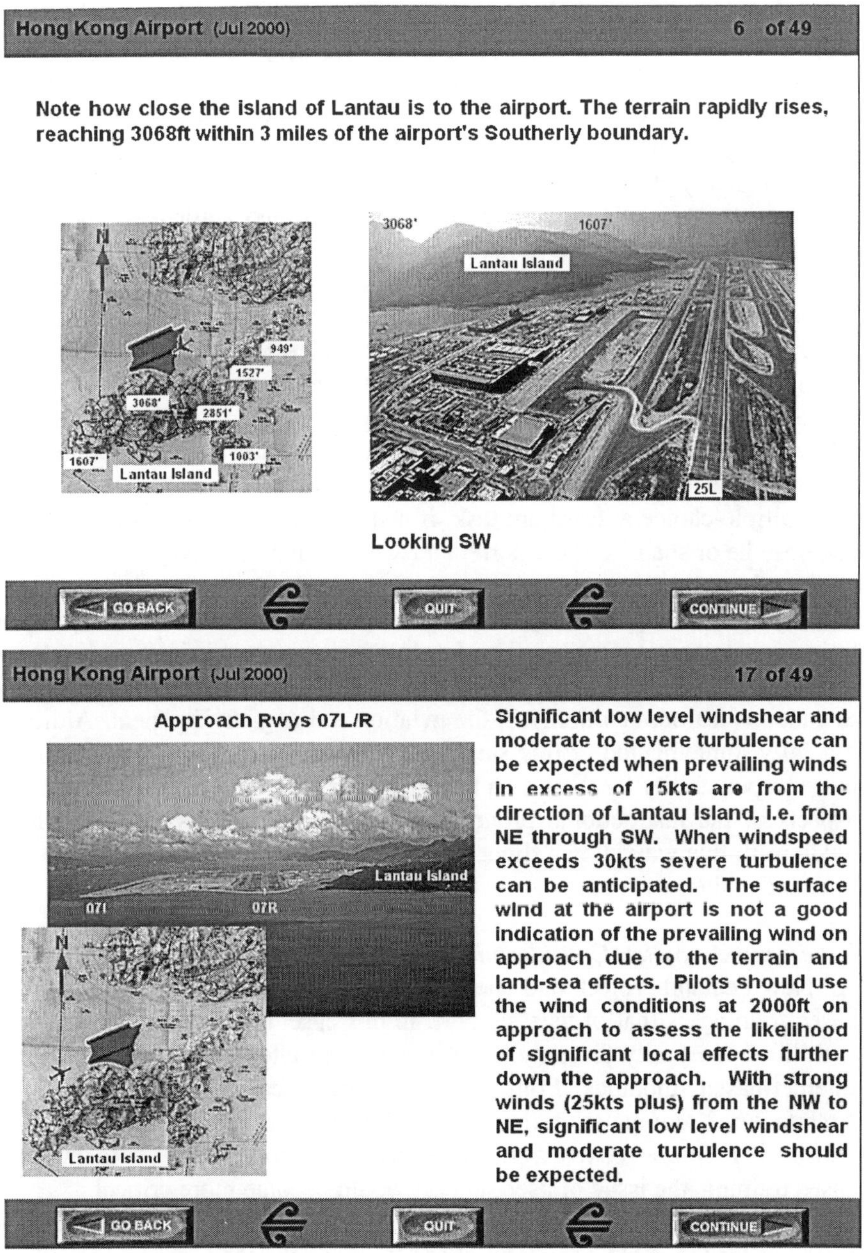

Figure 13.1 Examples of Internet-based airport familiarisation briefings

The briefings also include operational information and cover details of long final, short final and take-off, along with any operational restrictions relevant to particular airports. As employees work through each module, they are required to demonstrate adequate development of knowledge by successfully completing a series of multiple-choice assessment questions that are strategically placed throughout the module.

From the perspective of instructional design and adult learning theory, each module is constructed to take advantage of the non-linear capabilities of Internet-based multimedia delivery and provide for instruction that is to some degree individualised. If the assessment procedure identifies learning deficiencies — by the student providing an incorrect answer — the student is taken back to the relevant section of the module for further study. This process is similar to the "mastery" learning approach, whereby students are allowed a reasonable number of attempts at the assessment procedure until they meet the required standard (Keller, 1968). However, due to the nature of the multiple-choice assessment task, if a student provides a second incorrect response he or she is told the correct answer and not allowed further attempts on the relevant section.

As discussed previously, an important aspect of adult education and training is learner autonomy. The Internet-based airport briefings developed by Air New Zealand provide a good example of how the Internet can be used to enhance learner autonomy in the aviation training environment. Although not fully implemented at this stage, the ultimate aim is to give flight crew more responsibility for their own learning by providing access to the required aspects of the instructional program. Using the Internet as an instructional medium in this way allows flight crew to undertake their study in their own time and at their own pace.

Issues and Areas for Consideration An important issue highlighted by the Air New Zealand project, and one that is common to many forms of Internet-based training, is that of user-support. In this case, flight crews are provided with the opportunity to study from home as an alternative to access to the Internet-based modules from the Air New Zealand "Standard Client Architecture" computer terminals located at Air New Zealand bases. However, if future moves were made towards a complete shift to Internet-based training, the issue of user-support would become more critical.

Even within a large organisation, only limited technical support can be provided for the flight crew who chose to undertake their airport procedures training using the Internet from their home computers. If users have difficulty with the configuration of software or the installation of a necessary plug in on their home computers, most organisations are unable to provide detailed support.

One possible step towards a solution to this ongoing problem might be through standardisation, whereby a training provider will specify the required hardware, operating system and software necessary for successful operation of the Internet-based instructional materials (Jones, Kirkup, & Kirkwood, 1993). Air New Zealand makes recommendations for the minimum requirements in relation to the computer specifications for running the Internet-based instructional modules. However, an ongoing issue for any training provider relates to the responsibility for purchasing and maintaining "personal" equipment used for what is essentially the maintenance of work-related competencies.

Another issue that is highlighted by this case study is the need to clearly define the instructional purpose served by Internet-based instruction within the context of an integrated training program. It is apparent that Internet-based instruction has certain strengths and weaknesses that must be taken into account in the overall instructional design. For instance, while a practical and problem-based approach has been stressed as a key concept in adult learning, the common predilection towards the use of multiple-choice testing in the Internet-based training environment has the tendency to promote learning more aligned with the passive transfer of knowledge, or may encourage a surface approach to learning.

While multiple-choice testing is a particularly robust means of assessing knowledge-related learning outcomes, it is not always an effective technique for assessing skill-based behaviours such as problem solving (Gronlund & Linn, 2000). Furthermore, with the expectation of multiple-choice testing, students can commonly develop only declarative forms of knowledge whereby they are able to accurately replicate the information presented in the instructional material, but lack a sufficiently complex understanding of the material to effectively use it in the operational environment.

Within the context of flight crew acquiring and maintaining route and aerodrome proficiency qualifications, Air New Zealand uses Internet-based instruction specifically to streamline the process of preliminary knowledge acquisition. The Internet-based modules are, therefore, seen as a small element of a larger program in which the simulator and line training syllabus are designed to build the operational competencies for effective crew performance. This approach clearly defines the role of the Internet-based instructional materials, and uses the medium in a way that takes advantage of its strengths as an effective means of presenting information in a multimedia format.

Case Study Two: Massey University School of Aviation Internet-Based University Study

Project context: Training requirements and instructional objectives The delivery of flexible and distance learning within the higher education sector has been a focus of considerable recent growth. Most universities now offer some portions of their degrees through a distance learning option, and the Internet has become a common medium used to expand the range of educational delivery (Bates, 1995). The School of Aviation at Massey University has been a pioneer in the development of Internet-based aviation distance learning within its Bachelor of Aviation degree. As part of a university that has a long history of traditional paper-based distance learning programs, the School of Aviation has recently led the shift towards utilising Internet-based delivery as a core element of its distance education programs.

The aim of the Bachelor of Aviation degree is to assist students with the development of a wide range of knowledge and skills relating to professional practice in the international aviation industry. Not only concerned with the development of operational competencies, the degree program is designed to provide students with integrated academic, technical and practical knowledge, along with generic skills, such as high-level problem solving and decision-making skills (Hunt, 2000). Accordingly, the choice to include elements of Internet-based delivery in this academic environment has been driven by the need to explore alternative delivery modes that can be used to achieve these enhanced learning outcomes.

Furthermore, using Internet-based delivery in conjunction with more traditional delivery modes also provides a mechanism for the globalisation of the Bachelor of Aviation degree program. As a global telecommunications network, the Internet allows for access to study materials at any time and any place. Therefore, with the use of appropriate instructional design, students are able to engage in academic studies whether they are located on campus or overseas.

Technical Details and Instructional Processes The Internet-based course materials are developed using Macromedia Dreamweaver and Macromedia Flash software packages. These packages enable both the development and integration of multimedia content into individual modules of study. Once the course materials are developed, they are placed within the framework of a WebCT database. WebCT provides a flexible Internet-based learning environment which facilitates the organisation of multimedia course materials, enables the integration of on-line discussion and collaboration into the Internet-based courses, and also provides a number of tools for assessment, evaluation and student management. The use of this purpose-built software package enables the creation of learning environments in which a number of different forms of Internet-based learning can be integrated as elements of an overall instructional design.

The typical format adopted by the School of Aviation in the development of its distance learning curricula is that of multimodal delivery. Multimodal delivery involves the use of a variety of media as elements of an overall educational program. Furthermore, as a framework for instructional design, multimodal delivery uses what can be termed a complementary approach to information technology use in higher education. The complementary approach seeks to take advantage of the particular strengths of different instructional delivery modes in creating an enhanced educational program (Thomas, 2000). For instance, a print-based resource or perhaps a face-to-face practical session might be used to complement Internet-based course materials. The underlying assumption that drives the complementary approach is that the learning environment is a powerful mediator of the approaches to learning that students adopt, and the outcomes that they achieve. Accordingly, this approach to instructional design involves not using technologies for types of learning activities they cannot yet adequately support, and seeks to promote balance and diversity throughout the curriculum. The diagram in Figure 13.2 illustrates how both innovative Internet-based and more traditional face-to-face instructional elements can be integrated to produce an effective instructional design.

One of the major objectives of the complementary approach to Internet-based instructional design is the development of frameworks for students' knowledge construction and skill development. To achieve this objective, instructional materials are created such that the learner is guided through a developmental progression that begins with the introduction of prerequisite knowledge. Subsequent instructional materials and activities then assist students to progressively build on this foundation knowledge and create increasingly complex knowledge structures and skills.

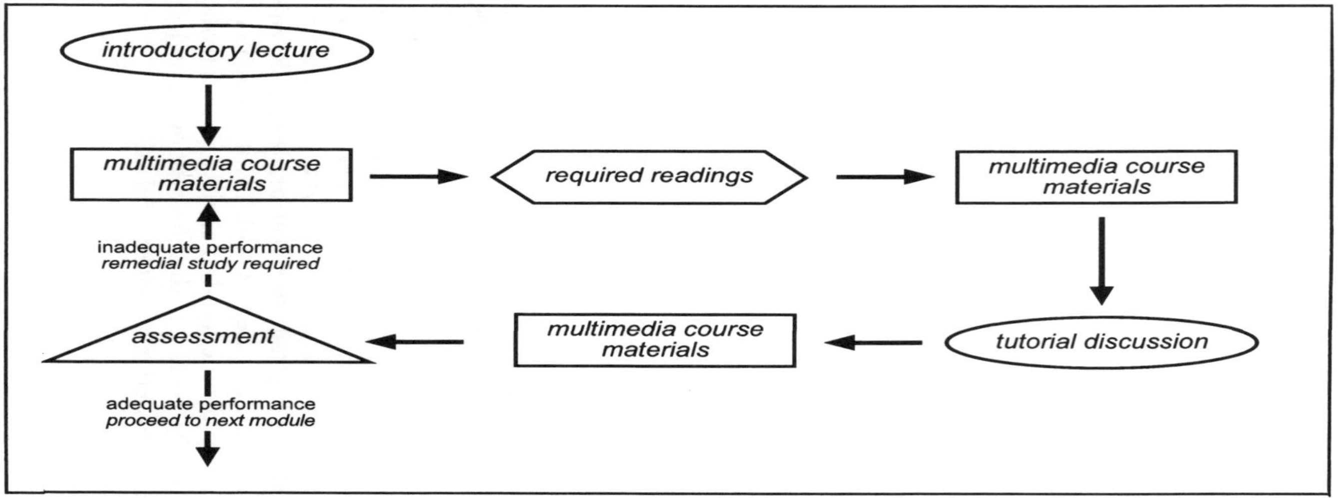

Figure 13.2 An example of a complementary approach to Internet-based instructional design

Another essential element of this approach is the alignment of appropriate learning activities with the desired learning outcomes. As Biggs (1999) argues, the learning activities and assessment tasks embedded within a curriculum are powerful determinants of student learning outcomes. Therefore, by aligning the desired learning outcome with appropriate instructional processes, the resulting curriculum can be both effective and authentic. Figure 13.3 shows two screens from a typical Internet-based course.

From the perspective of adult learning theory, the Internet-based distance learning programs at Massey University are designed to promote the three major aspects of andragogy. Firstly, learner autonomy is supported through the provision of learning self-paced learning materials. The students are provided with some structure to guide their studies, such as a scheduled lecture series or assessment due date. However, these events are used only as indicators to ensure progression through the materials and students are otherwise able to progress through the course of study at their own pace.

Secondly, the alignment of learning and assessment tasks ensures the development of a practical and problem-based curriculum. Throughout the course of study students engage in activities that require the use of newly acquired knowledge in solving authentic problems. Finally, the use of complementary face-to-face sessions and Internet-based communication tools are used to promote the ongoing use of internal and interactive dialogue in the learning process. Through the process of dialogue, students are encouraged to articulate their emergent understanding. Accordingly, they must engage with the academic subject matter at a deeper level. However, the use of Internet-based tools for online student interaction presents an area in need of further consideration.

Issues and Areas for Consideration

The social dynamics of computer-mediated communication presents an area for careful consideration, and the existing research presents some contradictory evidence in relation to the use of Internet-based communication tools in the educational setting. Harasim (1990), a pioneer of new information and communication technology use in education, introduced the educational world to the potential of Internet-based communication tools. She suggested that the new technologies would provide powerful new environments for learning which could enhance social and intellectual connections. She states that "like face-to-face education, online education supports interactive group communication. Historically, the social, affective, and cognitive benefits of peer interaction and collaboration have been available only in face-to-face learning. The introduction of online education opens unprecedented opportunities for educational interactivity" (p. 42).

During the last decade it has become widely accepted that information and communication technologies promote the development of communities of learning, which in turn promote learning characterised by student interaction, reflection, collaboration and cooperation towards the social construction of knowledge. However, there is little empirical support for these assumptions and even though the use of Internet-based communication tools has been increasing exponentially over the last decade, there is a considerable amount of contradictory evidence as to its potential in the educational setting (Walther, 1992).

As early as the mid-1970s it was hypothesised that new forms of telecommunications media vary in their degree of social presence, and therefore, have effects on group interaction (Short, Williams, & Christie, 1976). In relation to Internet-based communication tools, the asynchronous, text-based, and remote nature of computer-mediated communication strips communication to its very basis, and removes many supposedly extraneous elements. Further, it is often suggested that the lack of non-verbal cues produces a depersonalised form of interaction (Ruberg et al., 1996).

As Taylor (1998) has also found, introverted students are more likely to benefit from computer-mediated communication than extroverted students, as introverted students find it easier to express themselves in the depersonalised forum (see Chapter 4 for a description of personality types). Recent studies have suggested that Internet-based communication tools may be limited in their ability to support the rich forms of student interaction that are essential predicates to enhanced academic learning (Thomas, 2000). Accordingly, this remains an area where further empirical investigation is required.

Another issue raised by the Massey University School of Aviation case study relates to the control and validation of qualifications within a strict regulatory environment. For university aviation programs, in which both academic and practical elements of the curriculum contribute to the issue of qualifications and licences, it is essential that strict control be kept in relation to the authentication of student work. To this end, students' access to course materials and assessment is restricted with usernames and passwords and student access is logged to ensure they are covering the required syllabus.

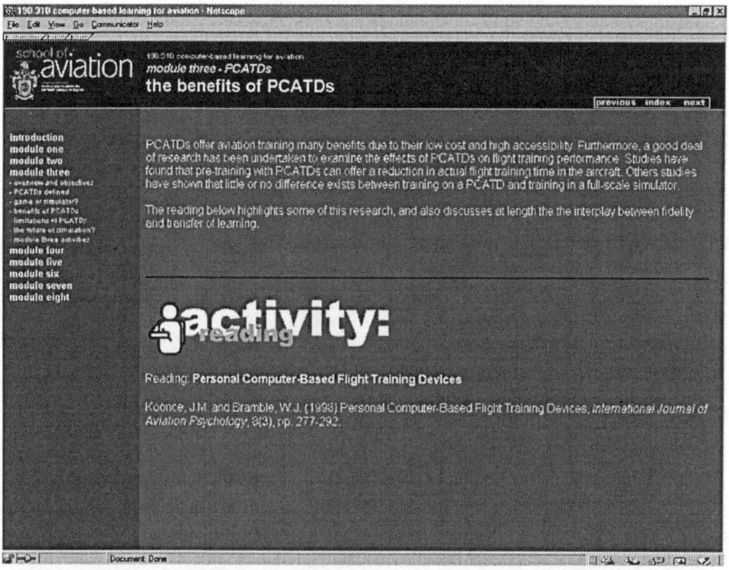

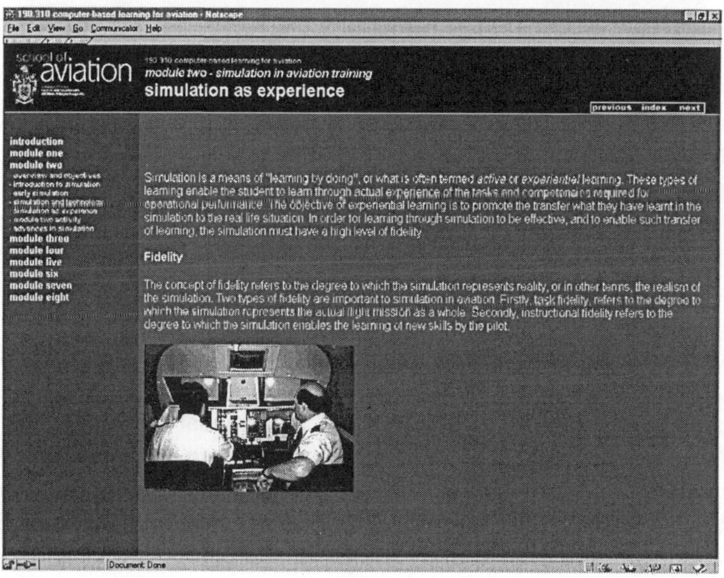

Figure 13.3 Examples of Internet-based course delivery at Massey University

However, even by restricting access and monitoring use of Internet-based materials the university has no means of unconditionally verifying the identity of students who participate in Internet-based instruction. Accordingly, it remains necessary to have in place assessment techniques in which the students' identity can be verified without question.

Conclusion: Promoting Enhanced Learning through Internet-Based Delivery

While the Internet obviously holds great potential to enhance education and training, there is also the need for caution as we move swiftly to adopt what are essentially radically new forms of computer-based learning. The Internet, like any other instructional medium, has unique characteristics, and its own set of strengths and weaknesses. From the preceding examination of the current research and practice in the field of Internet-based education and training, it is apparent that the Internet offers a great deal in relation to the presentation of multimedia course materials. Unfortunately, there is a trend to focus on the innovation involved in Internet-based delivery at the expense of achieving enhanced learning outcomes for the participants involved (Windschitl, 1998).

An important principle which must guide our adoption of Internet-based instruction is that the Internet is not a true educational technology. Indeed, at this point in the development of Internet-based educational delivery it is timely to be reminded of the origins of educational technology. As what was essentially an approach to instructional design, educational technology was first defined as a well disciplined and systematic approach to education and training, characterised by explicitness, sophisticated analysis, synthesis and the utilisation of optimal decision-making procedures (Davies, 1972). Unfortunately, this perspective of educational technology is now almost completely lost, and current definitions of educational technology focus on the use of computer software and hardware in the learning process.

By definition, the Internet is an information and communication technology, designed specifically for the efficient global transfer of information stored in a digital format. To this end, the use of Internet-based applications for the purposes of instruction can easily take for granted the need to adopt the appropriate principles of effective instruction design. Recent research has highlighted the failure of many Internet-based instructional materials to promote effective learning outcomes. As Landa

(1998) suggests, Internet-based materials have the propensity to actually inhibit the development of higher-order transferable cognitive skills (p. 55). This is largely because the medium lends itself to the presentation of information in a highly structured and condensed format. Accordingly, instructional materials can adopt an overtly didactic mode, teaching the solutions to problems rather than promoting the use of students' skills in problem solving, or promoting the development of socially negotiated structures of personal meaning. Further, Internet-based instruction frequently rewards the reproduction of information rather than deeper levels of cognitive engagement through the promotion of analysis, evaluation and personal reflection. This sentiment has been echoed by Doherty (1998), who states that "we are in danger of forcing the Internet into becoming a passive surfing technology. Instructors are presenting more and more traditional educational material on the Internet, with little consideration of the Internet's unique features. These features, if properly identified and utilised, can make learning an active experience" (p. 61).

High-quality instructional design and the use of a complementary approach to Internet-based instruction are important strategies that can be used to avoid the promotion of poor learning outcomes. Indeed, perhaps one of the most valuable benefits of shifting aviation education and training programs to Internet-based delivery will be a renewed focus on the fundamental principles of good instructional design. Through a process of needs analysis, instructional program review and curriculum redevelopment, the integration of Internet-based instruction has the potential to enhance both teaching and learning.

Although the last decade has been a period of radical change in relation to the instructional use of computers, there remains a considerable deficit in our understanding of the most effective use of Internet-based instruction. Particular focal points for future research include the analysis of the social dynamics of Internet-based communication, the further development of Internet-based assessment techniques along with the creation of instructional design tools to aid the development of Internet-based instruction that embodies the major principles of andragogy.

Recommended Readings

Bates, A.W. (1995). *Technology, open learning and distance education.* London, UK: Routledge.

Hunt, G.J.F. (Ed.). (1997). *Designing instruction for human factors training in aviation*. Aldershot, UK: Ashgate.

Laurillard, D. (1993). *Rethinking university teaching: A framework for the effective use of educational technology*. London: Routledge.

References

Barnard, J. (1997). The World Wide Web in higher education: The promise of virtual universities and online libraries. *Educational Technology, 37*(3), 30-35.

Bates, A.W. (1995). *Technology, open learning and distance education*. London: Routledge.

Biggs, J.B. (1999). *Teaching for quality learning at university*. Buckingham, UK: Society for Research into Higher Education and Open University Press.

Boshier, R. (1990). Socio-psychological factors in electronic networking. *International Journal of Lifelong Education, 9*(1), 49-64.

Burbules, N.C., & Callister, T.J. (1996). Knowledge at the crossroads: Some alternative futures of hypertext learning environments. *Educational Theory, 46*(1), 23-50.

Davies, I.K. (1972). The nature of educational technology. In I.K. Davies & J. Hartley (Eds.), *Contributions to an educational technology* (pp. 1-14). London: Butterworths.

Doherty, A. (1998). The internet: Destined to become a passive surfing technology. *Educational Technology, 38*(5), 61-63.

Earp, R.J. (2000, January/February). The evolution of distance learning. *Journal of Professional Aviation Training*, 12-15.

Fry, E.B. (1961). *Teaching machines and programmed instruction*. New York: McGraw-Hill.

Glaser, R. (1965). Towards a behavioural science base for instructional design. In R. Glaser (Ed.), *Teaching machines and programmed learning* (Vol. II, pp. 771-809). Washington, DC: National Education Association of the United States.

Gronlund, N.E., & Linn, R.L. (2000). *Measurement and assessment in teaching* (8th ed.). Upper Saddle River, NJ: Merrill.

Harasim, L.M. (1990). Online education: An environment for collaboration and intellectual amplification. In L.M. Harasim (Ed.), *Online education: perspectives on a new environment* (pp. 39-64). New York: Praeger.

Henley, I., Anderson, P., & Wiggins, M. (1999). Integrating human factors education in general aviation: Issues and teaching strategies. In D. O'Hare (Ed.), *Human performance in general aviation*. Aldershot, UK: Ashgate.

Hunt, G.J.F. (2000). System of regulating aircrew licensing in need of surveillance-based approach. *ICAO Journal, 55*(8), 18-20, 27.

Jacobs, G. (1992). Hypermedia and discovery-based learning: A historical perspective. *British Journal of Educational Technology, 23*(2), 113-121.

Jones, A., Kirkup, G., & Kirkwood, A. (1993). *Personal computers for distance education*. New York: St. Martin's Press.

Kearsley, G., & Shneiderman, B. (1998). Engagement theory: A framework for technology-based learning. *Educational Technology, 38*(5), 20-23.

Keller, F. (1968). "Goodbye teacher... ". *Journal of Applied Behavior Analysis, 1,* 79-89.

Knowles, M. (1983). Andragogy: An emerging technology for adult learning. In M. Tight (Ed.), *Education for adults: Adult learning and education* (pp. 53-70). London: Croom Helm.

Land, S.M., & Hannafin, M.J. (1996). A conceptual framework for the development of theories-in-action with open-ended learning environments. *Educational Technology Research and Development, 44*(3), 37-53.

Landa, L.N. (1998). Methodological drawbacks of current educational software and suggestions based on algo-heuristic theory. *Educational Technology, 38*(5), 54-56.

Laurillard, D. (1988). Computers and the emancipation of students: Giving control back to the learner. In P. Ramsden (Ed.), *Improving learning: New perspectives* (pp. 215-233). London: Kogan Page.

Laurillard, D. (1993). *Rethinking university teaching: A framework for the effective use of educational technology.* London: Routledge.

Laurillard, D. (1999). A conversational framework for individual learning applied to the "learning organisation" and the "learning society". *Systems Research and Behavioural Science, 16,* 113-122.

Mezirow, J. (1981). A critical theory of adult learning. *Adult Education, 32*(1), 3-24.

Park, O. (1991). Hypermedia: Functional features and research issues. *Educational Technology, 31*(8), 24-31.

Reushle, S.E. (1995). Design considerations and features in the development of hypermedia courseware. *Distance Education, 16*(1), 141-156.

Roselli, T. (1991). Control of user disorientation in hypertext systems. *Educational Technology, 31*(12), 42-46.

Rowntree, D. (1992). *Exploring open and distance learning*. London: Kogan Page.

Ruberg, L.F., Moore, D.M., & Taylor, C.D. (1996). Student participation. interaction, and regulation in a computer-mediated communication environment: A qualitative study. *Journal of Educational Computer Research, 14*(3), 243-268.

Short, J., Williams, E., & Christie, B. (1976). *The social psychology of telecommunications*. New York: Wiley.

Skinner, B.F. (1968). *The technology of teaching*. New York, USA: Meredith.

Starr, R.M. (1997). Delivering instruction on the World Wide Web: Overview and basic design principles. *Educational Technology, 37*(3), 7-15.

Taylor, J. (1998). Using asynchronous computer-conferencing to encourage interaction in seminar discussions. In R. Hazemi, S. Hailes, & S. Wilbur (Eds.), *The digital university: Reinventing the academy* (pp. 219-232). Berlin, Germany: Springer.

Thomas, M.J.W. (2000). *Virtual learning environments: The impact of communication and information technologies on a sustainable higher education.* Unpublished doctoral dissertation, University of Adelaide, Adelaide, Australia.

Walters, B. (1999, December). CBT enters a new dimension. *Civil Aviation Training,* 46-50.

Walther, J.B. (1992). Interpersonal effects in computer-mediated interaction: A relational perspective. *Communication Research, 19*(1), 52-90.

Warschauer, M. (1997). Computer-mediated collaborative learning: Theory and practice. *Modern Language Journal, 81*(iv), 470-481.

Wilson, J.R. (2000, April). Training set to go on-line. *Interavia,* 22-26.

Windschitl, M. (1998). The WWW and classroom research: What path should we take? *Educational Researcher, 27*(1), 28-33.

14 On-the-Job Training in Air Traffic Control

Kay H. Ellis and Patricia McDonell

Introduction

Air traffic control (ATC) is a complex interdependent system, but despite advances in technology that have changed the way information is derived, transmitted and displayed, it remains highly dependent on the individual and collective skills of the people who work within the system. Consequently, there is a great responsibility for organisations to train those people in the most effective and efficient manner and to the highest possible standard. Moreover, in the current climate of ATC personnel shortages, efficient training programs are essential.

On-the-job training (OJT) is the chief means by which controllers acquire the knowledge and skills to perform ATC functions safely and effectively. However, little research has been conducted on this aspect of ATC training, and how OJT is best achieved is neither completely understood nor agreed upon. OJT remains essentially an apprenticeship-based system where the novice learns the necessary skills by observation and imitation of competent practice, and absorbs the essential knowledge, attitudes and values by means of a process similar to osmosis. The practice of selecting and training OJT instructors has been based mainly on the premise that those who can do it can teach it. Most instructors have scant understanding of the cognitively demanding tasks that they perform routinely and automatically, and of how to impart these skills effectively and meaningfully to others. The relatively rapid introduction of computer-dependent and automated systems does not appear to have impacted greatly on the basic philosophy regarding OJT or the methods by which it is accomplished.

This chapter will examine the type of skills needed for effective ATC, and the ways they are acquired — in the classroom, in the simulator and on the job. It will also consider the skills and attributes of instructors and students essential for effective training. Issues relating to the problems that instructors and students experience during training will be addressed, including suggestions for further research in a number of areas. Reference is made to the Australian air traffic control environment due to the authors' training and experience.

The Air Traffic Control System

The primary function of ATC is to keep aircraft separated from one another. A secondary function is to make more efficient use of airspace by expediting traffic flow (Isaac & Ruitenberg, 1999; Nolan, 1999). However, in the increasingly commercially driven aviation environment, the cost effectiveness of airline operations is highly dependant on the manner in which ATC is effected, such that the secondary function has become a significant driving factor in the development and use of ATC equipment, procedures and practices. The organisations that provide ATC services and the people who work in them are also subject, for varying reasons, to the same economic pressures as airlines.

Controllers provide ATC services by monitoring aircraft, via radar and/or radio, and predicting their future positions. Early ATC systems comprised little more than a controller with a radio. Aircraft progressed along designated air routes from one airport to another. The position of aircraft was ascertained by regular radio contact between the aircraft and the controller. Paper flight-progress strips were used to aid the controller's memory and to cue the controller to perform required actions. A method of strip manipulation was developed to signify the position of each aircraft in relation to others. This method, procedural control, is still used today in areas where there is minimal potential for conflicts and where resources are constrained (Nolan, 1999).

The advent of radar for aviation use heralded a new era of ATC systems. Within the area of radar coverage, the position of aircraft could now be accurately monitored without recourse to radio communication. The need for aircraft reports was reduced, but some were still required. Although radar control enabled more efficient use of the airspace than procedural control allowed, all aircraft appeared as 'green blips' on the screen. Misidentification was a potential problem because of increased traffic within the controller's area of responsibility. More advanced ATC systems and increased use of aircraft transponders provided the controller with an increasing amount of information on each aircraft. Advances in technology permitted maps to be included on radar screens, as well as information about weather and equipment status. The design of controller workstations was also made more ergonomic. Thoughout this period, paper flight-progress strips were retained.

The new automated ATC systems currently being deployed throughout the world incorporate state-of-the-art, computer-based representations for aircraft position display, touch sensitive screens for information retrieval and mode selection, keyboards for data input and target acquisition, and automated messaging between ATC agencies. Satellite-based automated messaging systems reduce the need for radio contact between the aircraft and the controller in some areas. In remote areas where radar services and continuous radio communications are unavailable, procedural control is being assisted by computer displays derived from aircraft position reports, both verbal and automated. In some systems, paper flight-progress strips are being replaced by automated displays. Controllers transferring from earlier-generation ATC systems can face considerable change. Additionally, the new automated systems have the potential to radically alter the type and mode of delivery of ATC services (Hopkin, 1995).

The International Civil Aviation Organization (ICAO) has defined the types of services that ATC may provide, although different rules and regulations apply within different countries. Although there may be some difference in methodology and position titles across — and sometimes within — countries, the basic services provided are reasonably standard. The guidelines provided by ICAO are generally used for the development of rules and regulations, and ATC personnel are required legally to abide by them. The Australian ATC system is essentially ICAO compliant.

The Air Traffic Controller

"Visualize an environment populated by individuals who have allegedly been selected on the basis of their self-confidence, intelligence, judgment, problem-solving ability, motor skills, alertness, clarity of expression, perception, extraordinary good health, and other equally positive qualities" (Spring, 1998, p. 10). Although some elements of this definition might seem slightly exaggerated, air traffic controllers must remember vast amounts of complex information. They appear to be able to think of and perform many tasks at once, while maintaining a mental picture of the air traffic under their jurisdiction and in the airspace beyond, both in the present and into the future. Furthermore, they are required to withstand the rigours of what is generally accepted as a stressful occupation. Selection tests for prospective controllers traditionally screen for applicants who possess mental arithmetic skills, spatial awareness and general rational abilities, attributes that are broadly accepted as indicators of successful job performance. However, the intricacies of the air traffic controller's thought processes largely remain a mystery.

Expertise in Air Traffic Control

ATC expertise is often thought of in terms of the controller's performance of a multitude of tasks. The successful outcomes of the controller's actions in monitoring, organising and directing the flow of aircraft under their control determine the level of proficiency of an air traffic controller. The emphasis has been on the outward manifestations of the controller's thought processes — their actions. Traditional behavioural task analysis (TA) of an air traffic controller's work has been conducted to determine the types of activities performed, but this methodology does not allow for sufficient examination of the mental processes that precede these activities.

Cognitive Task Analysis (CTA), which focuses on the cognitive processes that underpin human performance, has been used successfully in other domains, but it is only recently that a systematic analysis of the cognitive elements of a controller's job has been undertaken. According to Seamster, Redding, and Kaempf (1997) "CTA identifies the cognitive structures (e.g., knowledge-base organization and representational skills) and processes (e.g., attention, problem-solving, and decision making) underlying job expertise" (p. 4). They maintain that CTA is the ideal mechanism to analyse many jobs in the aviation environment, such as that of the air traffic controller, as it describes the cognitive skills required for successful task performance, including interaction with the automated systems that are now part of most ATC environments.

Cognitive Work Analysis (CWA) is another strategy suitable for the analysis of ATC functions. Vicente (1995) argues that CWA is a complementary tool to both TA and CTA, as it focuses on "the functional structure of the work domain" (p. 536). It can appraise a series of tasks as a whole rather than as individual activities. It is also an important factor in complex systems where events rarely experienced by operators and not anticipated by designers have the potential to threaten system safety (p. 536). A critical issue is whether such complex systems are optimally designed to provide sufficient and unambiguous feedback to the operator to allow for an appropriate and timely response to abnormal situations. As air traffic controllers progressively move towards what Roske-Hofstrand and Murphy (1998) term "supervisory control" (p. 106), there has been insufficient research to indicate how this role shift will impact on a controller's skills. According to Garland, Stein, and Muller (1999), such automation will mean that "human problem solving and other cognitive processes will change or become additional complicating factors" (p. 487).

The Organisation and Structure of ATC Knowledge

Air traffic controllers are required to accumulate a vast amount of knowledge, often described as declarative (factual) and procedural (concerned with application). However, it is the efficient organisation of such knowledge — including a rapid means of recovering it when required — that is a key determining factor in expert performance (Ericsson & Smith, 1991; Redding, Cannon, & Seamster, 1992). Performances differ in that, as an individual progresses from novice to expert, there is increasing interconnection and chunking of information which makes the accessibility and subsequent application of that information more efficient (Glaser, 1987; Laurillard, 1993; Merriam & Caffarella, 1999).

Seamster et al. (1997) cite recent research indicating that knowledge structures and information processing skills change as learning occurs and expertise develops. In particular, many tasks that initially required attention are increasingly performed in a relatively automatic manner and without apparent effort. Additionally, a substantial amount of practice has contributed to the development of such expertise. These findings help explain why experts often find it difficult to explain many of their actions to novices: they no longer need to think consciously about *how* they perform many routine tasks.

Air Picture or Mental Model

The way in which individuals represent tasks is regarded as one of the most significant differences between novices and experts. Redding et al. (1992) suggest the use of a mental model or schema as a structure for both job performance and training. The mental model is not a new concept, and its use has been advocated in a wide variety of educational and instructional settings. Cornford (1999) succinctly summarises the view of many researchers: "A mental model or schema which embodies the components, processes, correct sequencing of components and processes, and the time relations between components and processes needs to exist in memory before a skill can be performed" (p. 267). It is this notion of timing and correct sequencing that is particularly pertinent to ATC, and in this domain is often called an air picture.

The air picture is an essential element in the development of situation awareness, that is, "the perception of the elements in the environment within a volume of time and space, the comprehension of their meaning, and the projection of their status in the near future" (Endsley, 1988 cited in Endsley & Smolensky, 1998, p. 116). Endsley and Smolensky further explain that, while novices often have to work hard to maintain this air picture, experts are more likely to use pattern matching with existing examples in long-term memory, thus freeing up working memory.

The Role of Memory

There has been a great deal of research on the types of memory necessary for effective ATC, that is, long-term memory, short-term or working memory, and prospective memory. Long-term memory contains the rules, procedures, standards, airspace charts, aircraft performance characteristics, as well as cognitive tasks involved in the organisation of information, problem solving and decision making (Garland et al., 1999). However, it is the controller's working memory that "directly or indirectly influences every aspect of his or her ability to control traffic" (p. 487) and where the changing air picture is most likely maintained. Unfortunately, working memory is also "heavily constrained and limited by such time-dependent processes as attention, capacity, and forgetting" (p. 456).

Prospective memory, or remembering to perform an intended action at some time in the future, is another dimension of working memory which is particularly susceptible to failure (Reason, 1990). A controller will often flag an intended action as one to be performed at a later time, often after another action has been performed, or as part of a future sequence. Memory cues, such as writing on a flight-progress strip or 'cocking' the strip half way out of the strip bay, are useful but may not be available in the fully-automated environment where flight data is presented on a computer display.

Intentional forgetting is a necessary part of ATC, as it frees up working memory for new information. However, rarely is guidance provided to the student about when, how and what to forget. It is another of the lessons learnt by trial and error, and remains largely beyond the control of the individual. Research may be able to provide some answers as to how controllers could effectively manage what they want and need to forget.

Errors

Garland et al. (1999) assert that "the volatility of information in working memory is potentially the greatest contributor to operational errors in ATC tactical operations" (p. 456). The complex ATC environment presents controllers with a high level of information that must be processed and updated on a continuing basis, and retained in working memory, in order to maintain situational awareness. In their research, Endsley and Rodgers (1997) found that the allocation of attention to prioritised information had the potential to lead to errors stemming from the remainder of the information being ignored, and that a lack of situational awareness can precipitate errors.

Research by McDonell (2000) categorised error types within the Australian military ATC tower environment against the four error types proposed by Reason (1990). A statistically significant proportion of these errors was identified as either mistakes or lapses. Mistakes are based on incorrect assessments (Reason, 1990), thus they arise from a lack of available information, either within the workplace or within the mind of the person involved (Reason, 1997). The availability of pertinent information is the basis of decision making. However, in complex environments, Endsley (1995) suggests that attention cannot be afforded to all information. Research into the type of information required, coupled with the optimum presentation format, could lead to a reduction in the frequency of mistakes within the ATC domain. McDonell (2000) speculates that the strategy required for reducing memory lapses within ATC is more problematic than for mistakes. As lapses are unintentional and memory-based (Reason, 1990) the development and teaching of strategies to increase memory potential may lead to some improvement in this area.

It is important to note that the consequences of error are not always detrimental. Rather, errors can be responsible for much learning and development. Learning environments — be they classrooms or simulators — where errors are observable and reversible could be described as 'safe'. Within these 'safe' environments, errors can be made deliberately to gauge their effect. When operating in real-time environments such as OJT within ATC, such "error-based learning" may be inappropriate, as the consequences can be immediate and catastrophic.

Problem Solving

Problem solving is regarded as an activity that is central to gaining vocational expertise. However, problem-solving skills are generally not transferable from one environment to another, but are context specific (Cornford, 1999; Gabrys, Weiner, & Lesgold, 1993). Hunt (1997) explains that people understand problems based on their domain knowledge and the way in which that knowledge is organised. A significant portion of research has focused on identifying the differences between the problem-solving skills employed by novices and experts, with the explicit aim of devising teaching and learning strategies aimed at developing those skills.

The most important difference seems to be the way in which the two groups categorise problems. Novices tend to see problems in terms of their surface features (Sweller, 1989, cited in Billett, 1998), while experts see problems in terms of their underlying principles (Glaser, 1987, cited in Laurillard, 1993). Experts often concentrate on finding the best representation of a problem (Cornford, 1999; Larkin, 1983, cited in Gabrys, et al., 1993), and selecting appropriate solutions. This method is especially important in novel situations, such as non-routine problem solving.

Teamwork

Recognition of the value of teamwork in problem solving is the basis for the concept of what has become known as Crew Resource Management (CRM). A team can be described as a group of individuals bound by a common purpose or goal, where the members operate interdependently (Isaac & Ruitenberg, 1999). Such interdependency exists within the ATC domain where safety and efficiency are paramount. To operate efficiently, all members of the team must work together, anticipating and understanding the needs of other team members. To aid safety, all members of the team must monitor the work of the team, and ensure that no member becomes overwhelmed or distracted. An essential element of teamwork is the development between team members of shared mental models, that is, "a set of task knowledge and organised expectations for performance, from which accurate and timely predictions can be drawn" (Bowers, Blickensderfer & Morgan, 1998, pp. 232-233).

Such predictions are essential in an environment where the monitoring of communications among team members is the chief means by which information about planned actions is gleaned. Understanding the requirements of other team members allows for a reduction in the need for verbal communication, coupled with an increase in performance (Bowers et al., 1998; Roske-Hofstrand & Murphy, 1998). Any breakdown in teamwork, caused either by poor communication or poor coordination, can lead to errors. CRM training endeavours to develop team situational awareness where the pooling of resources leads to improved overall performance (Endsley, 1995). CRM training has been included in the training of controllers for a number of years to assist in the development of efficient and effective teamwork.

Team skills include adaptability, coordination, cooperation, decision making, leadership and communication (O'Neil, Wang, Chung, & Herl, 2000). An individual within the team can perform in a manner to support coordination, but the concept of coordination requires the activity of more than one individual (Schiflett & Elliott, 2000). Isaac and Ruitenberg (1999) posit that the performance of a team is highly dependent on the skills of the individual members. Teams can include individuals of differing levels of expertise and experience, operating under differing workloads (Schiflett & Elliott, 2000) and teams may be either permanent or transitory. A student must develop the skills to adapt to these changing team dynamics (see Chapters 10 & 11).

Members of an effective team coordinate activity, communicate within the team and respond to changing conditions (O'Neil et al., 2000). There are many ways to successfully complete a team task and O'Neil et al. suggest that "interactions among individuals within a team and across task types is far more complex and less well understood that first thought" (p. 274). With the reduced use of both flight-progress strips and radio communications resulting from the introduction of automated ATC systems, the workings of a controller are becoming less and less obvious to other team members in the workplace. The impact this may have on ATC teamwork is an area where further research is needed, and needed quickly.

Air Traffic Control Training

The Purpose of Training

Authors have expressed differing views regarding the purpose of ATC training. Fisher and Kulick (1998) assert that it should provide "the knowledge and skills required to achieve and maintain successful job performance" (p. 273). Hopkin (1995) maintains that "the purpose of air traffic control training is to produce sufficient qualified controllers to meet staffing requirements, all of whom possess and can apply the requisite knowledge and skills to be safe and efficient and are motivated to remain as controllers" (p. 153). Given the present global shortage, the need to satisfy the demand for air traffic controllers is certainly a topical issue. However, it is debatable whether the current training addresses, or should be designed to address, the issue of staff retention.

Comprehensive versus Streamed Training

In Australia, ATC is provided by both military and civil controllers. To satisfy organisational and individual needs, military personnel, including air traffic controllers, are periodically relocated. This provides the organisation an opportunity to broaden its experience base and maintain sufficient numbers of qualified personnel to meet national commitments for military operations and exercises. Organisational needs determine the kind of training provided to military air traffic controllers. In order that they can be utilised in almost any ATC position, their basic training encompasses all aspects of ATC: procedural, radar and tower control. The usual career progression is from tower to approach radar control.

This process contrasts with the training and employment of civil air traffic controllers within Australia, where traditionally the tower controllers are the most experienced radar controllers who are subsequently cross-trained in tower. Generally, civil ATC training is streamed, and controllers may spend their entire working lives in one facility. This approach is less time-consuming and cheaper, but results in less flexibility for the controller and the organisation. As Hopkin (1995) notes: "National policies and needs differ on the best balance between fully comprehensive and narrowly specialized air traffic control training, and no single balance would be ideal everywhere" (p. 157). The Australian military and civil systems could be viewed as complementary, as many controllers move from military to civil ATC, in the same way that many military pilots take up jobs with civil airlines. Both groups may do so for financial reasons and greater locational stability but, in the process, they may sacrifice job diversity and operational opportunities.

A Model for ATC Training

There is no universal model for ATC training and, as long as broadly defined international requirements for performance standards are met, it is the responsibility of individual ATC providers to specify training philosophies and methodologies (Hopkin, 1995). Although ATC training is changing in some countries to combine knowledge acquisition with skill development, many ATC providers remain wedded to historical training methodology. Under this system, ATC training initially takes place in a college setting.

The extensive and complex body of knowledge required within the domain is acquired traditionally by means of an academic block at the start of training, followed by non-integrated and integrated simulation exercises. On successful completion of the course, the student is assigned to an ATC unit to undertake OJT. There, a further period of classroom instruction in local and emergency procedures precedes training in the nominated ATC position under the close supervision of a controller qualified in that position.

Fisher and Kulick (1998) acknowledge the difficulties in sequencing the content of ATC training. "There is no absolute sequence for performing air traffic tasks. Therefore, the order in which tasks are performed cannot be used to guide the sequencing of learning" (p. 287).

The traditional emphasis on behaviours and outcomes in ATC is reflected in OJT in the predominant emphasis on teaching the student, through demonstration and explanation, a vast array of control actions or behaviours that will, hopefully, result in desirable air traffic management outcomes. There has been little attention paid to the underlying thought processes that must inform these outcomes. Consequently, there is often no structure to the knowledge acquisition process that the student can identify.

Seamster et al. (1997) suggest that CTA can assist in the sequencing of knowledge and skills to promote effective learning. They further advocate that learning objectives should be organised around the concept of the mental model or representation. This mental representation "guides and organizes all learning activities and is progressively elaborated throughout training" (Seamster et al., p. 11). Training for and achieving the first rating in the primarily visual environment of the control tower could also enhance the development of an air picture.

Hopkin (1995) notes that, while there is agreement that OJT is valuable, there is less agreement on those aspects of ATC training that require OJT and those that could be accomplished in other ways. Computer-based learning, part-task training, non-integrated simulation and full simulation are used at various stages of training prior to commencing OJT in the live environment. The computer-based learning environment is ideal for acquiring and testing declarative knowledge, as well as simple exercises involving part-task training. For example, in the control tower environment, still photographs or video can be used by the student to learn the phraseology required for clearing aircraft to land and take-off. Non-integrated simulation is used to practise aspects of air traffic management without co-ordinating with other ATC agencies. Full simulation imitates the ATC environment where all aspects of control and co-ordination, as well as team work, are required. In all of these training settings, the instructor or student is able to select learning tasks and exercises, to work at a slower than real-time pace, to freeze action for review purposes, and to repeat certain sequences. These are all important precursors to operating in a live ATC work situation.

Simulation versus Live Training

ICAO notes that "skill acquisition in aviation has traditionally been achieved on the job or in the course of high-fidelity simulation" (1998, p. 2-1-18). However, in 1995, Hopkin asserted that "full fidelity and realism are unattainable in air traffic control simulation" (p. 163). There have been considerable developments in simulation as a result of advanced computer technology, but tower simulation remains a basic representation of the environment, and cannot be used to develop a realistic concept of visual judgement. It is only in the last 15 years that sophisticated simulation has been available at Australian ATC training institutions, and in the last five years at operational centres. Prior to this, all OJT was achieved in the live radar or tower environment.

Planning Training Sessions

The detailed planning of training sessions is a routine activity in most learning environments, but one that is extremely difficult to achieve in an ATC OJT context. During an OJT session, the instructor has virtually no control over the type, frequency or complexity of air traffic that occurs. Actual traffic sequences tend to drive the pace and direction of a training session. This presents less of a problem to the student who has already achieved a rating at another location or in another ATC element. Military air traffic controllers tend to undergo more OJT due to regular postings to different locations, and frequent interchange between tower and radar control duties.

Controller Confidence during Training

Whilst it may be argued that learning to control in real traffic scenarios can help the student to concurrently build a wide range of skills, real time ATC can be a daunting prospect. From the *ab initio* student's viewpoint, this can be likened to randomly picking up pieces of a jigsaw and trying to put them into a large blank space within vaguely defined borders with few reference points. There is the potential for the student to be so overwhelmed that they lose confidence, and thus are unable to function at a level already demonstrated in a simulated environment during initial training.

Anecdotal evidence suggest that the roller-coaster ride of self-confidence seems to be an accepted part of the controller's training. Most students are advised that everyone goes through it, and that the situation will improve with a rating and more experience. However, this advice is less than helpful to *ab initio* students who often have continuing doubts as to whether they possess "the right stuff" to be an air traffic controller. Fully-fledged controllers seem to be members of an elite fraternity, and students can feel daunted by the difference between what they know now and what they need to know to belong to that exclusive club. As much as controllers work closely together, at times it seems to the novice as though they are a loose collection of strong-willed egotistic individuals rather than a cohesive team.

Team Training

Hopkin (1995) acknowledges that "team training is an important feature of on-the-job training ... (and) training must include working harmoniously and effectively as a team member" (p. 168). Teamwork in relation to OJT is dependant on the behaviour of all those involved in the enterprise — not simply the student and the instructor. Without the presence of good team skills in the work environment, a student cannot develop such skills. However, apart from the emphasis on the development of team skills during CRM training, little attention is given to team-focused concepts and activities during OJT.

Instructors usually discuss the importance of teamwork required for effective coordination between ATC elements, but specific exercises designed to highlight and strengthen team relationships are lacking in most OJT situations. Schiflett and Elliott (2000) conjecture that team performance can be assessed against task outcome, organisational issues, team performance, individual performance and individual differences. Thus, they argue, team training should be based on these five elements, and incorporated into all phases of training.

The relevance of attitudes as an element of team performance is supported by Stout, Salas, and Fowlkes (1997). In a team training study they found that attitudinal change by individuals within the team resulted in improved team performance. Furthermore, the culture of the work environment will determine the level and effectiveness of teamwork. Learners are influenced both directly and indirectly by the people with whom they work and the cultural cues provided by the workplace setting. Indirect guidance occurs through observation of the performance of co-workers, how they operate in the physical environment and how they work together.

Billet (1998) acknowledges that inappropriate workplace learning in terms of knowledge, attitudes and values may occur through contact with less expert workers and the prevailing organisational culture. Expert co-workers may be reluctant to share their knowledge due to notions of privilege and status. However, these features exist in many workplaces and do not outweigh the benefits of situated learning.

As Hopkin (1995) notes, "on-the-job training usually presents the first opportunity to imbue professional standards, norms and attitudes", where the student also learns about "the expectations of other controllers, and the criteria for judging a good controller" (p. 167). *Ab initio* students will be experiencing for the first time "the stress and responsibilities of real air traffic control" (p. 167), which can be a difficult time both for students and those working alongside them, and will provide some indication of whether or not students are able to cope with the pressure of performing in a dynamic operational environment. The fact that the student is effectively operating on the licence of the instructor could make the inexperienced instructor feel anxious, as well as introduce a complex set of interactive obligations and responsibilities for both the instructor and the student.

The ATC Instructor

Once controllers have achieved competency or full performance status, it is assumed that they possess the attributes required to train less expert controllers. This reflects the attitude that has been the historical basis for most workplace training — those who are competent will be able to impart their knowledge and skills to others. Moreover, operational or staffing requirements may necessitate controllers becoming OJT instructors before they are ready or willing. This can result in motivational problems for some controllers who are directed to undertake OJT instruction.

In the past, preparation for OJT instructional duties was minimal with no formalised training available, or even considered necessary. It is only relatively recently that specific training for OJT instructors has been devised. Current courses concentrate on the basic principles of adult learning, as well as general teaching and assessment strategies, and provide advice about how to deal with certain personality types as part of the instructor-student relationship. On the military instructor course, the tower simulator is used as the venue for observing the OJT instructor's ability to direct and modify difficult and different types of student behaviour.

Hopkin (1995) asserts that "on-the-job instructors must combine the functions of instructing, assessing, monitoring progress, and assuming control if necessary. The instructor should combine the skills of a good controller, a good teacher and a good assessor" (p. 166). However, in reality, most controllers who become OJT instructors receive minimal training in the areas of teaching strategies and assessment. They do not usually have the same instructional background or training as instructors at formal ATC training institutions.

According to Fisher and Kulick (1998), controllers are generally not given sufficient training in how to communicate expert strategies to novices. They note that "experts tend to complete routine steps and make many decisions without thinking about them consciously. Novice controllers may need explicit instruction on those parts of the job that have become automatic to the expert controller" (p. 294). Seamster et al. (1997) explain that "often, the knowledge required for good performance has become intuitive" (p. 20), making it difficult for the experienced controller to describe the specific thought processes involved in the performance of many routine tasks, especially when the instructor has not been trained to recognise, or think consciously about, these thought processes.

Hopkin (1995) also states that judgement is a critical aspect of OJT. Instructors must judge — that is, make a subjective assessment based mainly on their own experience and on their observations of other novice controllers being trained — how proficient a student should be at a particular stage of training. Judging when to assume control can prove difficult, especially for less experienced OJT instructors, resulting in their taking over too soon or too frequently, or not soon enough, when students get out of their depth. Just as students can suffer from lack of confidence, so too can instructors regarding their ability to allow the student enough freedom to act and intervening only at the last possible moment. Isaac and Ruitenberg (1999) believe that the "unique" skill is "only gifted to a few controllers and should be sought and revered by management" (p. 128). Anecdotal evidence suggests that this is the case. However, for the sake of efficiency and economy, management needs to be able to supplement the pool of naturally gifted instructors with training using simulated exercises that provide guidance about when and how to intervene.

OJT instructors can also be likened to mentors, who are commonly used in educational settings (see Chapter 2). As Daloz (1998) explains, mentors are "interpreters of the environment" who also "model expected behaviour ... speak the language of the new world and understand its peculiarities" (p. 355). Although Daloz was referring specifically to unfamiliar academic environments, his observations are pertinent to the unique, and initially bewildering, ATC environment. As Isaac and Ruitenberg (1999) note, ATC personnel perform tasks that are seen in few other workplaces.

Instructor Workload and Performance

Although there are many models and definitions of workload, one that is useful in the ATC environment is provided by Stein (1998). He defines "the experience of workload (as) based on the amount of effort, both physical and psychological, expended in response to system demands (taskload) and also in accordance with the operator's internal standard of performance" (p. 157). This reflects a concern with the operator's subjective experience of workload, rather than any external measurement. According to Hopkin (1995), workload depends on who performs the task, with task demands varying significantly among individual controllers. He also contends that a controller's workload can be greatly increased by the presence of colleagues who lack experience or credibility, due to the perceived need to monitor the colleague's performance.

The authors believe that supervising a student during OJT also involves an increase in the instructor's workload. This is supported by research conducted by Ellis (1998) on the nature of instructor errors during OJT in the Australian military control tower environment and by investigations done by the Australian Bureau of Air Safety Investigation (1994, 1996) and Ratner (1987, 1982; also see Maurino, Reason, Johnston, & Lee, 1995). As part of the research conducted by Ellis (1998), participants were asked to rate and describe their workload during OJT. Although the sample of controllers interviewed was small, all instructors believed that their workload increased during training, with most assessing it as "double" their workload compared with when they were not training. Comments supporting their assessment of increased workload included "deciding how far to let a student go ... revising decisions ... prompting ... coaxing … anticipating the student's next actions" (pp. 13-14). Other factors relating to the OJT instructor's role which added to their psychological workload included "putting on a front (of unconcern) for the student's benefit" and "having more responsibility" (p. 14).

Whilst students are routinely questioned and counselled about their progress and well-being during OJT, instructors' needs are invariably neither considered nor attended to. OJT instructors will often consult their immediate supervisors and other controllers regarding the students' progress and their interaction with other team members. However, instructors may be unwilling to discuss their own feelings of anxiety or stress for fear of appearing weak or unable to cope. The observations made by the OJT instructors who were interviewed suggest that there may be more complex psychological demands placed on them than have been considered in the past. Future investigation is needed regarding the nature of workload during OJT with a view to devising strategies for dealing with this and other issues raised during the research.

The ATC Student

ATC students comprise two types: *ab initio* students who have never held a rating in an ATC element, and air traffic controllers who have held ratings and endorsements at one or more locations and who are undergoing training at a different location and/or in a new element of ATC. Initial OJT is invariably the most demanding phase of an air traffic controller's training, with the first rating usually being the most difficult to achieve. For the *ab initio* student, there is a vast difference between conducting exercises in a simulator and operating in the live environment of an ATC facility, coordinating with other ATC elements, talking to real pilots and managing actual traffic scenarios. Furthermore, as Hopkin (1995) notes, "on-the-job training reveals that real problems do not always conform with the neat categories in which they are taught, and their solutions can make some of the initial training seem simplistic" (p. 167).

Often, graduates from ATC courses are geographically separated from one another during OJT and will not have the opportunity to compare training experiences or discuss problems with their former course mates. They may find themselves training concurrently with other more experienced controllers who have newly arrived at a location, but *ab initio* students often feel reticent to discuss their training difficulties with those at the opposite end of the training spectrum. In any case, more experienced controllers are unlikely to be grappling with the same concerns as *ab initio* students, and will probably have forgotten how problematic their initial training was.

Experienced air traffic controllers who move to a new ATC facility or element tend to direct their own learning to a certain extent by asking relevant questions at appropriate times. They already have much of the required declarative and procedural knowledge, as well as appropriate knowledge structures, in long-term memory and will be able to perform many tasks automatically. Experienced air traffic controllers can self-diagnose their needs at different stages in their training, and refer to their own experience to monitor and direct their decisions and actions. They invariably become more proficient with each rating and endorsement, and therefore need less time and effort to attain the required proficiency standard. Equally, the cognitive and psychological demands placed on the instructor are reduced when training experienced air traffic controllers.

Conclusion

Four main issues face those conducting OJT in ATC. First, there needs to be an acknowledgment that OJT is a difficult and complex practice encompassing more than the traditional "watch me and do as I do" approach. Second, a more detailed and academically-informed understanding is required of the cognitive processes involved in both the teaching and learning of ATC tasks during OJT. Third, an understanding is required of how the ATC environment is affected by increased system automation, and the flow-on effect this will have on training.

More research is required to determine what impact increasing automation in ATC will have on future selection requirements and testing procedures. Finally, greater resources need to be allocated to more extensive and intensive training program for OJT instructors. It is time to rethink the traditional approach to OJT, based largely on historical practice, in favour of one that is more informed academically.

References

Billet, S. (1998). Understanding workplace learning: Cognitive and sociocultural perspectives. In D. Boud (Ed.), *Current issues and new agendas in workplace learning* (pp. 47-68). Leabrook: NCVER Ltd.

Bowers, C.A., Blickensderfer, E.L., & Morgan, B.B., Jr. (1998). Air traffic control specialist team coordination. In M.W. Smolensky & E.S. Stein (Eds.), *Human factors in air traffic control* (pp. 215-236). San Diego, CA: Academic Press.

Bureau of Air Safety Investigation. (1994). *A systematic investigation of airmiss occurrences 1991-1992*. Canberra, ACT: Author.

Bureau of Air Safety Investigation. (1996, May). *Follow-up investigation of systematic factors underlying air safety occurrences in the Brisbane Air Traffic Services Centre*. Canberra, ACT: Author.

Cornford, I.R. (1999). Skill learning and the development of expertise. In J.A. Athanasou (Ed.), *Adult educational psychology* (pp. 263-290). Katoomba, Australia: Social Science Press.

Daloz, L.A.P. (1998). Mentorship. In M.W. Galbraith (Ed.), *Adult learning methods: A guide for effective instruction* (2nd ed., pp. 353-372). Malabar, FL: Krieger.

Ellis, K.H. (1998). *The nature of training officer errors during on-the-job training in the military air traffic control tower environment.* Unpublished research project, University of Western Sydney, Macarthur, Australia.

Endsley, M.R. (1995). Towards a theory of situational awareness in dynamic systems. *Human Factors, 37*(1), 32-64.

Endsley, M.R., & Rodgers, M.D. (1997). *Distribution of attention, situation awareness, and workload in a passive air traffic control task: Implications for operational errors and automation* (DOT/FAA/AM-97/13). Washington, DC: Office of Aviation Medicine.

Endsley, M.R., & Smolensky, M.W. (1998). Situation awareness in air traffic control: the picture. In M.W. Smolensky & E.S. Stein (Eds.), *Human factors in air traffic control* (pp. 115-154). San Diego, CA: Academic Press.

Ericsson, K.A., & Smith, J. (1991). Prospects and limits of the empirical study of expertise: An introduction. In K.A. Ericsson & J. Smith (Eds.), *Toward a general theory of expertise: Prospects and limits* (pp. 1-38). New York: Cambridge University Press.

Fisher, S.G., & Kulick, I. (1998). Air traffic controller training: A new model. In M.W. Smolensky & E.S. Stein (Eds.), *Human factors in air traffic control* (pp. 273-298). San Diego, CA: Academic Press.

Gabrys, G., Weiner, A., & Lesgold, A. (1993). Learning by problem solving in a coached apprenticeship system. In M. Rabinowitz (Ed.), *Cognitive science foundations of instruction* (pp. 119-147). Hillsdale, NJ: Lawrence Erlbaum.

Garland, D.J., Stein, E.S., & Muller, J.K. (1999). Air traffic controller memory: Capabilities, limitations, and volatility. In D.J. Garland, J.A. Wise, & V.D.

Hopkin (Eds.), *Handbook of aviation human factors* (pp. 455-489). Mahwah, NJ: Lawrence Erlbaum.

Glaser, R. (Ed.). (1987). *Advances in instructional psychology* (Vol. 3). Hillsdale NJ: Lawrence Erlbaum.

Hitchcock, L. (1998). Air traffic control simulation: capabilities. In M.W. Smolensky & E.S. Stein (Eds.), *Human factors in air traffic control* (pp. 327-340). San Diego, CA: Academic Press.

Hopkin, V.D. (1995). *Human factors in air traffic control.* London: Taylor & Francis.

Hunt, L.M. (1997). Influences on the learning process and learning outcome: Practical implications for the instructor. In G.J.F. Hunt (Ed.), *Designing instruction for human factors training in aviation* (pp. 17-27). Aldershot, UK: Ashgate.

International Civil Aviation Organisation (1998). *Human factors training manual* (DOC 9683-AN/950). Montreal, CA: Author.

Isaac, A.R. (1997). Teaching for visualizing in air traffic control. In G.J.F. Hunt (Ed.), *Designing instruction for human factors training in aviation* (pp. 226-235). Aldershot, UK: Ashgate.

Isaac, A.R., & Ruitenberg, B. (1999). *Air traffic control: Human performance factors*. Aldershot, UK: Ashgate.

Kantowitz, B.H., & Casper, P.A. (1988). Human workload in aviation. In E.L. Weiner & D.C. Nagel (Eds.), *Human factors in aviation* (pp. 157-187). San Diego, CA: Academic Press.

Laurillard, D. (1993). *Rethinking university teaching. A framework for the effective use of educational technology.* London: Routledge.

Maurino, D.E., Reason, J., Johnston, N., & Lee, R.B. (1995). *Beyond aviation human factors*. Aldershot, UK: Avebury Aviation.

McDonell, P. (2000, November). *Human error types in Royal Australian Air Force air traffic control tower incidents.* Poster session presented at the Fifth Australian Aviation Psychology Symposium, Manly, Australia.

Merriam, S.B. & Caffarella, R.S. (1999). *Learning in adulthood: A comprehensive guide.* San Francisco: Jossey-Bass.

Nolan, M.S. (1999). Air traffic control. In D.J. Garland, J.A. Wise, & V.D. Hopkin (Eds.), *Handbook of aviation human factors* (pp. 431-454). Mahwah, NJ: Lawrence Erlbaum.

O'Neil, H.F., Jr., Wang, S., Chung, G., & Herl, H.E. (2000). Assessment of teamwork skills using computer-based teamwork simulations. In H.F. O'Neil, Jr. & D.H. Andrews (Eds.), *Aircrew training and assessment* (pp. 245-276). Mahwah, NJ: Lawrence Erlbaum.

Ratner, R. (1987). *Review of the Air Traffic Services system*. Canberra, ACT: Department of Aviation.

Ratner, R. (1992). 1992 *review of the Australian Air Traffic Services system.* Canberra, ACT: Civil Aviation Authority.

Reason, J. (1990). *Human error.* Cambridge, UK: Cambridge University Press.

Reason, J. (1997). *Managing the risks of organizational accidents.* Hants, UK: Ashgate.

Redding, R.E., Cannon, J.R., & Seamster, T.L. (1992). Expertise in air traffic control (ATC): What is it, and how can we train for it? *Proceedings of the 36th Annual Meeting of the Human Factors Society* (pp. 1326-1330). Santa Monica, CA: Human Factors Society.

Roske-Hofstrand, R.J., & Murphy. E.D. (1998). Human information processing in air traffic control. In M.W. Smolensky & E.S. Stein (Eds.), *Human factors in air traffic control* (pp. 65-114). San Diego, CA: Academic Press.

Schiflett, S.G., & Elliott, L.R. (2000). Synthetic team training environments: Application to command and control aircrews. In H.F. O'Neil, Jr. & D.H. Andrews (Eds.), *Aircrew training and assessment* (pp. 311-335). Mahwah, NJ: Lawrence Erlbaum.

Seamster, T.L., Redding, R.E., & Kaempf, G.L. (1997). *Applied cognitive task analysis in aviation.* Aldershot, UK: Ashgate.

Spring, E. (1998). One air traffic control specialist's perspective of air traffic control human factors. In M.W. Smolensky & E.S. Stein (Eds.), *Human factors in air traffic control* (pp. 1-15). San Diego, CA: Academic Press.

Stein, E.S. (1998). Human operator workload in air traffic control. In M.W. Smolensky & E.S. Stein (Eds.), *Human factors in air traffic control* (pp. 155-183). San Diego, CA: Academic Press.

Stout, R.J., Salas, E., & Fowlkes, J.E. (1997). Enhancing teamwork in complex environments through team training. *Group Dynamics: Theory Research and Practice, 1,* 169-182.

Vicente, K.J. (1995). Task analysis, cognitive task analysis, cognitive work analysis: What's the difference? *Proceedings of the Human Factors and Ergonomics Society 39th Annual Meeting* (pp. 534-537). San Diego, CA: Human Factors and Ergonomics Society.

Wickens, C.D., Mavor, A.S., & McGee, J.P. (Eds.). (1997). *Flight to the future: Human factors in air traffic control.* Washington, DC: National Academy Press.

15 Assessment and Evaluation

Ann Reich, Jayne Bye and Irene Henley

Introduction

Assessment is an ongoing, everyday activity for all of us. In many aspects of our lives we collect evidence and make judgements, for example, consumer decisions when shopping, interpersonal assessments when dealing with people, self-diagnosis of an ache or injury, and the decision about what to wear to work in order to have the desired effect at a meeting. In all cases, we collect some information or evidence about the activity being assessed, compare the evidence to some criteria, make a judgement as to whether the evidence meets the criteria, and then make a decision.

On the other hand, assessment also takes place in specialised contexts where assessors are highly trained, assessment strategies and procedures are codified, and the rules of evidence are documented. Examples of such contexts would include the legal system, the education system, medical diagnosis, systems of quality standards, and drug testing systems for the Olympic Games.

Similarly, evaluation is a term used to describe a process that also occurs in many contexts. The public sector, private industry and community groups all undergo evaluation. Evaluation occurs anywhere programs or activities are implemented to achieve specific outcomes or change. This feature of evaluation contexts is clearly evident in vocational education and training where programs are implemented to achieve changes in participants' knowledge, skills and attitudes. Program evaluation functions to provide feedback to trainers, participants, and managers about the success (or otherwise) of training initiatives. Therefore, evaluation is a crucial step in the cycle of program development and course planning described in Chapter 7. It is the key to quality training.

This final chapter focuses on assessment and evaluation making specific reference to aviation educational settings where appropriate. The first part of the chapter distinguishes between assessment and evaluation in education and training fields and then explores key aspects of assessment. These include the types of assessment used during the training cycle with special attention devoted to competency-based assessment because of its prominence within most aviation training contexts. The principles that are critical to assessment (i.e., validity, flexibility, reliability, fairness, authenticity and credibility) are highlighted and the types of evidence that need to be collected as the basis of reliable and valid decisions are described.

The second part of the chapter focuses on approaches to evaluating educational and training programs. The importance of the process of determining the 'value' or utility of an educational program to the aviation training context will be the underpinning message in this second section, where it will be suggested that effective evaluation is central to quality training. A range of types and models of evaluation will be presented, however, it must be noted that this overview can only begin to touch upon the vast amount of literature and theorising this process has generated. The specific examples chosen for elaboration here have been selected for their apparent appropriateness to the aviation training context. As well as the theories of evaluation, the practical skills of evaluation planning and data collection are also addressed. Useful guidelines for the preparation of questionnaires, surveys and other information-collection techniques will be presented as will key approaches to the presentation and dissemination of evaluation findings.

Assessment and Evaluation — Definitions

An important part of providing training and education is assessing trainees and evaluating the success of the program. The terms assessment and evaluation are often used differently by many writers in the education and training fields. In some cases, evaluation is the term used to describe the processes of assessing student progress as well as the value of the course. However, in this chapter a distinction is made between the terms 'assessment' and 'evaluation'. According to Athanasou (1995), "educational assessment judges student learning or performance and takes place against a set of explicit standards or other criteria; it may involve testing or measurement. Evaluation, as its name implies, is the process of determining the 'value' or utility of any educational program, policy or initiative" (p. 76).

What Is Assessment?

As mentioned in the introduction, assessment is an activity that we undertake in our daily lives. Harris, Guthrie, Hobart, and Lundberg (1995) note that in addition to our own assessment of situations, people and things, we also are personally assessed "by a plethora of individuals and in a wide variety of social, educational and work contexts. Much of this assessment is random, informal and subjective. It may be valuable or, alternatively, worthless, misleading and harmful. Its worth depends on both people and the processes they use" (pp. 159-160). Harris et al. add that a key component of these "assessment activities" is the collection of information or evidence about the activities and/or their outcomes. This evidence is then weighed against the criteria or standards of performance and a judgement is made as to whether the criteria or standards have been met. In most cases, we make a selection of a "sample of evidence" on which to make our judgement. However, assessment serves different purposes and the purpose will determine the type of assessment used. Assessment may be used:

- to let the student know how he or she is going (formative assessment);
- to help teachers or trainers assess the student's learning needs (diagnostic assessment);
- to determine whether learning outcomes have been achieved, for the purpose of formal recognition of training (summative assessment); and
- to determine whether a person has achieved as yet unrecognised standards of competency, for the purpose of gaining entry to or credit in recognised courses (assessment for recognition of prior learning). (Vocational Education, Employment and Training Advisory Committee [VEETAC], 1993, cited in Smith & Keating, 1997, p. 153)

Types or Forms of Assessment

Generally, there are five types or forms of assessment that can be used to make judgements regarding student knowledge, skills and progress. As mentioned above, these include diagnostic assessment, recognition of prior learning and/or experience, formative assessment, summative assessment, and holistic assessment. These are not mutually exclusive and in practice they will often be used together.

Diagnostic Assessment

Diagnostic assessment is used in assisting learners to determine their training needs. In assessing student needs, a diagnostic assessment indicates the performance gap between current competencies and the aim of the learning process (Rumsey, 1994). Many organisations have developed forms of skill or competency audits as the basis of this type of assessment. In some cases, it has become part of performance appraisal systems which assist in determining an employee's staff development needs, or performance management systems which focus more on the gaps in an employee's performance.

Recognition of Prior Learning (RPL)

Recognition of Prior Learning (RPL) has been defined as "the acknowledgment of skills and knowledge held as a result of formal training, work experience and/or life experience" (Harris et al., 1995, p. 262). RPL is a key part of the national vocational education and training systems in Australia and it is now a requirement that all accredited courses must include a process for RPL. Recognition of prior learning is also an important means of workplace learning being acknowledged and recognised in formal qualifications. It is perceived as being of benefit to both learners and employers. For the learners, there are the benefits of not having to undertake or "time-serve" in courses and not having to study subject material in which they are already competent. The benefits for employers are that they could more adequately target their training dollar.

Formative Assessment

Formative assessment occurs progressively throughout a program. It is used to provide feedback to trainees on their progress during training. As Rumsey (1994) suggests, "it is called 'formative' assessment since it helps students *form* the knowledge, skills and understanding that will eventually be needed to demonstrate the required competence" (p. 13). The emphasis of formative assessment is on providing feedback to learners during the learning process. Therefore, the results are not necessarily recorded. It is also used to assist learners assess their own achievements and to provide guidance about their strengths and weaknesses, with advice from teachers about how they can improve their skills and knowledge (Kingsland, 1993; Woods, 1994).

Examples of formative assessments include self-assessment activities at the end of each section of a module, 'practices' at practical tasks and debriefings after practical lessons, informal questioning by the trainer, and informal observations and feedback on the job. Essentially, formative assessments provide feedback to both teacher and students about progress and can also be a means of quality control.

Summative Assessment

Summative assessment is the most common form of assessment. It is usually undertaken at the end of a segment of learning. It is the formal process of collecting sufficient evidence to make a judgement that a person is competent in relation to a particular set of criteria. In a competency-based assessment system, the competency standards are used as the criteria. As part of a formal accreditation process, the purpose of summative assessment "is to produce certification of individual attainment of competencies in order for the individual to be eligible for further training or employment at an appropriate level" (Harris et al., 1995, p. 283). For example, in the aviation context, summative assessments are used to ascertain competency for the issue of licences and ratings. That is, the final test is a display of applied knowledge and skills and comprises the assessment upon which the formal qualification is based. In essence, summative assessments are used to assign grades or certify skills and abilities, to predict success in future courses, to determine the starting point in future courses, to compare results in different groups, and to provide feedback to students at the end of a course. Essentially, there are two main approaches to summative assessment: norm-referenced assessment and criterion-referenced assessment.

Norm-Referenced Assessment Norm-referenced assessment is commonly used in educational contexts. The passing standard is set in relation to the particular group of candidates undertaking the assessment task at that time. For example, universities who use a normative approach may define the percentage of students who will be granted a certain grade, such as only 10 percent of the group gaining a distinction or a high distinction, 30 percent being awarded a credit, 50 percent granted a pass and the rest receiving a fail result. Typically, the norm-referenced assessment has been based on the bell curve distribution. The necessary knowledge, understanding and skill to gain a certain grade will vary each year depending on the performance of the students in that particular year (McCurry, 1992). In short, in norm-referenced tests, the results of an individual are compared to the performance of others within the group.

Criterion-Referenced Assessment Criterion-referenced assessment requires the performance of students to be assessed in relation to a set of agreed criteria or standards. As Wolf (1993, cited in Harris et al., 1995) indicates,

> The crucial idea underpinning criterion-referenced assessment is that one should look at the substance of what someone can do, in effect, compare a candidate's performance with some independent, freestanding definition of what should be achieved, and report on the performance in relation to this criterion. (p. 160)

As most would be aware, criterion-referenced assessment is preferable in many aviation settings since competence or the ability to meet performance criteria is sought when assessing skilled performance. However, although criterion-referenced assessment is used in many aviation contexts, for example in flight tests and in air traffic control assessments, it is not the assessment mode used in theory examinations for pilot licensing purposes or in assessing the teaching ability of instructors. However, a popular form of criterion-referenced assessment is the recently adopted competency-based assessment.

Competency-Based Assessment Aviation training in several countries has recently embraced competency-based assessment. Undeniably, the core of competency-based approaches has been vocational and professional education and training because this is where the concept of competence can be given a coherent meaning and can be most successfully "operationalised" (Wolf, 1995). According to the Manpower Services Commission (1985) in the United Kingdom, competence simply means "the ability to perform the activities within an occupation" (cited in Wolf, 1995, p. 31). Further, Wolf explains that

> Competency-based assessment is a form of assessment that is derived from the specification of a set of outcomes; that so clearly states both the outcomes — general and specific — that assessors, students and interested third parties can all make reasonably objective judgements with respect to student achievement or non achievement of these outcomes; and that certifies student progress on the basis of

> demonstrated achievement of these outcomes. Assessments are not tied to time served in formal educational settings. (p. 5)

As indicated in this definition there are three main elements in competency-based assessment. There is an emphasis on the outcomes of performance; on the specification of these outcomes to make them transparent and clear to the person being assessed, the trainer, the assessor and others, such as employers; and the distinction and separation of assessment processes from a particular educational institution or learning program. In the Australian context, competency-based assessment has been defined by the Vocational Education, Employment and Training Advisory Committee (VEETAC) as, "the process of collecting evidence and making judgements on the nature and extent of progress towards the performance requirements set out in a standard, or a learning outcome, and, at the appropriate point, making the judgement as to whether competency has been achieved" (VEETAC, 1993, cited in Harris et al., 1995, p. 160).

The defined competency standards, developed and endorsed by national industry bodies, become the criteria against which performance at a particular level is assessed. In civil aviation, the agency responsible for establishing the performance criteria usually is the national civil aviation regulatory body, such as the Australian Civil Aviation Safety Authority, Transport Canada, or the American Federal Aviation Administration.

There has been much debate in the literature around competency-based assessment since its implementation as part of new vocational education and training systems in Australia and a number of Western countries. Wolf (1995) notes that the case for adopting competency-based assessment can more convincingly be made when referring to vocational education and training than when considering the requirements of liberal education or the development of higher level competencies, such as critical thinking, professional judgement, and decision making. One of the main reasons for this is that competencies that relate to the psychomotor (skill) domain and, to some extent, the cognitive (knowledge) domain are easier to observe and measure than those within the affective (attitudes, values, beliefs) domain of learning or those classified as higher level competencies. According to Kemp (1985), "when we turn to the affective domain – attitudes, feelings, and appreciations – we find difficulty in specifying objectives in clearly observable and measurable terms. Most often we indicate these objectives only indirectly by inferring from what can be observed" (p. 85). Lehrer (1997) agrees, but he states that while assessment in the affective domain may be difficult to carry out quantitatively, it is not impossible. He stresses, however, that the traditional assessment tools used in aviation, the written exam or the practical test, may not be appropriate to assess competencies in the affective domain or higher level competencies.

Hager, Gonzci, and Athanasou (1992) also raise some important issues in relation to assessing competence. This includes the criticism that it can only assess trivial and superficial skills and not the complexity of work, the issue of inference of competence, the issue of validity, the assessment of knowledge and attitudes and values in competency-based assessment, and the issue of professional judgement and objectivity.

Within the aviation context, Telfer (1994) highlights the issues facing the industry in applying training systems based on competency-based training. He states that the Systems Approach to Training (SAT), favoured by most air forces, has on-the-job performance as its ultimate criterion. However, he stresses that one of the main weaknesses within civil aviation lies in our ability to assess competencies. Telfer explains that "in part, this weakness currently comes from the tendency to use tests that are economic rather than effective. This places unfortunate emphasis on so-called objective tests (multiple-choice or true/false) at the expense of other means of demonstrating [competency]" (p. 352). He suggests that other forms of assessment, such as simulator scenarios, decision-making role-playing and critical incidents, may be more suitable since they closely resemble the actual workplace task.

Research conducted by Henley (1995a, 1995b) in both Canada and Australia also underlines the difficulties raised in assessing aviation theory subjects because the assessment of these subjects is primarily norm-referenced rather than based on criterion or competency standards. Henley (1995b) also highlights the problems encountered in assessing the teaching ability of flight instructors due to the lack of clearly specified criteria for higher level competencies and because of the perceived lack of assessor credibility. Similarly, Given (1993) raises the challenge of assessing additional competencies gained during training, stating that,

> the current emphasis in training theory is on output (what a person can do at the conclusion of a training program, or as the result of work experiences) as opposed to input (what a person has been exposed to in formal training programs). There is recognition that people develop a range of competencies, often without any formal training, and this results in a multi-skilled workforce which is able to perform a broad range of

> activities. By way of illustration, in the air force's military culture, technical airmen [sic] develop management skills in addition to their technical skills as they progress through the ranks. The question of assessing such competencies remains. (pp. 195-196)

These authors highlight the complexity of assessing performance for complex work, such as in aviation settings. Some researchers and educationists are advocating for a more holistic assessment process which includes multiple components, such as knowledge, understanding, problem solving, technical skills, attitudes, and ethics (Hawke, 1996; Wood, 1994).

Holistic Assessment

The use of holistic assessment processes has been raised by many researchers and writers as a way to overcome some of the perceived shortcomings of a narrow, atomistic approach to assessment sometimes associated with competency-based approaches (see Gonzci, Hager, & Athanasou, 1993; Harris et al., 1995; Hawke, 1996; Kingsland, 1993; Smith & Keating, 1997). Rumsey (1994) defines holistic assessment as covering "in an integrated way, multiple elements and/or units from relevant competency standards. [It] seeks to combine knowledge, understanding, problem solving, technical skills, attitudes and ethics into assessment tasks" (p. 12).

Rather than develop separate assessment tools to assess each unit of competency, a holistic approach seeks to integrate the assessment of knowledge, skills and values in one assessment activity. According to Hawke (1996), a holistic assessment approach aims to assess knowledge and values in the context of practice rather than as isolated or segregated activities. This is also in line with the objectives of problem-based learning discussed in Chapter 11.

To illustrate a holistic approach to assessment, the example provided by Hager, Athanasou, and Gonzci (1994) of some integrated methods of assessment for pilot competencies is reproduced here (see Table 15.1). This example highlights the way different learning outcomes have been grouped under categories and then outlines a range of possible forms of assessment used, such as observation, oral questioning, simulation and written tests.

Some of the methods used to assess Canadian flight instructors can also be used to illustrate a holistic assessment approach. The Canadian assessment design of the initial flight instructor rating is tripartite. Firstly, aspiring flight instructors must successfully complete a written examination covering administrative issues, the theory of flight, and instructional principles. This written examination is a traditional, standardised test using multiple-choice items. Secondly, candidates for the flight instructor rating must demonstrate their ability to present ground instruction for air exercises and conduct effective pre-flight briefings during a practical classroom examination with an inspector. This segment requires discerning observational skills and a considerable amount of flexibility on the part of the assessor because as the practice teaching proceeds, the assessor must be alert to suitable opportunities for questioning. According to Hawke (1996), "these need to be moments when relevant information, understanding or analysis can be sought and yet which do not disrupt or misdirect the flow of the activity" (p. 13).

Table 15.1 Example of an integrated approach in assessing pilot competencies

CATEGORIES	METHODS OF ASSESSMENT
Observable technical performance	Minimum 1500 hours, six month simulator check Six month route check Licence renewal Six month asymmetric check (i.e., engine failure) 90 day requirements for non-directional beacon VHF omni-directional radio range and instrument landing system 90 day requirements for take-offs and landings both day and night 5 hours instrument flight time Medical check In-flight or in-simulator assessment by check captain
Problem solving	Simulator In-flight problems assessed by check captain
Knowledge	Aircraft systems — oral questioning Civil Aviation examinations (meteorology 70%; navigation 70%; flight planning 70%; air legislation 80%)
Attitudes, Work habits	Oral questioning on company policy and requirements

Source: Hager et al., 1994, p. 18.

At the end of this practical teaching segment, potential instructors must also demonstrate their knowledge of flight tests standards for the private and commercial pilot licences through oral questioning. Thirdly, following the ground presentation, the candidates must pass a practical flight test with an inspector acting as a student (role playing) to show that they can perform, explain and demonstrate any of the flight manoeuvres for the private and commercial pilot licences. After the practical flight test, the instructor candidate must conduct a post-flight debriefing covering the essential elements of the practical lesson and provide student feedback for future improvement. Again, the assessor must take advantage of suitable opportunities to use perceptive questioning to assess the candidate's level of understanding and appreciation of values and ethical or safety issues.

The fact that multiple approaches are used to assess flight instructors in a holistic manner has definite advantages. The limitations of one approach are counter-balanced by the strengths of another and, as such, the fairness of the assessment should be increased (Andrews & Barnes, 1990; Hawke, 1996; Henley, 1995b). In spite of the potential benefits of the multiple approaches, the majority of flight instructors interviewed by Henley (1995b) noted that there were major shortcomings in each of the assessment approaches used to determine the competencies of flight instructors. However, before discussing the weaknesses that were identified by flight instructors, it is important to examine the principles of assessment because an understanding of these principles will help provide insight into the problem areas highlighted in Henley's research.

Principles of Assessment

Assessment systems must be seen to be rigorous, credible and defensible. As part of the Australian national vocational education and training system, National Assessment Principles have been endorsed. Of particular relevance are four key technical assessment principles: validity, reliability, flexibility, and fairness. According to the Australian National Training Authority (ANTA, 1998), "these technical principles of assessment must be addressed in the conduct of an assessment, in the development of assessment tools, and in the design establishment and management of the assessment system" (p. 17). This means that when planning assessments, trainers need to ensure that:

- the evidence proves that the person has the skills in the relevant competency standard not in other areas (*Validity*);

- another assessor would also make the same decision (*Reliability*);
- the assessment is either on- or off-the-job and at a mutually convenient time and situation (*Flexibility*);
- the method of collecting evidence didn't make it more difficult than the level of the competency standard (*Fairness*) (ANTA, 1999, p. 2.28).

To this list, Hawke (1996) adds authenticity as another important principle of holistic assessment. Research conducted by Henley (1995a, 1995b) also revealed the importance of credibility when assessments rely on the judgement of the assessor.

Validity

Validity is concerned with whether an assessment assesses what is claimed to be assessed. A straightforward example of this concept is outlined by Cropley (1995).

> As a test of ability to count we might hand students a box containing red, blue and green marbles and ask them to count how many of each colour there are. If a person is colour blind, the assessment will depend on colour perception as well as counting ability. In that case, the assessment would not be valid because it did not properly assess what it claimed to assess. (p. 28)

It is, therefore, very important that the methods of assessment match the aspect being assessed. Further, as assessment is basically a sampling process, the sampling needs to be robust enough to ensure the assessment process chosen provides the evidence to make valid decisions. In this sense, assessment needs to be holistic, "in that validity is a function of the extent to which the whole assessment process reflects the competencies to be acquired" (Harris et al., 1995, p. 168).

With particular reference to competency-based assessment, ANTA (1998) highlights factors that can enhance the validity of assessments when some or all of the factors below are applied:

- the assessment focuses on the appropriate areas of competence and skills;
- a sufficient range of the performance of the person being assessed is sampled;
- the assessment tasks resemble those encountered in the workplace;

- evidence of performance is obtained after the assessment to support predictive validity;
- evidence is gathered of transfer to new situations other than that used for assessment;
- the assessment procedure documents the links to workplace performance;
- multiple approaches to assessment are used; and
- the assessor can demonstrate how evidence of competency discriminates between unlike competencies and reinforces like competencies (p. 18).

Key questions we can ask ourselves in relation to the validity of an assessment tool are:

1. Does the tool measure the candidate's ability to meet the performance of the competency?
2. Does it reflect workplace realities?

Reliability

The methods and procedures of assessment used are reliable if the standards and their levels are interpreted consistently from learner to learner and over time. Reliability is concerned with whether an assessment measures consistently between repeated situations. It refers to the assessment instruments, the assessment strategy and the interpretation of evidence. As Hager et al. (1994) suggest, "without reliable assessments there can be no comparability of credentials. High quality competency standards are fundamental to reliability" (p. 14). Reliability is concerned with questions such as:

- Would other assessors agree with my assessment of the learner's performance?
- Would I myself assess the learner's performance in the same way if I assessed it again?
- Would the performance be different if it was done at other times or in other places?

A practical way to address these questions is to have an assessment guide which provides clear and consistent guidance to all assessors on the way in which the assessment criteria are to be used to reach an overall judgement regarding performance. An assessment guide can also be a valuable tool to both students and trainers, because the standards of performance and the assessment criteria are clearly spelled out. ANTA (1998) suggests the following should be implemented to enhance the evidence of reliability:

- comparing the results of two or more assessors (moderation);
- collecting evidence via a number of different assessment methods;
- collecting evidence across different locations and times;
- providing clear and careful instructions when an employee is requested to monitor his or her own or others' behaviour (documentation);
- specifying clearly competencies to be attained (documentation);
- detailing clearly items on self/peer/supervisor assessment (documentation);
- taking care in the construction of assessment tools (systematic procedures); and
- reviewing the training of assessors (systematic procedures) (pp. 18-19).

Key questions we can ask ourselves in relation to the reliability of an assessment tool are:

1. Does it consistently measure what it is supposed to measure?
2. Does it use a range of methods?
3. Does it generate sufficient evidence?
4. Is the evidence collected over a range of contexts and time?

Flexibility

According to Harris et al. (1995) and ANTA (1998), a range of approaches should be used which are appropriate to a range of delivery modes, sites and learner needs. Flexibility is concerned with making appropriate modifications to assessment procedures and methods to suit the particular needs of learners and adjusting the way that assessment is undertaken to suit the variety of contexts in which learning might be taking place. For example, flexible assessment might include the replacing of assessment instruments that depend on language or literacy ability with questioning, interviews or some other more user-friendly approach.

Further options for flexible assessment would include self-assessment, peer assessment, the use of negotiated, individual learning contracts, and negotiated partnerships with workplace supervisors and assessors for on-the-job assessment. Allowing assessment on demand can also provide flexibility. Learners could be given the opportunity to request assessment when they consider that they are ready. However, consideration needs to be given to the validity and reliability of any flexible assessment options. It must be stressed that flexibility is intended to apply to the process of assessment and not to the standards being assessed. ANTA (1998) states that to be flexible assessments should:

- cover both on- and off-the-job components of training where applicable;
- provide for the recognition of competencies no matter how, where or when they have been acquired;
- draw on a range of methods and be appropriate to the context, task and person;
- be made accessible to the person(s) being assessed so that they can proceed readily from one competency standard to another (p. 19).

A key question we can ask ourselves in relation to the flexibility of an assessment tool is: Can the procedures be adapted to suit the circumstances and needs of the enterprise, organisation or candidates?

Fairness

The assessment process must be equitable, accessible, transparent, participatory and open to reasoned questioning. It should not disadvantage particular learners. Fairness is concerned with equal access and opportunity to succeed, regardless of age, gender, ethnicity, disability, language and literacy levels or employment status. To achieve fairness, an assessment process should be well documented and assessment tasks should not go beyond the purpose of the course or, for vocational courses, what would reasonably be expected in the workplace. ANTA (1998) states that, to be fair, an assessment should:

- help the person being assessed understand clearly what is expected and what form the assessment will take;
- ensure that the assessment procedure supports the identification of training needs;
- be equitable to all groups of people being assessed (make reasonable adjustments to the methods used for collecting evidence depending on the characteristics of the person(s) being assessed);

- have criteria for judging performance that are made clear to all those seeking assessment;
- involve a participatory approach to assessment that is agreed to by the assessor and the person being assessed;
- provide opportunities that allow the person(s) being assessed to challenge assessments; and
- make provisions for reassessment (p. 19).

Key questions we can ask ourselves in relation to the fairness of an assessment tool are:

1. Does the tool include allowable adjustments that take into account: age, ethnicity, gender, language difficulties, disability, social and educational background? Is it free of bias?
2. Does the assessment procedure use multiple assessment methods to allow multiple checking of knowledge and skills?
3. Have there been many grievances and disputes?

Harris et al. (1995) raise some further issues about these four principles in competency-based assessment systems. As assessment processes are essentially a sampling process, the more holistic the assessment process is the more likely that the assessment will be robust enough to be a valid assessment of the competencies. Although, as Harris et al. state,

> Validity is an important constituent in a competency-based assessment system: reliability is less so because of the constraints which an undue concern for a reproducible process may have on the judgements of competence. An undue concern for reliability can very much affect the flexibility in approach often needed to make valid judgements. (p. 168)

The principle of fairness has a number of aspects. A fair assessment needs to be transparent in all aspects of the conduct of the assessment: the planning, conduct and reporting of the assessment. It needs to be transparent not just to the assessor but to the designers, teachers, learners, employers and to any other end-users (Harris et al., 1995). The nature and extent of feedback given to learners is of particular importance to a fair assessment process. As Harris et al. state, "if anything, the competency-based assessment system must attempt a greater level of transparency in its recording and reporting systems than many traditional systems, which are predominantly quantitative" (p. 169).

Authenticity

According to Hawke (1996) and Wiggins (1989, 1992), the design of a holistic assessment should closely approximate real workplace activities. Hawke stresses that "*authenticity* of the assessment is one of its defining features" (p. 12). The reason being that if students can perform competently in a simulated, work-related assessment, chances are they will be able to transfer these competencies to the real world of work. "If an assessment can capture the complexities of the real work environment and provide opportunities for [learners] to develop processes to integrate and reflect on the skills involved then they have a better chance of being able to transfer learning from one context to another" (Hawke, 1996, p. 14). This can present a challenge since educational systems may not have the capability to simulate the workplace, and simulated work-related assessment can be costly and time consuming. However, within the aviation context, the use of simulators can provide authenticity. As Lintern (1995) argues, optimal training benefit in aviation accrues in a situated learning environment, such as Line-Oriented Flight Training (LOFT). This same type of environment that duplicates as nearly as possible the situation in which the learning will later need to be applied can also be used effectively to assess skilled performance.

Credibility

When an assessment is entirely dependent upon ratings based on an examiner's subjective impressions, as is the case during a flight instructor's ground briefing, credibility becomes an important issue. For example, while observing a flight instructor teach selected ground briefings, the examiner is called upon to make a value judgement about the competence of the instructor. The examiner gains information about the candidate's teaching ability mostly by impressionistic means and then, by virtue of his or her expertise and experience, renders a judgement that constitutes the assessment. Much of the credibility of the assessment is invested in the assessor as a recognised expert in aviation education and as an assessor (Henley, 1995b). This type of assessment based on professional judgement has certain advantages. The assessment can be quickly done because the assessor is usually an expert who has extensive experience in the area and can draw upon it without being limited or artificially constrained by measurement instruments. Another strength of the assessment based on expert judgement is that all of the variables involved in the educational process can be brought into consideration. Notwithstanding these advantages, assessments based on the judgement of an assessor can have definite shortcomings, as was revealed by Henley's (1995a, 1995b) research into the quality of the development and evaluation of flight instructors in Canada and Australia.

Henley (1995b) found that despite the fact that multiple approaches were used to assess flight instructors, the majority of flight instructors in both Canada and Australia did not consider that the flight instructor test was a valid and reliable assessment of their teaching competence. This was due to the fact that one or more of the principles of assessment described above were deemed missing or lacking in some way.

Firstly, most interviewees felt that the validity and the reliability of standardised, multiple-choice written test used to assess flight instructors' knowledge of teaching principles was questionable because some questions were ambiguous and poorly worded. Further, most respondents felt that the multiple-choice test was not a fair and valid assessment tool because it was limited in scope and concentrated on a narrow band of knowledge and facts that are easily measurable. It did not assess higher level competencies, such as decision making, critical thinking and judgement or essential teaching skills and attitudes, such as anticipation, flexibility, reflection-in-action, and empathy. These competencies are not included in standardised tests because they are mainly "intangible" and cannot be easily measured using such measurement instruments.

Secondly, the majority of flight instructors indicated that they did not feel that the ground presentation (micro-teaching) was a true and valid indication of their teaching ability because it was a narrow sample of their teaching skills. Further, many also felt ill at ease with "role playing" as a means of assessing their teaching ability and the majority questioned the credibility of the examiners, who they felt were not recognised expert educators and/or examiners.

The reliability and the objectivity of judgements based on subjective impressions can be questionable, even in situations where instructors and trainees were willing to concede the expertise and the integrity of the assessor. In the case of the flight instructor test, for example, because the assessment process is implicit, it hides from scrutiny both the data and the criteria or standards used to assess teaching performance (Henley, 1995b).

The reliability and objectivity problem is further accentuated by a notable lack of meaningful and operational guidelines to judge teaching effectiveness. However, by far the most objectionable weakness identified by the majority of the flight instructors interviewed by Henley (1995b) in Canada and especially in Australia was the fact that very few examiners in the aviation industry have received any specialised training on how to teach and how to conduct assessments. Therefore, the credibility and expertise of the assessors was seen to be lacking.

Furthermore, two assumptions present a basic limitation in assessing a skill like teaching. First is the assumption that variables — such as a person's ability to instruct, his or her attitudes, his or her ability to make decisions or render judgements, to improvise, to communicate effectively, to work with students, to understand their needs and their apprehensions, and to be patient with them and show concern and empathy, none of which involve production absolutes — can be reduced to a single figure on a rating form.

The second assumption is that the skills and attitudes needed for effective teaching can be assessed in a few hours. The quality of instructional skills has to do with nuances, with details, with the subtle and unique factors that make a difference beyond the points on a rating scale and, according to most instructors, it cannot be accurately and authentically determined in a few hours using role playing. In such situations, much of the face validity is invested in the assessor as a recognised expert in the field being evaluated. However, while the great majority of examiners were considered experts in aviation and in some cases in flight training, they were not recognised by instructors as assessor or as educators *per se* (Henley, 1995b).

Finally, the great majority of flight instructors who participated in Henley's survey indicated that the practical flight test was a fair assessment of their flight skills but not of their teaching skills. They noted that standards of performance for the flight manoeuvres were clear. However, what was expected of an instructor in a teaching role, the assessment approach itself and the standards to be met, varied greatly from region to region and from one examiner to the next. The assessment of an instructor's effectiveness in teaching in-flight manoeuvres and in analysing a student's learning problems were seen as lacking because of the use of unrealistic 'role playing', the uncertainty regarding the assessment criteria used, the lack of assessor credibility, and the limited amount of time devoted to assessing the complexities of teaching in a dynamic environment. Most flight instructors felt that "role playing lacks realism, is confusing [and] is inflexible" (cited in Henley, 1995b, p. 146). Further, many instructors said that they prepared "for the 'personality' of the examiner more than for the task" (cited in Henley, 1995b, p. 154) because the test was "purely subjective [and restricted] to the model presented by the examiner" (cited in Henley, 1995b, p. 148).

This situation, whereby flight instructors are assessed by those they consider to be in a poor position to be assessors, understandably can lead to feelings of resentment and frustration toward the assessment process. Clearly, it can be seen that even when a more holistic approach is used, if the principles of assessment are not met, the merit of the assessment can be compromised. Most instructors suggested that the assessment would be fairer and more valid if they were assessed while teaching "real students" and the assessment included a sample of their teaching skills in actual learning situations and on a variety of occasions (Henley, 1995a, 1995b).

The Evaluation Process

As stated in the introduction, the term evaluation is used in many different contexts and it has many different forms. The focus here will be on how evaluation is used in a training context. Therefore, the main concern will be how evaluation can add to the effectiveness of education and training programs. In fact, it will be seen that evaluation activities are continually occurring in this context, at all levels of decision making and participation, from management, program planners, trainers, potential employers to participants. Scott (1992) concisely describes this ongoing process:

> Evaluation is the driving force behind the whole process of teaching and learning. For example, administrators and industry managers allocate funds to one training program or provider not another, they prefer one learning approach to another, they give priority to one lot of trainees over another, and they emphasise one cluster of content not another. They continuously make judgements about the worth of particular programs and the relative competence of particular trainers.
>
> Teachers/trainers evaluate what is likely to work when preparing for a teaching-learning session. They make judgements about how well things are going once they start teaching. They evaluate how successful the session has been after it is over.... It is most important to recognise that this sort of evaluation is going on all the time.... (p. 56)

The centrality of this process to the training context is further underlined by Field, Newman, Morris, and Burg (1985) who link evaluation to "both personal and organisational accountability, which in turn are closely tied to things like funding and accreditation" (p. 53).

Definitions of Evaluation

Examining a range of definitions of evaluation sheds some light on how the process is conceptualised in the field of education and training. The following definition from Pithers (1998) encapsulates some key features of evaluation.

> Evaluation can be defined as a reasonably systematic, continuing process which examines the value of training from the design of the training program, through on-the-job outcomes, the benefits of the training program to the enterprise for whom the training was developed and provided. (p. 204)

This definition stresses the range of features of training that can be subjected to evaluation. This issue of what can be evaluated and whose concerns drive evaluation procedures will be discussed in some detail below.

Another definition by Beatty, Benefield, and Linhart (1991) focuses on the fact that there are many stakeholders involved in any evaluation and that its outcomes can impact on all of them. According to Beatty et al., "evaluation is the mutual, interactive, and systematic process of identifying, gathering, and analysing information in order to make decisions which would enhance learning for all persons involved" (p. 164).

A key feature of some useful definitions is the focus on the "valuing" aspects of evaluation as Athanasou's (1995) succinct definition reveals: "Evaluation, as its name implies, is the process of determining the 'value' or utility of any educational program, policy or initiative" (p. 76). Scott (1992) elaborates upon this theme of value to get to what he perceives as the essence of the process.

> It is most important to be clear on what the process of *evaluation* involves. At the heart of the term evaluation is the word "value". Making judgements of worth (attributing value) is ultimately a question,

> therefore, of the application of subjective perceptions of what is "good" and "bad". (p. 56)

One thing this range of definitions reveals is that making these types of important judgements about training is clearly going to be a complex process. Firstly, evaluation has implications. For example, if a training program is evaluated negatively, steps must be taken to either re-think, re-design or cut the program. These may be decisions that impact upon budget allocations, employment of training staff and resources. This may be one of the reasons (apart from the time and money thorough evaluation requires) that, as Brookfield (1986) suggests, *systematic* evaluations are rarely carried out.

However, it must be argued that despite the difficulties and complexities of evaluation, it is a crucial aspect of providing quality training, especially in today's training context where program designers are increasingly accountable for trainees and workers meeting competency standards. Evaluation is one way to assess how your training "measures up".

Purposes of Evaluation

Evaluation is a process that can be shaped to serve a variety of purposes. As seen in Chapter 7, planning and implementing a training course involves a complex series of decision-making steps. In the evaluation process, the outcomes of any or all of these decisions can be evaluated. The following series of questions adapted from Rae (1997) provide a very useful overview of the many purposes evaluation can serve and the various aspects of training on which it can focus.

- *Content of training* — Is it relevant and in step with training needs? Is it up to date?
- *Method of training* — Were the methods used the most appropriate ones for the subject and for the learning styles of the participants?
- *Amount of learning* — What was the material of the course? Was it new to the learner or merely the mixture of previously covered material? If not new, was it useful as confirmatory or revision material?
- *Trainer skills* — Did the trainer have the necessary attitude and skill to present the material in a way that encouraged learning?

- *Length and pace of training* — Given the material essential to learning, was the learning event of the appropriate length and pace? Were some aspects laboured and others glossed over?
- *Objectives* — Did the training satisfy its declared objectives? Was the learner given the opportunity to try to satisfy any personal objectives? Was this need welcomed? Were personal objectives actually satisfied?
- *Omissions* — Were any essential aspects omitted from the learning event? Was any material included that was not essential to the learning?
- *Learning transfer* — How much of the learning is likely to be put into action in the workplace? If it is to be a limited amount only or none, why is this? What factors will deter or assist transfer of the learning?
- *Accommodation* — If course accommodation is within the control of the trainer, or is relevant to the type of training event, he or she may wish to ask whether the training milieu was suitable. Was the accommodation acceptable? Were the meals satisfactory?
- *Relevance* — Was this course, seminar, conference, workshop, tutorial, coaching, assignment or project, etc. the most appropriate means of presenting a learning opportunity or achieving the learning outcomes?
- *Application of learning* — Which aspects within the workplace now include elements that are a direct result of the learning event? Which new aspects of work have been introduced as a result of the learning? Which aspects of previous work have been replaced or modified as a result of the learning? Which aspects of learning have not been applied? Why not?
- *Efficiency* — How much more efficient and/or effective is the learner in his or her work as a result of the training? Why or Why not? The question could also be posed to the learner's boss and his or her subordinates, but, as in so many cases, it will be highly subjective evidence and the questioning will need to be extensive (adapted from pp. 11-12).

Types of Evaluation

Traditionally, broad categories have been used to describe different types of evaluation. The following terms are widely used in the evaluation literature and it is important to have an understanding of their different uses.

Formal and Informal

Formal evaluation is characterised by the systematic collection of a range of evidence relating to the training activity, which is then measured against the evaluation criteria. This type of evaluation is often reported on in writing and distributed to relevant stakeholders for consideration. Alternatively, informal evaluation "lacks systematic procedures and formally collected evidence" (Worthen, Sanders, & Fitzpatrick, 1997, p. 7) and often relies on the experience and instincts of those involved. However, this does not mean it is less useful. When those involved are knowledgable, experienced and fair, the outcome can be very sound (Worthen et al., 1997).

Formative and Summative

As the term suggests, formative evaluations are carried out "during the developmental (i.e., formative) stages of a program" (Field et al., 1985, p. 54). The purpose of evaluation at this stage is to allow the evaluation data to be utilised in the program planning cycle so that the program is monitored from the outset of its implementation. In contrast, summative evaluation "refers to evaluation conducted after the program is fully functioning, for the purposes of justification or accountability" (Field et al., 1985, p. 54).

Worthen et al. (1997) summarise some of the key differences between these two types of evaluation.

> Note that the audiences and uses for formative and summative evaluation are very different. In formative evaluation, the audience is program personnel.... Summative evaluation audiences include potential consumers ... funding sources ... and supervisors and other officials, as well as program personnel. Formative evaluation leads to (or should lead to) decisions about program development (including modification, revision and the like). Summative evaluation leads to decisions concerning program continuation, termination, expansion, adoption and so on. (p. 15)

Internal and External

Due to the fact that evaluation is often utilised to serve a variety of purposes and can at times be a process charged with a range of divergent interests (both political and pragmatic), it is not surprising then that sometimes people who are external to the training program are called in to conduct the evaluation. In this case, the term external evaluation is used to describe the process (as distinct from when program employees conduct their own evaluation of their work, known as internal evaluation). Worthen et al. (1997) summarise the advantages and disadvantages connected with both internal and external evaluation roles. Internal evaluators normally will be more knowledgeable about the program than any outsider would be, but they may also be too close to the program and thus be unable to be completely objective. Seldom is there as much reason to question the objectivity of external evaluators (unless they have a particular axe to grind), and this dispassionate perspective can perhaps be the greatest asset of having an external evaluator. On the other hand, it is difficult for an external evaluator to ever learn as much about the program as the insider knows (p. 18).

The following discussion highlights two other types of evaluation commonly carried out in the training context: cost-benefit analysis and self-evaluation.

Cost-Benefit Analysis

This type of evaluation generally occurs once training has been completed and, while it can be carried out by training staff, it is often initiated by management and involves others in the workplace training setting such as accountants (Pithers, 1998). The aim of this type of evaluation is to "attach monetary values to training activities" (Newby, 1996, p. 217). Pithers (1998) underlines the importance of this process in the current workplace training setting and also the complexity of this type of evaluation.

> All training programs cost money to develop, present and assess, not just in terms of the resources required, but in terms of the people involved and their costed time out of their current job. The issue is whether, over time, the organisation is able to recover these costs and, in reality, have value added above that level. This is sometimes, although not always, the case. The difficulty in attempting this form of evaluation is that a monetary value has to be put onto training costs and training outcomes. This is not too difficult if training results in measurable changes to productive output, that is, if the product continues to sell or be valued by consumers. (p. 207)

However, not all training is so closely and immediately linked to profit ratios. Sometimes training has less direct benefits to a work environment. For example, team-building training may have the effect of reducing staff turnover and increasing productivity due to increased ability of the work team to carry out their work with less friction and greater efficiency. In this case, the evaluation strategy may have to be creatively designed so that lower recruitment costs and fewer losses of productivity due to employee attrition are clearly linked to the team-building training outcomes and, therefore, show up in the cost-benefit analysis.

Self-Evaluation

Gowing and Saunders (1992) highlight the fact that trainers these days often find themselves under strong pressure to evaluate their performance and the course structure in order to report to the organisation that provides the funding for the training program. In the light of this context where program designers and trainers are becoming more accountable for productivity through training and performance, Gowing and Saunders suggest that trainers can conduct effective self-evaluations.

Central to the process of good self-evaluation is the concept of reflection. This is a concept widely discussed in the literature of effective training practice and requires some elaboration here. It is generally agreed that there are two basic reflective processes central to reflective practice: *reflection-in-action* and *reflection-on-action* (Merriam & Caffarella, 1999, p. 235). The theorist most closely associated with describing these two key processes is Donald A. Schön, perhaps one of the most well-known theorists in this area. In his influential work, *The reflective practitioner: How professionals think in action*, first published in 1983, Schön describes the powerful insights to be gained about practical and professional knowledge through understanding reflection-in-action and reflection-on-action.

Reflection-in-action is a type of reflection about practice that occurs while the practitioner is working. It is the practical bases of decision making practitioners use when drawing on their own informal theories about practice, that is, the knowledge that they have gained by carrying out their practice as distinct from knowledge that might be gained from reading about formal theories of practice. Schön (1998) describes reflection-in-action as recognising that "sometimes we think about what we are doing. Phrases like 'thinking on your feet', 'keeping your wits about you', and 'learning by doing' suggest not only that we can think about doing but that we can think about doing something while doing it" (p. 54). This level of self-awareness in relation to actions can lead to increased understanding of how effective practice occurs. In this way, practitioners are said to be able to not only learn more about effective practice but also monitor and increase their repertoire of responses to the many and varied situations a work context can create (Foley, 1995). In this way, most practitioners are conducting continuous self-evaluations which feed back immediately into teaching practice.

Reflection-on-action is an activity that can occur after an event. Foley (1995) describes reflection-on-action as "a retrospective process, in which practitioners return to the experience, attend to their own and others' feelings about the experience, and re-evaluate the experience" (p. 47). This re-evaluation can lead to a questioning of what occurred and of the assumptions that may have guided decision making in the event and, sometimes, can result in a new perspective on a practice, issue or situation. If this type of reflection is carried out in a systematic way, it becomes a powerful self-evaluation tool (see Chapter 8 for a more detailed discussion on reflective practice).

Stages in Evaluation

So far in this chapter evaluation has been defined, its purposes examined and different types of evaluation have been outlined. The remaining section of this chapter is, therefore, focused on the practical aspects of evaluation — the "how to" of evaluation. This is an important focus for the chapter since it must be acknowledged that effective evaluation is a skilled process. Field et al. (1985) suggest that evaluation consists of five main stages which are outlined separately below:

1. establish criteria for evaluation;
2. gather evidence;
3. compare this evidence with the criteria;
4. make judgements about what this comparison reveals; and
5. report the results of the evaluation (pp. 53-54).

Establishing Criteria

This stage of evaluation is a twofold process. Firstly, it is important for the evaluator to gain a clear picture of what the program was designed to achieve

(Field et al., 1985). In the case of internally conducted evaluations, this should be a simple step since the people conducting the evaluation have had close involvement with the training program. However, if the evaluation is being conducted externally, this is a crucial step since it is impossible to assess the value of a program without having a clear understanding of its objectives. Secondly, the evaluator must decide which aspects or characteristics of the program are being evaluated. Some examples of commonly evaluated program characteristics are the quality, the suitability, the effectiveness, the efficiency, and the importance to the participants and the community.

Gathering Evidence

Knox (1986) suggests that "the core of an evaluation plan is the design and the procedures for data collection and analysis" (p. 168). Scott (1992) concurs with this emphasis, however, he notes a "tendency for people to use scant data when drawing conclusions about the worth of a particular individual or program" (p. 58). Knox (1986) suggests that a range of "quantitative and qualitative data [be] collected from representative samples of participants, instructors, administrators, and records in ways that minimise disruption of the ongoing program and encourage use of the findings" (p. 168). However, Field et al. (1985) note that many evaluations "fail to get higher than the level of learner satisfaction" (p. 57). The implications of this statement are such that many evaluations are limited in scope and depth and fail to consider the full range of training course objectives and the fact that there are often important unintended outcomes of a learning process.

Comparing Evidence to Criteria

The data or evidence collected might take a number of forms. It might be statistical information based on participant attrition or ratings scales of trainer performance. It could be financial calculations concerning cost effectiveness or resource use. It might also include qualitative data that allows individuals and groups to provide information about their responses to the training in written or verbal formats. However, as Field et al. (1985) note, this evidence "is of little or no use until it is compared with the evaluation criteria which have been established" (p. 57).

Making Judgements

This stage of the evaluation involves coming to some conclusions about what the evidence suggests when viewed in relation to the evaluation criteria. As outlined in the previous section on definitions of evaluation, making judgements about a training program has implications for the organisation and the people involved in making training decisions and carrying out the training. This can be a difficult process and caution must be exercised. Caffarella (1994) for example, warns that "unless the evaluation design is sophisticated enough to control for all variables that could affect the outcomes, program planners must be careful not to attribute success or lack of success to their educational program without considering other contextual variables" (p. 140). This statement emphasises the importance of thought and planning in the collection of evidence and careful analysis of the findings. On a practical note, Field et al. (1985) recommend that program evaluators make "careful, reasoned judgements ... on the basis of the best evidence available" (p. 58).

Reporting Results

If, as Knox (1986) contends, the purpose of conducting training program evaluation is to "make judgements for planning, improvement, or justification" (p. 169), then plans for reporting the findings of evaluation are an important part of the evaluation process. Careful reporting of findings and recommendations will facilitate the implementation of desired changes to training. Therefore, a sound evaluation plan should clearly indicate effective ways of reporting pertinent findings to each of the target audiences and suggest ways to encourage actual use of findings (Knox, 1986). Phillips (1997) suggests the following guidelines for effectively communicating program evaluation findings:

> *The communication must be timely.* Usually program results should be communicated as soon as they are known. From a practical standpoint, it may be best to delay the communication to a convenient time such as the next edition of the newsletter or general management meeting. Questions about the timing must be considered. Is the audience ready for the results in view of other events which may have happened? Are they expecting them? When is the best time to produce the maximum effect on the audience?

> *The communication should be targeted to specific audiences.* The communication will be more efficient when it is designed for a particular group. The message should be specifically tailored to the interests, needs, and expectations of the target audience.
>
> *The media should be carefully selected.* For particular groups, some media may be more effective than others. Face-to-face meetings may be better than special bulletins. A memo to top management may be more effective than the company newspaper. The proper method of communication can help improve the effectiveness of the process (pp. 337-338).

The process of conducting training evaluations is complex; however, if managed carefully, it can yield very useful results for a range of stakeholders. In concluding this chapter, a valuable set of guidelines for conducting an evaluation are proposed:

- strive for fairness, objectivity, and credibility even as you acknowledge that the evaluation process is a combination of objective descriptions and subjective judgements;
- explain evaluation plans and procedures to people likely to be affected by them;
- provide clear instructions with all data-collection instruments and procedures;
- recognise that evaluation can be disruptive, so be as organised and unobtrusive as possible, be accommodating and minimise disruptions;
- as a part of the evaluation process, try to develop commitment to use the findings;
- use pre-testing and pilot projects to refine data-collection instruments and procedures;
- look for strengths as well as weaknesses;
- be alert to unexpected assumptions, influences, viewpoints, and consequences;
- provide feedback from evaluation findings;
- help participants learn how to use evaluation procedures as part of their own learning activities;
- in reporting evaluation findings, indicate implications and alternative courses of action;
- anticipate consequences of plans, activities, and recommendations;

- be willing to do additional analysis; and
- express appreciation to those who cooperate with or contribute to the evaluation effort (adapted from Knox, 1986, p. 185).

Conclusion

This chapter has examined various aspects of assessment in educational settings. It has examined different types of assessment and in particular it has explored competency-based assessment. The chapter also highlighted the principles that are important in order to ensure effective assessment, such as validity, reliability, flexibility, fairness, authenticity and credibility. In addition, some guiding questions for ensuring these principles are incorporated into practice were suggested. Examples of the ways these principles have been applied (or the consequences of their lack of application) in aviation settings have also been discussed, highlighting the important implications for future assessment practice in the aviation industry and the need to address the education and assessment expertise of the examiners.

Evaluation is a term used to describe a process that occurs in many contexts. In the training context, where programs are designed to achieve measurable outcomes, program evaluation provides feedback to trainers, participants, and decision makers to guide current and future decisions about training. Evaluation must be seen, therefore, as a crucial aspect of providing quality training. In today's training context where program designers are increasingly accountable in terms of trainees or workers meeting competency standards and working within budgets, evaluation is one way to assess how a training program "measures up".

Evaluation was also examined through a range of broad categories that have been used to describe types of evaluation. These categories included formal and informal evaluation, formative and summative evaluation and internal and external forms of evaluation. Finally, the practical skills of evaluation were covered and a range of methods of data collection was discussed. Useful ideas for collecting important evaluation information were presented and easy-to-follow guidelines were suggested. The reporting to, and use of evaluation findings by, target audiences were also considered due to their importance to having program evaluation recommendations and findings implemented.

References

Andrews, T.E., & Barnes, S. (1990). Assessment of teaching. In W.R. Houston (Ed.), *Handbook of research on teacher education* (pp. 569-598). New York: Macmillan.

Athanasou, J. (1995). Issues in the evaluation of adult education. In G. Foley (Ed.), *Understanding adult education and training* (pp. 75-88). St. Leonards, Australia: Allen & Unwin.

Australian National Training Authority (ANTA). (1998). *Training package for assessment and workplace training BSZ98*. Melbourne, Australia: Author.

Australian National Training Authority (ANTA). (1999). *Training package for assessment and workplace training BSZ98: Facilitator's pack: Plan assessment; conduct assessment; review assessment and train small groups.* Melbourne, Australia: Author.

Beatty, P.T., Benefield, L.L., & Linhart, L.J. (1991). Evaluating the teaching and learning process. In M.W. Galbraith (Ed.), *Facilitating adult learning: A transactional process* (pp. 163-192). Malabar, FL: Krieger.

Brookfield, S. (1986). *Understanding and facilitating adult learning.* San Francisco: Jossey-Bass.

Caffarella, R.S. (1994). *Planning programs for adult learners: A practical guide for educators, trainers and staff developers.* San Fransisco: Jossey-Bass.

Cropley, M. (1995). Validity. In W. Hall (Ed.), *Key aspects of competency-based assessment* (pp. 27-45). Adelaide, Australia: National Centre for Vocational Education and Training.

Field, L., Newman, M., Morris, R., & Burg, J. (1985). *Program development implementation and evaluation: A resource guide.* Sydney, Australia: Sydney College of Advanced Education.

Foley, G. (1995). Teaching adults. In G. Foley (Ed.), *Understanding adult education and training* (pp. 31-53). Sydney, Australia: Allen & Unwin.

Given, K.C. (1993). The evaluation of training. In R.A. Tefler (Ed.), *Aviation Instruction and Training* (pp. 189-202). Aldershot, UK: Ashgate.

Gonzci, A., Hager, P., & Athanasou, J. (1993). *The development of competency-based assessment strategies for the professions.* Canberra: Australian Government Publishing Service.

Gowing, R., & Saunders, S. (1992). Self-evaluation. In A. Gonczi (Ed.), *Developing a competent workforce: Adult learning strategies for vocational educators and trainers* (pp. 286-303). Leabrook, Australia: National Centre for Vocational Education Research.

Hager, P., Athanasou, J., & Gonzci, A. (1994). *Assessment: Technical manual.* Canberra: Australian Government Publishing Service.

Hager, P., Gonzci, A., & Athanasou, J. (1992). General issues about assessment of competence. In P. Hager & C. Chappell (Eds.), *Competency-based standards:*

The challenge of assessment (pp. 4-15). Sydney, Australia: School of Adult & Vocational Eduction, University of Technology Sydney (UTS).

Harris, R., Guthrie, H., Hobart, B., & Lundberg, D. (1995). *Competency-based education and training*. Melbourne, Australia: MacMillan Education.

Hawke, G. (1996). *Integrating assessment of learning outcomes: An approach to holistic assessment*. St Leonards, Australia: Assessment Centre for Vocational Education.

Henley, I. (1995a). The flight instructor test: Is it a valid and reliable assessment of an instructor's teaching competence? In R.S. Jensen & L.A. Rakovan (Eds.), *Proceedings of the Eighth International Symposium on Aviation Psychology* (pp. 1082-1088). Columbus, OH: Ohio State University.

Henley, I. (1995b). *The quality of the development and evaluation of flight instructors in Canada and Australia*. Unpublished doctoral thesis, University of Newcastle, Newcastle, Australia.

Kemp, J.E. (1985). *The instructional design process*. New York: Harper & Row.

Kingsland, A. (1993). Problem-based learning: Efficient, affordable and stress-free implementation. In G. Ryan (Ed.), *Research and development in problem based learning* (pp. 311-319). Campbelltown, Australia: Australian Problem Based Learning Network.

Knox, A.B. (1986). *Helping adults learn.* San Fransisco: Jossey-Bass.

Lehrer, H.R. (1997). The evaluation of aviation curriculum in the affective domain: Some preliminary thoughts. In R.A. Telfer & P.J. Moore (Eds.), *Aviation training: Learners, instruction and organization* (pp. 180-194). Aldershot, UK: Avebury Aviation.

Lintern, G. (1995). Flight instruction: The challenge from situated cognition. *The International Journal of Aviation Psychology, 5*(4), 327-350.

McCurry, D. (1992). Assessing standards of competence. In A. Gonczi (Ed.), *Developing a competent workforce* (pp. 220-239). Adelaide, Australia: National Centre for Vocational Education Research.

Merriam, S.B., & Caffarella, R. (1999). *Learning in adulthood: A comprehensive guide.* San Fransisco: Jossey-Bass.

Newby, A.C. (1996). *Training evaluation handbook.* Aldershot, UK: Gower.

Phillips, J.J. (1997). *Handbook of training evaluation and measurement methods* (3rd ed.). Houston, TX: Gulf Publishing.

Pithers, R.T. (1998). *Improving learning through effective training.* Katoomba, Australia: Social Science Press.

Rae, L. (1997). *How to measure training effectiveness* (3rd ed.). Hampshire, UK: Gower.

Rumsey, D. (1994). *Assessment: Practical guide*. Canberra, Australia: Australian Government Publishing Service.

Schön, D. (1998). *The reflective practitioner: How professionals think in action.* London: Ashgate.

Scott, G. (1992). Challenges facing vocational teachers and trainers in the 1990s. In A. Gonczi (Ed.), *Developing a competent workforce: Adult learning strategies for vocational educators and trainers* (pp. 44-78). Leabrook, Australia: National Centre for Vocational Education Research.

Smith, E., & Keating, J. (1997). *Making sense of training reform and competency based training.* Wentworth Falls, Australia: Social Science Press.

Telfer, R. (1994). Improving aviation instruction. In N. Johnston, N. McDonald, & R. Fuller (Eds.), *Aviation psychology in practice* (pp. 340-357). Aldershot, UK: Avebury Technical.

Wiggins, G. (1989, May). A true test: Toward more authentic and equitable assessment. *Phi Delta Kappan, 70*(9), 703-713.

Wiggins, G. (1992, May). Creating tests worth taking. *Educational Leadership, 49*(8), 26-33.

Wolf, A. (1995). *Competence-based assessment.* Buckingham, UK: Open University Press.

Woods, D.R. (1994). *Problem-based learning: How to gain the most from PBL.* Waterdown, Canada: Donald R. Woods.

Worthen, B.R., Sanders, J.R., & Fitzpatrick, J.L. (1997). *Program evaluation: Alternative approaches and practical guidelines* (2nd ed.). New York: Longman.

Index